MODERN STEAM ENGINES

BY JOSHUA ROSE, M.E.

Originally published in 1887

Reprinted in 2003 by

The Astragal Press
Mendham, New Jersey

Originally published in 1887 by Henry Carey Baird & Co., Philadelphia
Reprinted in 1993 by Lindsay Publications, Inc.
This edition printed in 2003 by Astragal Press

International Standard Book Number 1-931626-15-4
Library of Congress Control Number 2003113283

Published by

THE ASTRAGAL PRESS
5 Cold Hill Road, Suite 12
P.O. Box 239
Mendham NJ 07945-0239

Cover design by Donald Kahn
Manufactured in the United States of America

MODERN STEAM ENGINES:

AN ELEMENTARY TREATISE UPON THE STEAM ENGINE, WRITTEN IN PLAIN LANGUAGE;
FOR USE IN THE WORKSHOP AS WELL AS IN THE DRAWING OFFICE.

GIVING

FULL EXPLANATIONS

OF

THE CONSTRUCTION OF MODERN STEAM ENGINES;

INCLUDING

DIAGRAMS SHOWING THEIR ACTUAL OPERATION;

TOGETHER WITH

COMPLETE BUT SIMPLE EXPLANATIONS OF THE OPERATIONS OF VARIOUS KINDS OF VALVES, VALVE MOTIONS, AND
LINK MOTIONS, ETC., THEREBY ENABLING THE ORDINARY ENGINEER TO CLEARLY UNDERSTAND THE
PRINCIPLES INVOLVED IN THEIR CONSTRUCTION AND USE, AND TO PLOT OUT
THEIR MOVEMENTS UPON THE DRAWING BOARD.

BY JOSHUA ROSE, M. E.,

Author of "The Complete Practical Machinist," "Mechanical Drawing Self-Taught," "The Pattern Maker's Assistant,"
"Modern Machine Shop Practice," "The Slide Valve."

Illustrated by Four Hundred and Twenty-two Engravings.

PHILADELPHIA:
HENRY CAREY BAIRD & CO.,
INDUSTRIAL PUBLISHERS, BOOKSELLERS AND IMPORTERS,
810 WALNUT STREET.
LONDON:
SAMPSON LOW, MARSTON, SEARLE & RIVINGTON,
CROWN BUILDINGS, 188 FLEET STREET.
1887.

INTRODUCTION.

This book is intended for those who desire to acquire a knowledge of the construction of Modern Steam Engines, and to thoroughly understand the distinguishing features of each class of engines and the action of their important parts.

It is written in the language of the workshop so as to make it useful to the practical Engineer.

The various kinds of slide-valve motions have been treated very fully, because in the construction of the valve motion, chiefly lies the distinguishing features of most engines.

The diagrams explaining the action of each valve motion, have been obtained by moving the engine throughout a revolution, and measuring the port openings, both for the admission and exhaust, at each inch of piston motion. These diagrams therefore represent the *actual* workings of the valves.

In the examples, engines of as nearly as possible the same size have been shown, so as to enable the reader to compare the action of the different valve motions, and to make this comparison still more complete, a diagram of the action of each class of valve motion is given, as well as diagrams of the same valve motion, under different conditions of eccentric position and valve travel.

The author has endeavored to omit nothing that is essential to those who may begin their studies of the steam engine from the pages of this book, and pains have been taken to render it easy to follow the text. To this end the following means have been employed.

When a certain mechanism is to be considered, the engravings first show it as a whole and explain its general action. It is then treated in detail and moved through its various positions, a separate engraving showing each new condition, and a diagram showing the action under each condition.

The engravings have been made large and are, in many cases, repeated, so as to render them easy to follow and avoid turning to back pages.

v

Each subject is complete in itself, hence some subjects are treated in repetition. This possesses no disadvantage because it saves turning to back pages when studying particular mechanisms or movements, while it serves the learner as a review. Thus the effect of the connecting-rod in varying the piston speed, is treated of in connection with Common Slide-valve engines, Adjustable Cut-off engines, Automatic Cut-off engines, Shifting Eccentrics, Diagrams for designing Valve Motions and Link Motions.

Again, the subject of diagrams for designing valve motions and for investigating the action of valve motions is treated in several different ways, each explaining the groundwork upon which such diagrams are based. This lays a solid foundation upon which the reader may afterwards proceed without difficulty, and it is hoped and believed that the work will be found full and clear in its treatment, and easy to follow.

JOSHUA ROSE.

New York City, January, 1886.

P. O. Box 3306.

CONTENTS.

CONTENTS.

CONTENTS

CONTENTS.

CONTENTS.

MODERN STEAM ENGINES.

CHAPTER I.

CLASSIFICATION OF STEAM ENGINES——THE COMMON SLIDE VALVE ENGINE.

The different forms in which the steam engine appears may be classified as follows:

The *high pressure engine*, in which the steam, at whatever pressure it may be used, exerts its pressure in one cylinder only, and is then exhausted into the atmosphere.

The *compound engine*, in which the steam, after having been used in one cylinder, passes to a second and, in some cases to a third cylinder where it is used expansively before being exhausted into the atmosphere. When the steam expands in a third cylinder, the engine is said to have a *triple expansion*.

. The *condensing engine*, in which the steam, instead of being exhausted into the atmosphere, is condensed, creating a vacuum (or, more properly, a partial vacuum) on one side of the piston, thus relieving it from the pressure of the atmosphere which would act to counteract the steam pressure on the other side.

The *compound condensing engine*, in which the steam is used first in a high pressure cylinder recieving steam from the boiler, and then in one or more low pressure cylinders and is finally condensed, forming a vacuum on the exhaust side of the piston.

In some one of these forms the steam engine appears

in each of its applications. Engines are, however, also designated to indicate the purposes for which they are used, as *marine* engines for steamships, *locomotives* for railways, *portable* engines for those intended to be moved from place to place, *stationary* engines, as those set upon permanent foundations as in factories.

They are further designated from especial features in the design, as *beam engines*, where the piston delivers its power to a beam, *side lever engines*, where the beam or beams are at the side of the cylinder instead of above or below it.

Oscillating engines, in which the cylinder oscilates upon journals or trunnions in order to avoid the use of guide bars, and thus economize space.

Direct acting engines, in which the piston rod is connected direct to the crank by means of a connecting rod.

Vertical engines or *horizontal engines*, according to whether the cylinder bore stands vertical or horizontal, and *inclined engines* when it stands in neither of those positions.

Traction engines are those employed to draw loads, without the use of rails, usually upon common roads or highways.

2

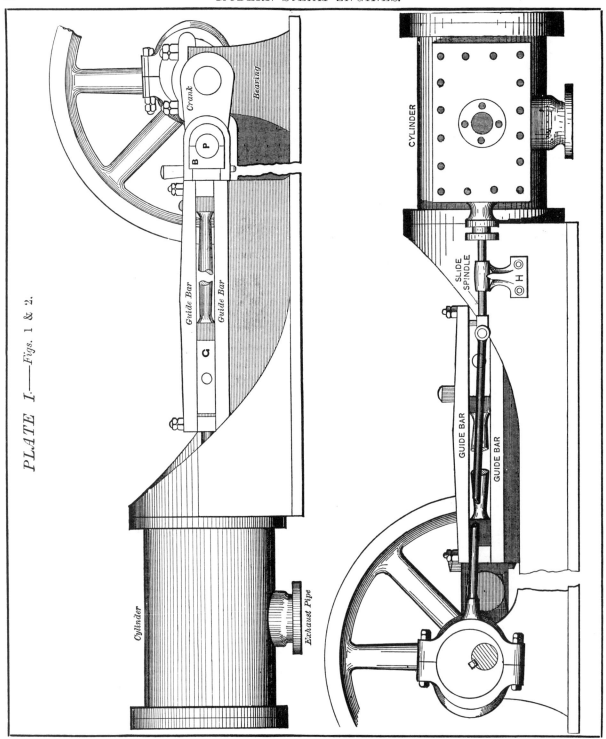

PLATE 1.—Figs. 1 & 2.

An inverted cylinder engine is one in which the cylinder stands vertical and the piston operates through the lower cylinder cover.

When, as in the case of most ocean going steamships, the cylinders stand in a line with the crank shaft, the engines are said to be *tandem* engines, or the cylinders arranged *tandem.*

Rotary engines are those in which the piston motion is in a circle around the piston rod or shaft, or it may revolve around the shaft in some curve. It is, however, attached direct to its rod or shaft, while *semi-rotary engines* are those in which the piston reciprocates in an arc or segment of a circle of which the shaft is the center.

Each of these kinds of engines, however, may be further designated by the peculiar design or operation of its parts; thus a simple slide valve engine is one in which the admission of steam is effected by a simple slide valve, or a common D valve, as it is sometimes called.

A *throttling engine* is one in which the speed of the engine is regulated by a fly-ball governor which partially closes or throttles the bore of the steam pipe, and thus causes the steam to enter the steam chest at a reduced pressure; an action that is termed *wire drawing* the steam.

A *cut-off engine* is one in which the steam supply to the cylinder is governed by a cut-off valve or valves. When the point in the piston stroke at which the cut-off valve will act is adjustable by hand, the engine is an *adjustable cut-off engine.* When the point of cut-off is governed by the engine itself, it is an *automatic cut-off engine.*

There are also *disc engines,* in which the pistons operate against a disc. *Multi—cylinder engines,* in which a number of cylinders are arranged either side by side, or with their bores radiating from the engine shaft, each taking steam at one end only.

THE COMMON SLIDE VALVE ENGINE.

The simplest form of high pressure, or non-condensing engine, is the common slide valve engine, whose construction is shown in Figs. 1, 2, 3 and 4, which represent a horizontal stationary engine. Fig. 1 is a

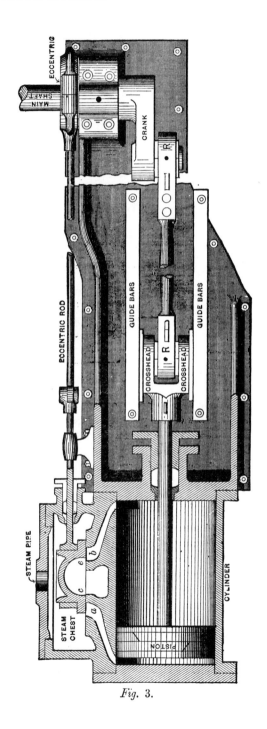

Fig. 3.

side view, showing the guide bars and crank with the crank end of the connecting rod. Fig. 2 is a view from the other side showing the eccentric and the slide spindle and its guide H. Fig. 3 is a top view of the engine, the cylinder and steam chest being shown cut

Fig. 5 represents (removed from the other parts of the engine) a cylinder C, steam-chest S, slide valve V and a valve rod, or spindle, R. The cylinder, C, is provided with three ports or openings, *a*, *b*, and *c*, the first and second of which are called the *steam ports*, while *c*

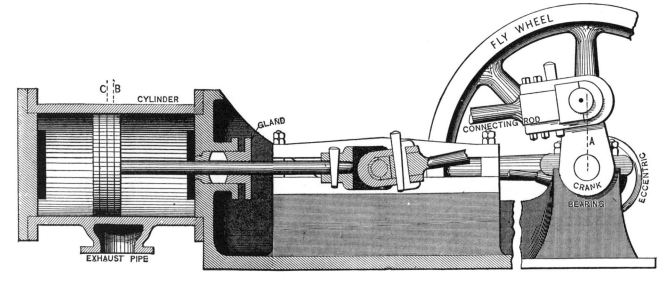

Fig. 4.

in half horizontally so as to expose the mechanism, and, as the names of the parts are marked upon them, there is no occasion to enumerate them. It may be noted, however, that the bed plate, the connecting rod and the eccentric rod are shown broken for convenience of illustration. In Fig. 3, the crank is shown on its dead center and the piston, therefore, at the end of the stroke, and it is obvious that the pressure exerted upon the piston by the steam would have no effect in moving the engine, because the crank, connecting rod and piston are in a straight line. Fig. 4, however, shows the parts in the position they would occupy when the crank was at its point of full power, and it is obvious that as soon as the crank has moved off its dead center the steam pressure upon the piston is in a direction to cause the crank to revolve and drive the fly-wheel whose momentum carries the crank past its dead center.

The action of the *slide valve* that governs the admission of the steam into, and its exhaust out of, the cylinder may be explained as follows:

is the *exhaust port*. The slide valve, V, fits closely to the face where these ports emerge into the steam chest and is traversed to and fro across them, the distance it

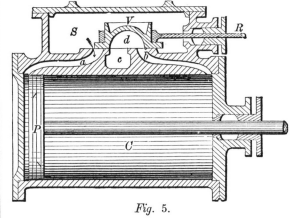

Fig. 5.

travels in one stroke being called the amount of the *valve travel*. It is operated by a rod R, or the slide valve *spindle* as it is termed, which receives motion

from the eccentric rod shown in Fig. 3. Now supposing the crank to be on its dead center and the piston in
the position it occupies in the figure and the valve will
have left port *d* open to the amount of the lead, the
steam passing from the chest V, through *d*, into the
cylinder. At this time, port *b* is acting as an exhaust
port, the steam that propelled the piston during the
previous stroke passing through the valve exhaust port
(or exhaust cavity, as it may more properly be termed)
and into the cylinder exhaust port *c*, whence (in a high
pressure engine) it passes into the atmosphere. Thus
while *a* is acting as a steam port, *b* is acting as an exhaust
port. But by the time the piston has reached the other
end of the cylinder, and has therefore completed its
stroke, the valve will have moved so as to leave *b* open
to the steam chest and leave communication between
port *a* and (through *d*) the cylinder exhaust port *c*.
Hence *a* and *b* act alternately as steam and exhaust
ports, and when either of them is admitting steam it is
called a *steam* port, while when it permits the steam
to escape out of the cylinder it is called an *exhaust*
port.

The action of a slide valve may, within certain
limits, be varied at will by altering its proportions and
the amount of its travel or sliding motion. It may be
designed so as to let the steam from the steam chest
pass into the cylinder during the whole of the piston
stroke, or so as to close the steam port before the piston
has traveled a full stroke, so that after the valve closes
the steam port, the steam already in the cylinder
expands and drives the piston for the remainder of the
stroke without using any more steam from the steam
chest. This action is called working steam *expansively.*
The point in the piston movement at which the valve
closes the steam port is called the *point of cut-off.* The
steam that enters the cylinder while the steam port is
open is *live steam*, while from the moment the steam
port closes and the steam in the cylinder begins to expand, it is no longer live steam but expansive steam.

Fig. 6 represents the construction of a valve to let
live steam follow the piston during full stroke; A and
C are the cylinder ports, and B is the cylinder exhaust
port; D is the valve exhaust cavity and E and F are
the *bridges*. The edges G and H of the valves are the
live steam edges, because it is their passage over the

respective ports A and C that admits the live steam to
or cuts it off from entering those ports. The inside
edges of the valve, as denoted by the arrows in port D,

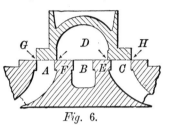

Fig. 6.

are the *exhaust-edges* of the valve, because it is their
passage over the ports A and C that opens or closes
them as exhaust ports. The total width of the valve,
from edge G to edge H, is in this case (to let live steam
follow full stroke) made to just cover the two ports A,
C, so that when the valve is in the middle of its travel
(as it is shown to be in the cut), no steam can get into
or out of the passages A, C. The width of the valve
exhaust port D (that is, the distance between its exhaust
edges) is, in this case, made to extend as nearly across
the two bridges E, F, as is compatible with covering
them sufficiently to prevent the passage of steam from
either C or A into D and E, when (as in the cut) the
valve is in the middle of its travel. But if the valve
be moved from this position, in either direction, both
ports will be put in action, one as a steam and the other
as an exhaust port. Thus in Fig. 7, the valve having

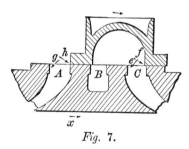

Fig. 7.

traveled in the direction of the arrow, port A is open,
so that steam may enter the cylinder driving the piston
in the direction of the arrow *x*. During this valve movement, the edge *e* of the port C has, in conjunction with
edge *f* of the valve, afforded an opening for the steam
in C to escape, and the edge *e* is, therefore, called the
exhaust edge of the port. On the other hand, edge *g*

of port A has, in conjunction with edge *h* of the valve, afforded ingress to steam at A.   Hence edge *g* is the *steam* edge of the port.

By the time the piston has arrived at the end of the stroke, denoted by *x*, the valve will have traveled back to the position shown in Fig. 6.   When the piston has made one-half its return stroke, the valve will be in the position shown in Fig. 8; both piston and valve moving

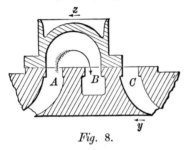

Fig. 8.

in the same direction, as denoted by arrows *y* and *z*. At this time, port C will be full open as a steam port, and port A full open as an exhaust port, the steam passing from A into B, as denoted by the curved arrow. These respective valve and piston movements being repeated, the piston is driven to and fro in the cylinder by steam on one side of it, while that which drove it on the previous stroke is exhausted, through B, into the atmosphere.

STEAM LAP.

To enable a slide valve to cut off the steam supply to the cylinder before the piston has completed its stroke, and thus cause the steam, admitted before the cut-off occurred, to work expansively during the remainder of the piston stroke, what is termed *steam lap* is given to the valve, as shown in Fig. 9.   Here the steam edges,

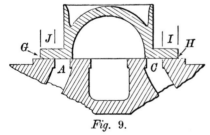

Fig. 9.

G and H, of the valve, instead of barely covering the ports A and C, as in Fig. 6, are prolonged over them

by the distances I and J, respectively, and the amount to which this prolongation or overlapping extends (the valve being in the middle of its travel) is called the *steam lap.*   The measurement of this lap is designated in terms of its length on each side.   Thus if I, and J measure an inch each, the valve has an inch of steam lap on each end.

The *lip* of the valve means that part of the flange, at each end of the valve, that covers the steam port and extends beyond it; thus, in Fig. 9, the lip, at one end, is from the steam edge H to the left-hand edge of the port C, and at the other end of the valve it is that part from the edge G to the right-hand edge of port A.

The action of steam lap, in cutting off the steam supply, is shown in Fig. 10, in which the valve and

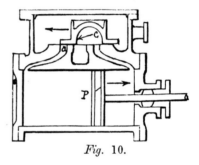

Fig. 10.

piston moving in the direction denoted by the arrows, the steam on the side P of the piston is enclosed by the walls of the cylinder, the face P of the piston, and the valve face at *a;* the resulting expansion of the steam continuing until the edge *c* of the valve meets the edge of the port *a*, at which time the piston will have nearly completed its stroke.

Now, whether a valve has a steam lap or not, it is evidently essential that when the piston is at either end of its stroke, the valve must be in a position to admit steam to one end of the cylinder and permit it to escape from the other end.   Furthermore it is found necessary (in order to prevent the parts from reversing their direction of motion with a *pound, knock*, or *thump*) to prevent all the steam from being exhausted, a certain proportion being enclosed in the cylinder before the piston reaches the end of its stroke, so that the piston, during the latter part of each stroke, has, on one side, a steam pressure forcing it ahead, while, on the other side, it is compressing some of the steam that would otherwise exhaust. The steam, thus prevented from escaping, acts as a cushion on the advancing piston and causes the reciprocating parts (as the piston and its rod, the crosshead and the connecting rod) to reverse their motions easily. This is termed *cushioning* or *compression*, and the amount necessary for the above purpose depends upon the speed of the engine and other considerations that will appear hereafter.

For the same purpose, and also to fill the steam port with live steam at a pressure nearly equal to that in the steam chest, the valve is given what is called *lead*, which will be explained presently.

The valve is operated by an eccentric which is driven by the crank shaft, axle, or main shaft of the engine.

The *throw* of an eccentric is the amount its center is distant from the axis of its bore or from that of the shaft by which it is revolved, and a straight line passing through these two centers is termed the *throw line*. Thus, in Fig. 11, A represents the center of the bore

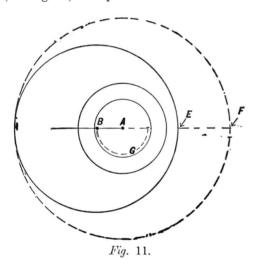

Fig. 11.

of an eccentric, and B the center of the eccentric, hence the radius from A to B is the throw of the eccentric—which equals one-half the travel of the valve. Thus if the eccentric be moved through one-half a revolution on its axis A, its center B would move in the dotted arc G, and the distance it would move its rod would be equal to the radius E, F, or twice the distance A, B.

The throw line A, B, is, for convenience, taken to represent the position of the eccentric, and the throw-line of the crank is taken to represent the position of the crank. When the eccentric throw-line moves in advance of the crank in the path of revolution, it is said to *lead* the crank, and it is obvious that, in this case, the crank and eccentric will, at the beginning of the piston stroke (the crank having moved past either of its dead centers) move in the same direction, whereas, when the eccentric follows the crank, it will (at the beginning of the piston stroke) move in the opposite direction to the crank, the word *direction* referring to right and left, and not to the path of revolution. In a simple slide valve engine, such as in Figs. 1, 2, 3, and 4, the eccentric leads the crank.

VALVE LEAD.

The *lead* of a valve is the amount to which it has opened the steam port when the piston is on the dead center. Lead is given to a valve by advancing the position of the eccentric with relation to the crank. If a valve has neither lap nor lead, the throw-line of the eccentric will stand at a right angle to the throw-line of the crank, as shown in Fig. 12, in which S represents a steam chest, V the valve broken away at the bottom to expose the cylinder ports, P the direction of crank revolution, A the throw-line of the crank, and B the throw-line of the eccentric. To give the valve lead, it would be necessary to move the eccentric forward on the shaft until the valve opened the port to the required amount, in which case its throw-line would stand at an obtuse angle to A instead of at a right angle.

To reverse the direction of crank revolution, all that is necessary is to move the eccentric on the shaft until its throw-line stands at E, or still further according to the amount of the lead required. When the eccentric rod is attached direct to the slide valve spindle, the eccentric throw-line always stands in advance of that of the crank, no matter in which direction the crank is to revolve. This is shown in Fig. 12, in which (motion being supposed to commence from the dead center) it is obvious that B will pass the other dead center in advance of the crank. The amount to which the eccentric throw-line is set forward, or in advance of the crank throw-line, is termed its *angular advance*, and is measured in degrees of angle; in the upper half of Fig. 12

it is shown at a right angle, or an angle of 90° to the crank.

The amount to which an eccentric requires angular advance increases in proportion as the valve is given steam lap and lead. Thus in the lower half of Fig. 12 is a valve having steam lap and lead. The eccentric throw-line is seen to be at an angle of 120° instead of at an angle of 90°, as in the upper half of the figure. The positions of the valves are, so far as the steam port at end S is concerned, the same in both figures, except

In Fig. 13 is shown a valve having exhaust lap, the width at P being less than that at Q, and the amount of exhaust lap being denoted by K or by L. The manner in which exhaust lap operates is shown in Fig. 14, where the piston, moving in the direction of *f*, is near the end of its stroke; the valve, moving in the direction of *d*, is about to open port *a* as an exhaust port. Suppose then that the valve cavity D has exhaust lap added, as denoted by the dotted line, then the valve will require to move still further before the port *a* opens to the

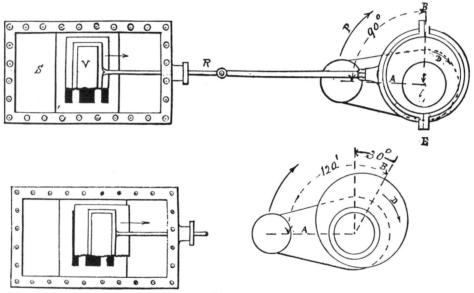

Fig. 12.

that, in the lower half of the figure, the valve is shown to have a slight lead.

The *exhaust lap* of a valve is the amount to which its exhaust cavity or port (when in its mid position over the cylinder ports) covers the cylinder port bridges.

exhaust, and during this valve movement a certain amount of piston movement will occur, and to this amount the steam will have been detained in the cylinder by reason of the exhaust lap.

Thus, then, the effect of lead and of exhaust lap is in

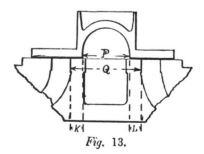

Fig. 13.

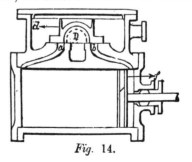

Fig. 14.

the same direction, *viz.*, to cause the steam port to be filled with steam by the time the piston reaches the end of its stroke. But the steam admitted by the valve lead is live steam drawn from the steam chest, while that enclosed by exhaust lap is saved from the exhaust, or, in other words, is a part of the steam admitted to the cylinder on the previous piston stroke. Both, however, obviously serve to arrest the piston motion at or towards the end of the stroke, and, therefore, cause the reciprocating parts of the engine, *i. e.*, the piston and its rod, the cross-head and the connecting rod, to reverse their direction of motion easily or without shock, which results because the pressure on the piston, due to the cushioning, reverses the direction of contact of the journals before the piston reverses its motion. Thus when the piston is pulling the connecting rod, its pressure is transferred through the half brass (on the cross-head and crank pin respectively) that is furthest from the cylinder, while when the piston is pushing the connecting rod, its pressure is transferred through the half brasses that are nearest to the cylinder. Now if the transfer of pressure or contact from one half brass to the other takes place at the extreme end of the stroke, and coincident with the admission of steam, the motion of the piston, cross-head and connecting rod will be reversed suddenly and violently, especially if there be any play or looseness between the connecting rod brasses and the crank pin or cross-head journals. But when cushioning is resorted to, either by means of lead or exhaust lap, the contact between the brasses and their journals is transferred from one brass to the other while the piston is moving in the one direction and is just completing its stroke, so that by the time it has completed it, the brass and journal contact is on the proper side for the ensuing stroke.

CLEARANCE.

Clearance in a valve is the amount to which its exhaust cavity is wider than the extreme width of the bridges, leaving all the ports open when the valve is in mid-position as at A, in Fig. 15. With no steam lap, clearance permits the steam to escape when the port is opened for admission, as at B. This may be avoided by giving more lap than clearance. Clearance reduces the compression, and prevents the engine from thump-

3

ing when the valve has a maximum of steam lap. In position C, for example, the dotted half circle represents no clearance, and it is seen that the compression for port

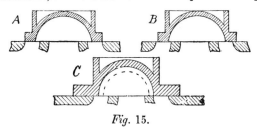

Fig. 15.

c, is delayed by the clearance during a portion of valve movement, represented by the distance between the edge of the valve cavity and the dotted half circle.

VALVE TRAVEL.

The travel of a valve is the amount of its motion over or across the steam ports, and is varied to suit the proportions of the valve. The manner in which a valve will admit steam to the cylinder, and the relation of such admission to the piston movement, is governed by the steam lap, the lead, and the travel of the valve, while the manner in which the steam will be exhausted from the cylinder and its relation to the piston movement is governed by the valve exhaust port, the lead, and the valve travel. Now the least variation in either the lap or travel of a valve has a marked effect upon the disposition of the steam to the cylinder, and the combination of varying proportions that may be given to a valve without varying the dimensions of the cylinder ports are so numerous that only the general effects of each of the elements (as steam lap, exhaust lap, travel and lead) will be at present considered.

In proportion ¡as steam lap is given to a valve its travel and, therefore, the eccentric throw must be increased in order that the port may open fully as a steam port. But the width of the bridges must be sufficient to prevent the valve from moving so far over the ports as to leave less opening at the cylinder port F, Fig. 16, than there is at the exhaust port (as B in the figure) or otherwise the cylinder exhaust port will be cramped, as in the figure where the valve is shown to have over-travel, causing the width at F to be less than that of port B. If the valve have steam lap and no overtravel, the lap permits a freer exhaust, as shown in Fig. 17 where the valve has steam lap equal to the width of the

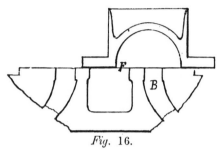

Fig. 16.

port, and is shown in the position it would occupy when the piston was at the end of a stroke, and ready to make the next in the direction of arrow G. The exhaust port B is here fully open, and this shows us that,

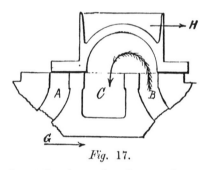

Fig. 17.

unless exhaust lap is employed, steam lap permits the steam to exhaust earlier than it otherwise would do. This is shown in Fig. 18, in which the valve has lap equal to half the width of the steam port, it being obvi-

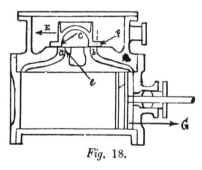

Fig. 18.

ous that the valve edge F must, in any case, meet the port edge *b*, at the dotted line, by the time the piston has arrived at the end of the stroke. In the absence

of steam lap, the exhaust edge *c* of the valve would be coincident with edge *e* of the port when the piston was at the end of its stroke; hence the port *d* would still be closed, whereas, having steam lap to one-half the port width, the exhaust is open to one-half that width, plus the amount of lead when the piston is at the end of its stroke. Since, however, the piston is, at this part of its movement, traveling very slowly, while the valve movement is at about its quickest, the steam is not exhausted much too early in the stroke unless the valve has a maximum of steam lap, or an amount more than equal to the steam port width; in which case the evil may be, to some extent, remedied by giving it exhaust lap.

THE IRREGULARITY OF THE PISTON MOTION.

The action of a slide valve, having an equal amount of steam lap for each cylinder steam port, is not the same for the two piston strokes occurring during a complete revolution of the crank.

This may be seen from Fig. 19, in which the piston is shown in the middle of the cylinder, and the crank at mid position, or half way between its points of dead center. Suppose now that we take the distance from the center A of the crank to the center of the cross-head, and we may mark an arc F, but if we take the same distance or radius, and from the center of the crank pin mark a second arc, it will be at E, showing that the piston will, with the crank in the position shown, be pulled forward to the amount of the distance between E and F on the line of centers of the engine (as the line passing through the center of the cylinder bore to the center of the crank shaft is called). This is due to the connecting rod which, during that part of the piston stroke in which it moves away from the line of centers, retards the motion of the piston, while, during that part of its motion in which its crank end is approaching the line of centers, it causes the piston motion to accelerate.

At all times, except when the crank is on a dead center, or dead point, the connecting rod is at an angle to the line of centers D, D, and the variation of piston speed, above referred to, is said to be due to the *angularity* of the connecting rod, meaning its angle to the line of centers of the engine. The angularity of the eccentric rod or, in other words, its movement out of a straight line, also causes the admission, point of cut-off and release of the steam to vary for the two piston strokes, but in a minute degree only; unless, indeed, the eccentric rod is unusually short in proportion to the amount of valve travel. As the cause is the same for both the connecting rod and the eccentric rod we may

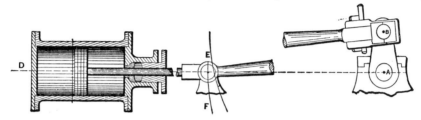

Fig. 19.

further explain it in connection with the connecting rod only.

Thus suppose that in Fig. 20, I represents the center of the cross-head journal when the piston is at half

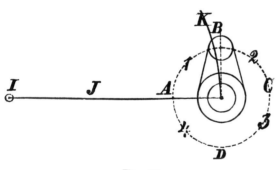

Fig. 20.

stroke and J the length of the connecting rod, and that the dotted line B represents the crank throw line when at half stroke. Let the dotted circle represent the path of the center of the crank pin and 1, 2, 3, 4, respective quarter-revolutions of the crank pin center. If, then, a pair of compasses be set from point I to the center of the crank shaft, and, from I as a center, an arc of a circle be struck it will be denoted by K, and the intersection of K with the dotted circle will be the location of the crank pin center when the piston is at half stroke. This is obvious because when the crank pin center is at A, the piston will be at one end, and when it is at B, the piston will be at the other end of its

stroke, and the point I will recede from the crank center, a distance equal to that from A to the crank center in one case, and to that from C to the crank center in the other case. Clearly, then, while the piston is making the first half of its stroke, the crank pin will move from its dead center at A to the point where the arc K cuts the dotted circle. While the piston is making its second half stroke, the crank pin center will require to move from K to C, and as the length of the eccentric rod is greater in proportion to the valve travel than the length of the connecting rod is to the piston stroke, therefore it moves at a more uniform speed than the piston does, and the points of cut-off, release, etc., of the steam is not timed equally for the two piston strokes.

The nature of the variation (considered with relation to a single piston stroke) will always be that the piston, starting from its farthest point from the crank, will travel more than half the length of its stroke while the crank makes its first quarter-revolution, and less than half its stroke while the crank makes its second quarter. But considered with relation to a full revolution of the crank, the piston will travel the least while the crank is making the half-revolution furthest from the piston, as from B, past C, to D.

Referring now to the motion imparted by the eccentric to its rod, and comparing it with that of the crank and piston movements, let A, B, C, D, in Fig. 21, represent the four quarters of the crank revolution, and E, F, G, H, the corresponding eccentric movements, the dead center A being that farthest from the cylinder, and it will be observed that while either the crank or the eccentric is moving from A to B, the linear motion produced by the rod will be retarded by the angularity of the rod to the line of linear motion, while when either of them is moving from B to C the angularity of the rod will cause its linear motion to be accelerated. From C to D the linear motion would also be accelerated, and from D to A retarded.

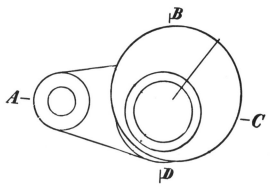

Fig. 21.

Considered, however, with relation to the half-revolution from B, past C, to D, the rod angularity would accelerate, while from D, past A, to B it would retard the linear motion of the rod, or, what is the same thing, of the piston or valve, as the case may be.

But when the eccentric rod connects direct to the valve spindle, its throw line will always be somewhere between B and C when the crank is at A, its exact location depending upon the amount of steam lap and the lead of the valve. It will be seen, therefore, that there is no uniformity between the variation of linear motion given to the piston and valve by their respective rods.

The amount of angular advance (which in the shop is sometimes termed eccentric lead or lead of eccentric) necessary to give to a valve a certain amount of lead varies with the throw of the eccentric. Suppose, for example, that a valve has no steam lap and that it be required to have say $\frac{1}{8}$ inch lead, then the amount of angular advance of eccentric, necessary to give this $\frac{1}{8}$ inch lead, will be less in proportion as the throw of the eccentric is greater. To demonstrate this let A, Fig. 22, represent an eccentric whose throw line may be moved from C to D, the inner circle representing the path of motion of the eccentric center when the throw line is at C and has no angular advance, while the dotted circle is the path of motion of the center when the throw line is at D and has an angular advance of 30°, and it is

clear that moving the throw line from C to D would increase the lead more if the path of the center of the eccentric was on the dotted arc than if it were on the inner circle.

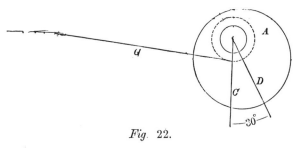

Fig. 22.

Again, the angular advance of an eccentric necessary, under any given amount of valve steam lap, to produce a given amount of lead will vary according to the position of eccentric with relation to the crank, and this varies with the construction of the engine.

It has been shown, in Fig. 12, that when the valve connects direct to the eccentric strap the throw line of the eccentric leads the crank. But when a rock shaft is employed, the throw line of the eccentric follows the crank, thus Fig. 23 represents an eccentric, rock shaft, and valve connection; C represents the crank, and D the eccentric throw line, E the eccentric rod, R the rock shaft, and S the valve spindle, the direction of crank movement being shown by the arrow. If the eccentric rod is attached direct to the valve spindle, without the intervention of a rock shaft, the eccentric throw line would require to stand at F, in which case with the same lap, lead, and travel of valve, the angular advance of the eccentric would be different, on account of the angularity of the eccentric rod. To demonstrate this, Fig. 24 represents two eccentrics, A and B; P represents the crank pin to move in the direction of the arrow; C represents a line at a right angle to the crank throw line, and D the throw lines of the two eccentrics, both standing at 30° angle from C. The circle N represents the path of the center of the eccentric, hence its diameter equals the travel of the valve. Now let the line E represent the center of the cylinder ports (that is the center of the cylinder exhaust port), and the diameter of half circle F (equal to the diameter of the circle N) will represent the travel of the valve. Let

G and H represent the respective eccentric rods of equal length, and attached direct to the valve. Then in moving eccentric A, so that its throw line moves from C to D, the valve will move from E to I, while in moving

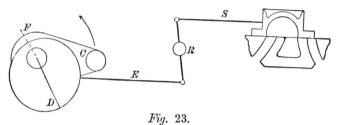

Fig. 23.

the throw line of B from C to D, the valve moves from E to J, which being a greater distance than from E to I, shows that though the two eccentrics have been moved an equal amount, the valve has *not* moved an equal amount. Suppose that the extreme diameter of the half circle F, or the two lines L, M, represent the steam edges of the steam ports, then there will be less lead at J than at I. To make the lead equal, the angular advance of eccentric A would require to be diminished.

the same eccentric, B being the position when the crank pin is on the dead center shown, and A the position when the crank is on the other dead center.

Then let E represent the center of the cylinder ports, and the amount of lead, given by the two eccentrics, varies as the difference in distance of I and J from the diameter of F, as before. To equalize the lead, the rod G may be shortened, but this being done, the travel of the valve would not be equal on each side of the cylinder ports.

The cause of this variation of lead (due to a given and equal degree of angular advance of eccentric) between a directly connected valve gear and one having a

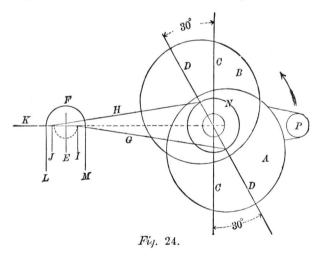

Fig. 24.

Now A occupies the position necessary when a rock shaft is used (the crank leading the eccentric) and B the position when the eccentric is attached direct, hence when a rock shaft is used less angular advance of eccentric is necessary to give a certain amount of lead, though the lap and travel of the valve remain the same.

But we may assume A and B to be two positions of

rock shaft, or is indirectly connected, may be further explained as follows:

In Fig. 25, let c and c′ represent the throw lines of the eccentrics, while the valve ends of the rods are at E, then in moving the throw lines, or the centers of the eccentrics, from c to d and from c′ to d′, the valve end of the rod H moves from E to J, and the valve end of rod G from E to I. Let the lines *d e* and *d′ e′* be perpendicular to the line *c c′* and parallel to the line K, and, of course, they are equal, each being the sine of 30°. Now let the eccentric centers, instead of moving

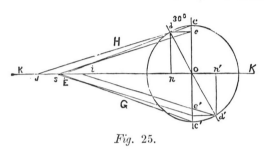

Fig. 25.

along the circumference of the circle, move the one from *c* to *e*, the other from *c′* to *e′*, then the valve ends of both rods will be at the point S, which is farther from the crank shaft than E. Then while one eccentric center moves from *e* to *d* the valve end of rod H moves from S to J, which equals the distance *e* to *d*, since *e d* is parallel to the line K, and the other eccentric center moving from *e′* to *d′* moved the valve end of rod G from S to *i* = the distance *e′* to *d′*, which equals distance *e* to *d*, therefore S *i* = S J, and E *i*, being less than S *i*, is less than E J.

DIAGRAMS OF STEAM DISTRIBUTION.

The manner in which a slide valve opens the ports for the admission and exhaust of the steam, and the general effect produced by various amounts of lap travel, etc., may be very clearly perceived from the following diagrams, more clearly, it is believed, than by any other form of diagram. In our first example suppose the proportions are as follows:

Length of piston stroke — —	24 inches.
Length of connecting rod — —	72 "
Width of steam port — —	1 "
" " bridge — — — —	$\frac{3}{4}$ "
" " cylinder exhaust port —	1 "
Steam lap of valve — — —	0 "
Exhaust lap of valve — — —	$\frac{1}{32}$ "
Lead of valve — — —	$\frac{1}{16}$ "
Width of valve cavity — — —	$2\frac{7}{16}$ "
Travel of valve — — —	2 "

The construction of the diagram is as follows: The line A B, Fig. 26, represents the full stroke of the piston. Line C C is drawn parallel to A B, and is dis-

tant from it to an amount equal to the full width of the cylinder steam port. Line D D is also parallel to A B, and distant from it to an amount equal to the full width of the cylinder steam port. Line A B is divided by

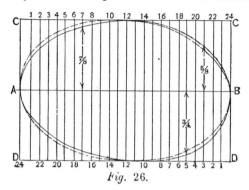

Fig. 26.

the vertical lines into as many equal divisions as there are inches in the piston stroke—in this case, 24 inches. The curved line on the upper half, that is between A B and C C, shows the port opening to admit steam through one cylinder port and the lower curved line shows the exhaust opening of this same port.

The dotted curve above A B shows the port opening for the admission of steam for the other cylinder steam port, and the dotted curve below A B shows the port opening during the exhaust of this second piston stroke. The manner of obtaining these curved lines was as follows: The engine fly wheel was moved around until the piston had moved an inch, the amount the port was open was measured, and this measurement was marked on line 1 above A B. The piston was then moved another inch, and the port opening again measured and marked on the second vertical line above A B, and so on throughout every inch of piston movement for that stroke. Thus at line 7, the piston had moved 7 inches from its dead center A and had left the steam port open $\frac{7}{8}$ of an inch, as is marked in the diagram; or, again, when the piston had moved 21 inches, the port was open $\frac{5}{8}$ inch as marked on the diagram. Through the points thus marked on the vertical lines the full curve, starting from A, passing up to C and ending at B, was marked, thus showing the port opening for every inch of piston movement and for the whole stroke. For the exhaust of this steam, the piston was moved an inch on its return stroke and the width of the same port, acting as an exhaust port, was measured and marked from the vertical line 1 and below A B; the piston was then moved another inch and the exhaust opening again measured, and so on throughout the whole stroke; thus when the piston had moved 4 inches on its return stroke the exhaust port had opened ¾ of an inch as marked on the diagram.

The admission and exhaust of the steam for the other steam port was obtained and marked on the diagram in the same way, but was marked in the dotted curves so as to distinguish it.

In these diagrams, therefore, the actual working of the valves is shown. From the diagram in Fig. 26 we perceive that the steam ports were not opened full until the piston had moved 12 inches on one stroke, and 11 inches on the other stroke, and that the exhaust port was not opened full until the piston had moved 9 inches on one stroke, and did not open quite full for the other stroke.

In the full line diagram, the port opened slower and closed slower for the live steam, but opened quicker and closed quicker for the exhaust steam. There was no expansion (that is to say, the steam followed full stroke) and no cushion, the steam exhausting during the entire stroke. To show the effect of steam lap, let ¾ inch of steam lap be added to this valve which will necessitate making the cylinder exhaust port 2 inches wide instead of 1 as before, widening the valve exhaust port from $2\frac{7}{16}$ inches to $3\frac{7}{16}$ inches and increasing the valve travel from 2 to $3\frac{1}{2}$ inches. In both cases, the travel of the valve is equal to twice the width of the steam port added to twice the amount of the steam lap, and the valve exhaust port has $\frac{1}{32}$ of an inch exhaust lap, while the cylinder exhaust port is equal in width to twice the width of the steam port, added to the amount of steam lap on the valve, these proportions being those which (disregarding the slight error or variation due to the angularity of the eccentric rod) just gives a valve travel sufficient to open the ports fully as steam and exhaust ports.

A diagram of the valve movement, under this new condition, is shown in Fig. 27, and we find the effect of adding the steam lap to be a much quicker port opening to admit steam and, in the full line diagram, a closure of the valve at $19\frac{1}{2}$ inches of piston movement, the

steam being shut in by the valve and working expansively during the next 3 inches of piston movement, the exhaust opened when the piston had traveled about $22\frac{1}{4}$ inches and, therefore, $2\frac{1}{4}$ inches before the piston had arrived at the end of its stroke. The exhaust was

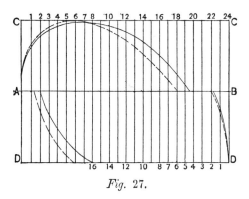

Fig. 27.

full open when the piston had moved ⅜ inch on its return stroke, remaining full open while the piston moved to its 16th inch of movement, and finally closing at the 22nd inch of piston movement, when it shut in the steam, giving 2 inches of cushion or compression.

Comparing one stroke with the other, there is $2\frac{1}{4}$ inches difference in the amount of the expansion, ⅜ inch in the commencement of the exhaust, or point of release, and ⅝ inch difference in the amount of compression.

These variations are due to the angularity of the connecting rod which has been already explained. But in the full line diagram the port has not opened fully, and this is due to the angularity of the eccentric rod.

We may now examine some of the means that may be employed to equalize these differences and to detain the steam longer in the cylinder. First, to prolong the point of release, let us take the same valve, under precisely the same conditions as for the last diagram, and give to it $\frac{5}{16}$ of an inch of exhaust lap, and the diagram of the port openings is given in Fig. 28.

Here we have, as a marked and entirely new feature, the circumstance that the exhaust port opens full and immediately begins to reclose, which is caused by the exhaust lap partly closing the cylinder exhaust port as shown in Fig. 29, at A, where the width is less than it is at B.

It is laid down by most of those who have investigated the subject that the port should permit about one-half more opening for the exhaust than for the live

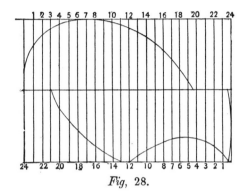

Fig, 28.

steam, because of the diminishing pressure and velocity of the steam in leaving the cylinder. Now if we add up the length of all the vertical lines, measured from the line A B, in Fig. 27, to the line of the upper curve, and divide by the number of vertical lines so meas-

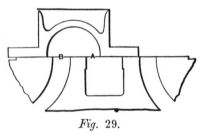

Fig. 29.

ured, we shall obtain the average port opening for the live steam, and, by a similar process, the average exhaust opening may be obtained, but a glance will show that in Fig. 28 there is but little if any difference, while actual measurement will show them practically equal. The exhaust could not, in this case, be free during the early part of the stroke, hence a back pressure would be induced which would more than off-set the advantage gained by the prolongation of the points of release which is shown by the diagrams to have taken place. It will be observed, however, that the addition of the exhaust lap has increased the cushion, or compression, from about 2 inches, in Fig. 27, to 3 inches in Fig. 28.

It is here to be observed that the defect, shown in

the exhaust, may be remedied by increasing the width of the cylinder exhaust port, which would also involve a corresponding increase in width of the valve exhaust port. Or the width of the bridges between the cylinder ports and the valve exhaust port might be widened, which would accomplish the same result.

The points of full port opening, port closure, point of release, and point of compression may be made more uniform for the two piston strokes, by giving to a valve an increase of travel, thus, in Fig. 30, is represented a

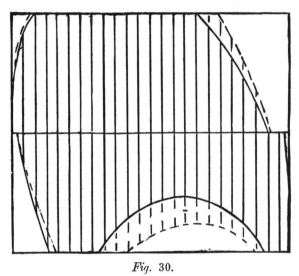

Fig. 30.

diagram of the port openings of a valve motion hav:ng the following elements :

Width	of	Steam ports	–	–	$1\frac{1}{4}$ inches
"	"	Bridges	–	–	1 "
"	"	Cylinder exhaust port	–		$2\frac{1}{2}$ "
"	"	Steam lap	–	–	$\frac{3}{4}$ '
"	"	Exhaust lap	–	–	$\frac{1}{32}$ '
		Travel of valve	–	–	$5\frac{3}{8}$ "

These dimensions represent the average employed upon American passenger locomotives having pistons 16 inches diameter and 24 inches stroke, the port openings and closures are here seen to be very nearly equal for both strokes, but the average exhaust area is less than the average steam area, on account of the valve traveling so far as to partially close the cylinder exhaust port, as shown in Fig. 31 at F, where the opening is less than at B.

The exhaust, however, will not be cramped providing the effective area of the cylinder exhaust port is equal to the area of the nozzle of the exhaust pipe, which, in American practice, is limited so as to produce sufficient draft for the combustion of the fuel in the firebox.

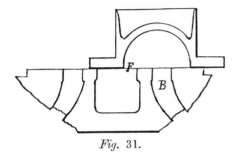

Fig. 31.

Fig. 32 represents a diagram of the port openings of a valve motion, having precisely the same dimensions except that there is $\frac{1}{8}$ inch more steam lap, and the valve travel is $4\frac{1}{2}$ inches only, instead of $5\frac{3}{8}$, the $4\frac{1}{2}$ inches being $\frac{1}{4}$ inch more than twice the width of steam port, plus twice the steam lap.

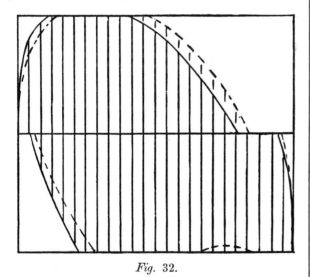

Fig. 32.

The main point of difference between these last two diagrams is that in the first, the average steam opening is 1.07 inches, average width of exhaust opening $\frac{15}{16}$ of an inch. In the last, there is an average steam port

4

opening of $\frac{9}{10}$ of an inch, and an average exhaust opening of $1\frac{12}{100}$ inches.

STEAM EXPANSION.

A slide valve of the form hitherto referred to will not serve to advantage to cut off the steam at a period earlier than at about three-quarters of the stroke, leaving the remaining quarter stroke to be made under expansive steam, because when the amount of steam lap is excessive, the exhaust occurs too early, and furthermore the admission and amount of expansion varies greatly in one stroke as compared to the other.

The object of using the steam expansively is to obtain more duty from it before it is exhausted into the atmosphere ; suppose, for example, that the stroke of a piston is 12 inches, and that after it has traveled 6 of these, the supply of steam to the piston is cut off by the valve, then all the work done by the steam, during the remaining 6 inches of piston stroke, is obtained without taking any more steam from the steam chest, or what is the same thing, from the boiler. The absolute power of a given cylinder is, of course, diminished in proportion as the steam is worked expansively, because the steam pressure decreases in the ratio that its volume is increased, thus suppose we have a cylinder 10 inches long, and that 6 inches of its length is filled with steam at a pressure of, say, 50 lbs. per square inch of piston area, and that, no more steam being admitted, the piston moves down the cylinder, then the steam pressure would diminish as the piston moved, the pressures at the end of each inch of piston motion being as marked in Fig. 33. When the piston had moved from the sixth to the seventh inch, the steam would occupy one-seventh more space, hence its pressure would be one-seventh less, therefore we reduce the 50 one-seventh, obtaining 42.86 as the pressure at the end of the seventh inch of piston motion. When the piston had moved from the seventh to the end of the eighth inch of its stroke, the steam would occupy one-eighth more space than it did at the end of the seventh, or what is the same thing, than it did at the beginning of the eighth inch of its motion ; hence the pressure of 42.86 lbs. would be reduced one-eighth, making it 37.51 lbs. at the end of the eighth inch of the piston stroke, and so on throughout the whole stroke. To obtain the average pressure of the steam throughout the whole stroke, we add together the pressure at the end of each inch of piston stroke and divide the sum so obtained by the

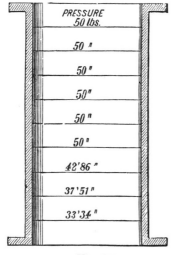

PRESSURE
50 lbs.
50 "
50 "
50 "
50 "
50 "
42'86 "
37'51 "
33'34 "

Fig. 33.

number of inches in the whole stroke which gives the average pressure. Thus

	LBS. PER SQ. IN.
Pressure at end of 1st in. of piston motion	50
" " " " 2nd " " " "	50
" " " " 3rd " " " "	50
" " " " 4th " " " "	50
" " " " 5th " " " "	50
" " " " 6th " " " "	50
" " " " 7th " " " "	42.86
" " " " 8th " " " "	37.51
" " " " 9th " " " "	33.34
" " " " 10th " " " "	30.01
	443.72

and 443.72 divided by 10 equals 44.37, hence the average pressure is 44.37 lbs. per square inch of area.

THE ALLEN VALVE.

If sufficient steam lap be given to a common D valve, (such as has thus far been considered) to enable it to cut off the steam earlier than at about five-eighths of the stroke, the throw of the eccentric requires to be so much increased, in order to obtain sufficient valve travel, that the angularity of the eccentric rod causes the points of cut-off to vary to an objectionable degree; this may be remedied by giving to the valve a different amount of steam lap for each steam port. There are, however, objections to this which will be explained hereafter.

An early point of cut-off, with a minimum of travel and a maximum amount of port opening may, however, be obtained by the use of Allen's valve, which we may now consider with relation to cutting off at some definite point, leaving its employment in connection with the link motion to be treated of in connection with that subject. The construction of the Allen valve is shown in Fig. 34, in which the valve is shown in mid position.

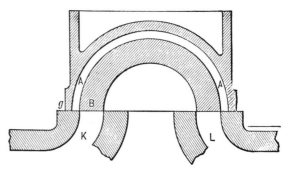

Fig. 34.

A A is a supplementary steam port, which acts to admit steam as well as the steam edge *g* of the valve.

Fig. 35 shows the valve in its position when the

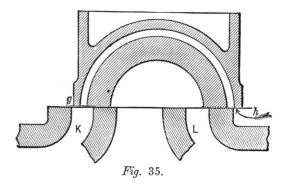

Fig. 35.

crank is on the dead center, and it is seen that when the valve moves to the right, steam will be admitted to port K by the edge *g* and, simultaneously, through the supplementary port as denoted by *h*.

In Fig. 36, the valve is shown at the end of its travel, the port K being closed to the amount of the thickness of metal at *c*, a circumstance that is taken into consideration in determining the width of the steam port.

The exhaust is effected in the same manner as in a simple slide valve, and independently of the supplemental port, as may be seen from the figure.

The steam lap of the valve is the distance from the inner edge (*f*, Fig. 37) of the supplementary port to the outer edge *g*, this being the amount the valve overlaps

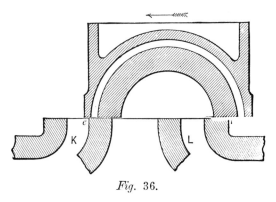

Fig. 36.

the steam port when in its mid position, as shown in Fig. 34.

The steps *a*, Fig. 37, in the valve seat, must be fair with the edge *d* of the supplementary port, when the edge *g* is fair with the outer edge, or steam edge, of port K, so that when the valve moves to open port K the steam will be admitted simultaneously through the

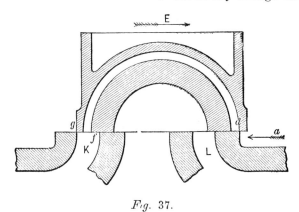

Fig. 37.

supplementary port and through the port opening left by the edge *g*, as the valve moves in the direction denoted by the arrow E.

The width of B, Fig. 34, must be sufficient to cover the steam port and keep it steam tight when the valve is in mid position, and for this purpose $\frac{1}{32}$ of an inch on each side will suffice, making B $\frac{1}{16}$ of an inch wider than the steam port.

The inner edges of the supplemental steam port must be $\frac{1}{32}$ of an inch wider than the extreme width of the outer edges of the cylinder steam ports, so as to isolate the ports when the valve is in mid position.

We now come to the width of the supplementary port, and it is obvious that the thickness at *e*, Fig. 36, must be enough to leave sufficient strength, and, since this thickness covers the port, as seen in the figure, it is desirable to leave it no more than its strength requires, thus leaving the supplementary port as wide as possible.

By the employment of the supplementary port the steam port is opened quicker, remains full open longer and closes quicker. On referring again to Fig. 35, it will be seen that if the valve were moved $\frac{1}{8}$ inch the port would be left open $\frac{1}{8}$ inch at *g* and $\frac{1}{8}$ inch at *h*, and when the valve had arrived at the position shown in Fig. 38 the port will be opened the amount at *g* and

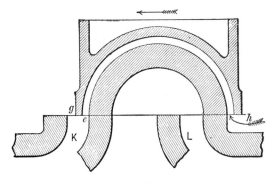

Fig. 38.

the full width of the supplementary port, as seen at *h*, *e*, whereas a common slide valve would have the opening at *g* only. As the valve motion continues, the port opening remains full because to whatever amount the supplementary port closes at *e*, the opening at *g* increases. Similarly after the valve has finished its stroke, and is returning to effect the cut-off, the full port opening will be maintained until the valve has traveled back to the position shown in the figure, the opening at g and at e being equal. Suppose, for example, the valve has traveled back as far as shown in Fig. 39,

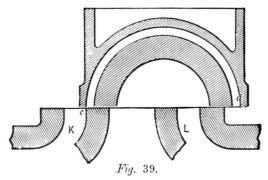

Fig. 39.

and to whatever amount g closes, the supplemental port will open, thus maintaining the full port opening. This will continue until the openings at g and e are of equal width, after which time the end d of the supplementary port will close as rapidly as the edge g does.

We have here considered the working of the Allen valve for late points of cut-off only, and its action may be seen by referring to Fig. 40 and 41, the former

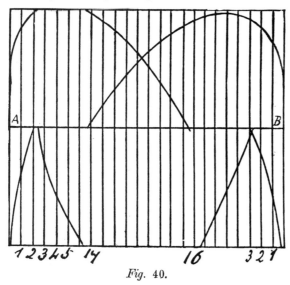

Fig. 40.

showing the port openings of a common slide valve, and the latter of the Allen valve. Both valves have $1\frac{3}{8}$ inches steam lap and $5\frac{1}{4}$ inches travel. The width of port for the common valve is $1\frac{1}{4}$, and that for the Allen

valve $1\frac{5}{8}$, the thickness c, Fig. 36, being $\frac{3}{8}$, leaving the effective steam port opening $1\frac{1}{4}$ inches; as an exhaust port, however, there is a full opening of $1\frac{5}{8}$ inches.

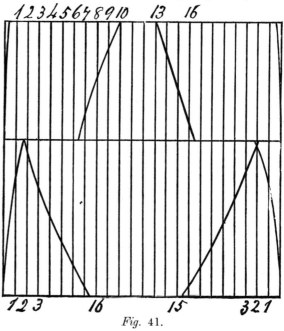

Fig. 41.

A comparison of the two diagrams gives us as follows for the forward stroke :

	Allen valve.	Common valve.
	IN.	IN.
Amount of piston motion before the port had opened full – –	$\frac{1}{2}$	$4\frac{1}{2}$
Inches of piston motion under a full port opening – – –	13	$6\frac{1}{4}$
Point of cut-off – – –	$16\frac{1}{2}$	$15\frac{1}{2}$
Point of release – – –	$21\frac{3}{4}$	21
Cushion – – –	$1\frac{1}{4}$	$2\frac{1}{2}$

It is seen, therefore, that the advantage is on the side of the Allen valve in every particular, and it is to be noted that we may add to the Allen valve sufficient exhaust lap to prolong the point of release to $22\frac{1}{4}$ inches, and thus keep the steam in during $1\frac{1}{4}$ inches more of piston stroke while having the same amount of cushion as the common valve. The two points of cut-off for the Allen valve are at $16\frac{1}{2}$ and at $17\frac{1}{2}$ inches respectively of piston motion, a variation of one inch only, while for the common valve we have the point of cut-off at $15\frac{1}{2}$ inches for one stroke and $17\frac{1}{4}$ for the other, a variation of 1¾ inches. The point of release varies in the case of the Allen valve ½ inch only, while in the case of the common valve it varies an inch.

If we give to the valve sufficient travel to close the supplementary port, as shown in Fig. 42, we may give

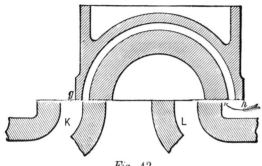

Fig. 42.

to the valve as much steam lap as the common slide valve and get a much better distribution of steam as has been shown, but we may give to the valve only sufficient travel to cause it, at the end of its stroke, to come to the position shown in Fig. 39, the opening at *g* being equal to the width of the supplementary port and thus enable the valve to cut-off at early points in the stroke without the employment of excessive steam lap and valve travel, while obtaining a better steam sup-

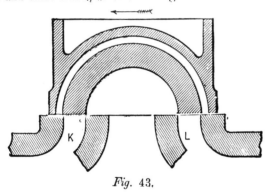

Fig. 43.

ply than with the common valve, and avoiding the irregularities in the points of cut-off of release and of compression that arises when the amount of steam lap is excessive, as 1¼ or 1½ times the width of the steam port.

Fig. 43 shows the valve in mid position, the lap

being so proportioned to the port width that when the valve is at the end of its stroke, as in Fig. 39, the opening at *g* is equal to the width of the supplementary port. To find the amount of lap necessary to accomplish this result and the width of the supplementary port, we subtract the thickness of metal between edge *g* and the supplementary port from the width of the steam port, and divide the remainder by two, and the sum so obtained will be the width of supplementary port necessary to give at *g* opening equal to the supplementary port. By adding the width of the supplementary port to the thickness of the metal at *g* we obtain the amount of steam lap. The correctness of this method will be seen on referring again to Fig. 38, where it is plain that the openings at *g* and *e* are equal, and when added to the thickness of *g* equal the width of the port. If, in order to effect an early point of cut-off, we were to increase the amount of steam lap, we must correspondingly increase the valve travel, and the opening at *g* will be diminished.

In Fig. 43 *a*, we have a diagram of the port openings

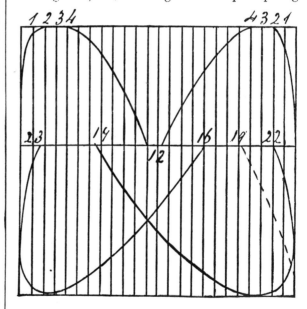

Fig. 43 a.

of an Allen valve, the steam port being 1⅝ inches, but as the thickness at *e*, Fig. 39, is ⅜ inch, the effective port width is 1¼ inches. The steam lap is, therefore,

$1\frac{1}{4}$ inches (width of port $1\frac{5}{8}$ less thickness at e $\frac{3}{8}$ $=1\frac{1}{4}$) and the travel is $4\frac{1}{2}$ inches (twice the width of the port $1\frac{5}{8}$ inches added to the amount of steam lap, $1\frac{1}{4}$ inches, $=4\frac{1}{4}$). Now the common valve, whose port openings were shown in Fig. 40, had $1\frac{3}{8}$ steam lap and yet the steam followed $15\frac{1}{2}$ inches on one stroke and $17\frac{1}{4}$ on the other, whereas in this case there is $\frac{1}{8}$ inch less lap and the valve cuts off at 11 inches on one stroke and $11\frac{1}{2}$ on the other, which occurs on account of the reduced travel. Furthermore, the admission is more full and the cut-off much sharper, evidencing the superior-

ity of the Allen valve for early cut-offs. The exhaust is opened very freely, and would begin at $19\frac{1}{4}$ inches on the forward stroke, but by adding $11\frac{1}{16}$ inches of exhaust lap, the point of release is at the 22nd inch for one stroke and at $22\frac{1}{2}$ inches for the other. The compression occurs at $17\frac{1}{2}$ inches on one stroke and 16 inches on the other, which is not objectionable considering the early part of cut-off, and, also, that the port opens to its full width of $1\frac{5}{8}$ inches when acting as an exhaust port.

——— o : o ———

CHAPTER II.

DIAGRAMS FOR DESIGNING VALVE MOTIONS OR MECHANISMS.

The action of a given slide-valve mechanism may be investigated, or the proportions of lap, lead, travel, etc., necessary to admit cut-off, and exhaust the steam at predetermined points in the piston strokes, may be found by a combination of lines forming a diagram, the principles involved in constructing such diagrams being as follows: The path of motion of the center of an eccentric describes a circle, whose diameter equals the travel the eccentric will give to the valve. Thus, in Fig. 44,

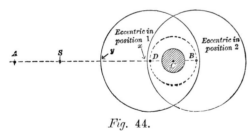

Fig. 44.

is shown an eccentric in two opposite positions. Its shaft center is at C. When the eccentric is in the position marked 1, its center is at D, and its throw is the distance from C to D. If the shaft revolves the eccentric one-half-revolution, the point D, revolving about the center C, will arrive at B, and the eccentric will stand in position 2. If the shaft makes another half-revolution, the point D will again reach the position it occupies in the figure, its whole path being denoted by the dotted circle D B.

That the diameter of this dotted circle is equal to the whole of the movement the eccentric is capable of moving its rod in a straight line, may be shown as follows:

Suppose the eccentric is in position 2, and the direction of rod motion is on the line A B, and that a pencil point be rested against the eccentric at x, then when the eccentric had made a half-revolution it would move the pencil along the dotted line from x to y, and from x to y is the same distance as from D to B. Or suppose that —the eccentric being in position 1—we set a pair of compasses to the length of the eccentric rod and, resting one point at D, we may mark the arc passing through the dot at A. Then resting one point at B we mark dot S, and from A to S is the amount the eccentric is capable of moving the valve, and is also equal to the distance from D to B. If we wish to find how much the eccentric has moved the valve while moving from one point to another in its path, we proceed as in Fig. 45, in which B represents the point the eccentric

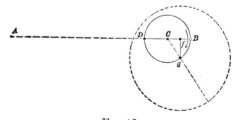

Fig. 45.

started from, and g the point it has arrived at, and to find how much it has moved its rod in a straight line,

33

all we have to do is to mark a line *f* at a right angle to the line A B, thus producing the point *f* on line A B, and from *f* to B is the distance the rod—or, what is the same thing, the valve—was moved on the line A B, while the eccentric center moved from B to *g*.

The path of the crank pin is obviously also a circle, whose diameter equals the stroke of the piston, and may, therefore, be taken to represent it, and if we leave the length of the connecting rod out of mind, or suppose it to be so great as not to cause the piston motion to be accelerated at one part and retarded at another part of its stroke, we might find the position of the piston, for any position of crank pin, by lines in the same way as we have done in the case of the eccentric. In Fig. 46., for

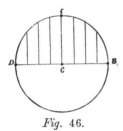

Fig. 46.

example, the circle represents the crank pin path, and *i* the position of the crank pin, the piston being at C. Similarly, throughout the whole figure, one end of each vertical line may be taken to represent the position of the piston, and the other end of the same line will represent the corresponding position of the crank pin. Now, as both the valve and crank pin motions may be represented on a circle, it will be obvious that we might use the same circle to represent the two, using separate dots to denote their relative positions. In Fig. 47,

the valve travel is 1 inch, and the circle will represent it full size, so that every $\frac{1}{16}$ inch on the diameter will represent $\frac{1}{16}$ inch valve travel. Now, suppose that the

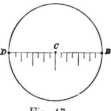

Fig. 47.

piston stroke is 16 inches, and the circle, being but 1 inch, is $\frac{1}{16}$ full size when considered to represent the crank pin path, and, being $\frac{1}{16}$ full size, every division on its diameter will represent an inch of piston movement; hence in using the same circle for both crank pin and eccentric path, all we require to consider is that its diameter represents both on a certain scale, or one full size and the other to a certain scale.

As soon, however, as the length of the connecting rod is considered, a new element is introduced, inasmuch as that the divisions or scale, while still serving to denote piston positions and valve movement, will not serve to denote crank pin positions. Thus, in Fig. 48, we have a crank pin circle, D, I, B, and an eccentric circle, *d b*. The crank pin is shown at G, and the corresponding position for the eccentric center is at *g*. To find the position of the valve we mark the dotted line *f*.

To find, however, the position of the piston we must mark the center line A B, and set a pair of compasses to represent the length of the connecting rod on the same scale as the crank pin circle D B is drawn. Thus, if

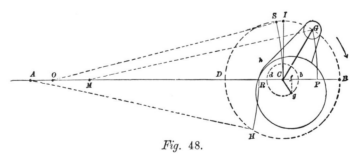

Fig. 48.

for example, we have a circle whose diameter (1 inch) is divided into sixteenths of an inch. Suppose, then, that

this circle is one-eighth the diameter of the actual crank circle, the compasses must be set to one-eighth the actual length of the connecting rod. Then one end of the compasses is rested at the crank pin center G and a mark is made at M, representing the center of the cross-head journal. The compasses are then rested at M, and an arc, G P', is drawn, the point P representing the position of the piston when the crank pin is at G. On the other hand, to find the position of the crank pin from that of the piston we proceed as follows:

Suppose the piston is at C, or at mid-stroke, then the compasses are set at C and point O is marked. Then from O as a center, arc C S is drawn, and S is the position of the crank pin when the piston is at C. The variation between the crank pin and piston position is, therefore, represented by the distance between point S and point I. Similarly, when the piston is at R the crank pin is at H, as is shown by the arc H R.

Now, let it be noted that since the crank and eccentric are fast upon the same shaft, and therefore revolve together, they may be represented by a two-armed lever or bell crank, as denoted in Fig. 49 by a thick line from G to C and from C to g, and that, this being the case, we are enabled from the position of one to find the position of the other, whether the crank pin and eccentric center circles are drawn to the same scale or not.

In Fig. 49, for example, it is supposed that the crank

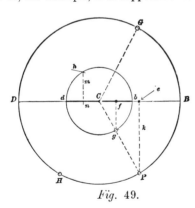

Fig. 49.

pin stands at G and the eccentric center at g, and it is required to find where the eccentric center will stand when the crank pin has moved to position H. To do this, we set a pair of compasses to the radius from G to g, and then, resting one point at H, mark the arc h, and where h cuts the inner circle is the new position for g. If we wish to find how much the valve has moved

5

from its mid-position over the ports, we drop a perpendicular m, and get point n, and from n to C is the amount or distance the valve has moved from mid-travel.

It now remains to show that the circle used for the path of the crank pin can just as well be used for the path of the center of the eccentric as not, and, therefore, that the inner circle can be dispensed with. Suppose, then, that the crank pin started from D and has arrived at G, and the center of the eccentric will be at g, and the valve will have moved three-quarters of its whole travel, or, in other words, the amount of its motion from the end of its travel is from d to f, which is three-quarters of its whole travel from d to b. Instead, however, of letting the inner circle represent the path of the center of the eccentric, suppose that the outer circle does so, and we may prolong the throw line of the eccentric so that it runs from C to P. The diameter of the outer circle will then represent the valve travel, and to find where the valve stands we draw a line K, giving us the point e, distant from D three-quarters of the diameter of circle D B; hence we find that the valve has traveled three-quarters of its stroke, whether measured on the inner or on the outer circle; hence the outer circle is a perfect substitute for the inner one.

We have, in this case, found the position of the eccentric center from that of the crank pin; but suppose that we have an engine whose piston travel is 12 inches, length of connecting rod 24 inches and valve travel 2 inches, and we may find the position of the piston for any given position of crank pin as follows:

In Fig. 50 is a circle of an inch diameter, and, there-

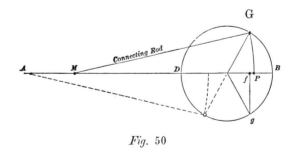

Fig. 50

fore, equal to one-twelfth the piston stroke, while it represents half the valve travel. Let the crank pin be at position G and the eccentric center at *g*, and setting a pair of compasses to a radius of 2 inches, so as to represent the length of the connecting rod on the same scale as the diameter of the circle represents the piston travel (that is, one-twelfth full size), we rest one point on the line A B (representing the center line of the engine), and mark the arc G P, and P is the position of the piston when the crank pin is at G and the valve is at *f*. Now, suppose the crank pin is at H, and we may find the corresponding piston position by a similar process, as is shown in the figure by dotted lines.

The effect of this variation between the position of the crank pin and that of the piston is shown in Fig. 51, in which it is supposed that the point of cut-off is

mechanism is so constructed that the cut-off occurs at equal points of crank movement, it will occur at unequal points of piston movement; and, conversely, if it occurs at equal points of piston motion, it will occur at unequal points of crank motion. The position of the valve is represented at the top of the figure on the small circle, *d b*, (this circle representing the full valve travel) at point *f*, the valve having moved from *b* to *f*, which is equal to the width of the port.

It is obvious that the angularity of the valve rod will vary the motion of the valve in the same manner as the angularity of the connecting rod varies that of the piston, but the length of the valve rod is usually so great in proportion to the valve travel that the error is

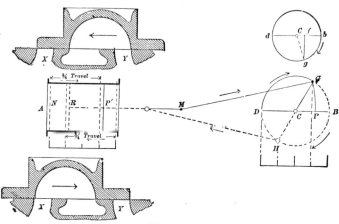

Fig. 51.

for one stroke at G and for the other at H. The valve position, when the crank pin is at G, must be as shown at the top of the engraving, the valve traveling as denoted by the arrow and port *x* being closed. The corresponding piston position is shown in the cylinder at P′, and also within the circle at P, the distance from N (representing the beginning of this stroke) to P′ in the cylinder being the same as from D to P measured across the circle, and it is seen that while the crank has moved three-fourths of its stroke the piston has moved more than three-fourths of *its* stroke.

Similarly, when the crank pin is at H, and has moved three-quarters of the distance from B to D, the piston will be at R, and has traveled less than three-fourths of its stroke, and it becomes evident that if the valve

too minute to have any practical or appreciable importance, and may therefore be discarded.

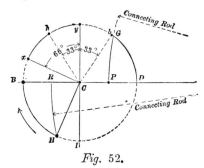

Fig. 52.

Now, suppose we have a valve mechanism in which the valve has neither lap nor lead, and that the circle shown in Fig. 52, represents the path of the center of the eccentric drawn full size, and also the path of the crank pin drawn to some scale, as, say, one-tenth full size, and as the valve has no lap, the throw line of the eccentric will stand at a right angle or angle of 90° to the crank. Suppose the crank stands at H C, and the eccentric will stand at x, its throw line being at x C, which is at a right angle to the crank. As the eccentric and crank are fast on the same shaft, the length of arc passed over by both will be equal in equal spaces of time; thus, while crank pin H moves to B, the eccentric center will move from x to y, arc $x\ y$ being equal in length to arc H B.

Now, suppose that instead of the valve having no lap, and the live steam following the piston during its full stroke, it is required that when the crank pin is at H the steam is to be cut off, and the remainder of the stroke is to be performed with the steam already in the cylinder acting expansively, and then the arc $x\ y$ must be passed over by the eccentric after the steam valve closes the steam port and cuts off the steam supply at the head end of the cylinder (the end furthest from the crank being termed the head end of the cylinder). Hence, during that part of the eccentric motion from x to y, the valve must have lap enough to keep the port closed at both its steam and exhaust edges, so that steam can neither get into nor out of the cylinder.

As a given amount of eccentric motion gives more motion to the valve in proportion as the throw line of the eccentric is nearer to its mid-position y, therefore, instead of leaving the eccentric (as at x) at an angle of 90° to the crank, or 66° from its mid-position, suppose it to be given an angular advance equal to one-half the arc $x\ y$, or 33°, in which case when the crank pin is at H, the eccentric will be at h. When, therefore, the crank has moved to B (at which time steam is to be admitted to the other steam port) the eccentric will have moved to b and will stand at 33° ahead of its mid-position y. Thus it will be seen that, although the length of arc $h\ b$, traveled over by the eccentric while the steam port is closed, still equals in length the arc H B, traveled over by the piston during the period of expansion, yet by setting the eccentric ahead, giving it the angular advance from x to h, its motion, during the period of expansion, has been brought equally on each

side of y, causing it to move the valve quicker and to give it a greater amount of motion while it is acting to cut off the steam for the port at one port, and traveling to open the other for admission.

Referring to Fig. 53, in which H represents the position of the crank pin at the point of cut-off, and h the position of the eccentric at the same instant, then from d to D represents the amount of valve travel that occurs after live steam is cut off, and from d to C represents the amount of valve lap required to cut off when the crank pin is at H and the piston is at R. Similarly for the other stroke, the crank pin starting from B, the dis-

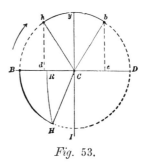

Fig. 53.

tance from C to e represents the amount of lap the valve must have to cause the steam to be cut off when the crank has arrived at a point exactly opposite to H. In this construction we have found at h the eccentric position for crank position H, and then moved them around the circle at the same distance apart, just as though we were moving a two-armed lever on the center C. But we may move the eccentric back from h to H, thereby making H represent both the crank and the eccentric at the same time, and, in this case, all the points and lines on the diagram, that relate to the eccentric and valve, would be turned back to the same amount. Therefore, if h, Fig. 53, becomes H, Fig. 54, then b, Fig. 53, will become B in Fig. 54, and the point of mid-eccentric travel y will be midway between H and B, or y, Fig. 54. So, likewise, the line y C, Fig. 53, will become y C in Fig. 54. Now the line on which the travel of the valve is shown must be at a right angle to the line y C, and is, therefore, shown in Fig. 54 by the line v V, and the lines H d and b e (corresponding to lines h d and b e in Fig 53) may be drawn, giving in their distance apart the total amount of lap, one-half, as *d* C, being for one port, and the other, C *e*, serving for the other port.

Thus the movements of piston, crank and valve may be shown on the diagram, the position of the piston for any position of crank being shown on the line, B C D,

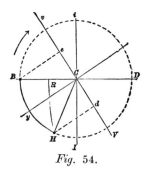

Fig. 54.

as before, while the positions of valve corresponding to those of the eccentric may be shown on the line V C *v*. But it must be borne in mind, that while B *i* D is the line of motion relating to the crank, line V *v* is that relating to the valve motion, and that this line stands at the same angle to the line B C D for piston travel that the eccentric is in advance of the crank. Thus, in Fig. 54, V C is at an angle of 123° from the crank H C, and hence represents an angular advance of eccentric amounting to 33°, because without angular advance the eccentric would stand at 90°, and 123° less 90° is 33°.

In Fig 55, we have a diagram constructed upon the

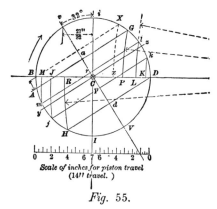

Scale of inches for piston travel
(14″ travel.)

Fig. 55.

foregoing principles, this particular form being that employed by Mr. J. W. Thompson, of the Buckeye Engine

Works. The circle is drawn to a diameter equal to the full travel of the valve, which is the most convenient, although it is obvious that the larger the circle, the more accurate are the results obtained. The piston stroke being, in this example, 14 inches, a scale, equal in length to the diameter of the circle, is constructed below it. Now, suppose a slide valve is to be designed which will cut off, say, at three-quarters stroke, and do so equally on both strokes, then set off on line B D, a point P three-fourths of the diameter of the circle distant from B, and set-off R three-fourths of the distance from D to B, then P and R, will represent the piston positions at the points of cut-off. With compasses set to represent the length of the connecting rod—in this case 2½ times the diameter of the circle—mark arcs P G and R H, locating at G and H the corresponding crank pin positions. From H draw the line H D, meeting the circle exactly at D, if the valve is to have no lead at the head end of the cylinder, (the head end being represented by end D of the line B C D).

Now, since the admission is to occur when the piston and crank are at the beginning of this, which we may term the forward stroke, and steam is to be cut off at H, then the crank must pass over that part of the circle lying between D and H, while the port at the head end is open.

Now, since the same point is taken to represent the eccentric and the crank, therefore when the eccentric is at D, the edge of the valve is moving away from the steam edge of the port to open it, but, by the time the crank has arrived at H, the edge of the valve will have first moved over the port, leaving it full open, and then moved back again to close it, just closing it at the time that the crank has reached point H; hence the line *v* V, drawn through the center C and at right angle to H D, will divide the arc D V H equally, and that part of the arc from D to V will represent the period during which the valve is traveling to open the port, while the part from V to H will represent the period during which the valve is traveling to close the port and effect the cut-off; hence V *v* represents a center line on which the valve is moving and on which the positions of the valve at any point of the stroke may be located and on which all laps and leads must be laid out. The point C, of course, still represents the center of the valve travel, and a line, drawn through C and at a right angle to V*v*, locates line *y z*, which, relatively to line V *v*, is the point of mid-travel of the valve.

To find the amount of valve lead necessary to be given to the valve at the crank end, in order to equalize the point of cut-off for the two piston strokes, draw the line G A parallel to line D H. The amount to which end A, of line G A, falls below B (that is, distance A B) is the required amount of lead. This will appear on applying similar reasoning to that already given when supposing that if, on the return stroke, the valve at the crank end opens when the piston and crank are at B, then—as motion occurs in the direction of the arrow—when the eccentric has reached its center line V *v*, and is at *v*, the valve will begin to close, and would finally close when the crank and eccentric are at the distance *v* B, from *v*, or at point X, corresponding to a piston position of *x*, which is too soon, as the piston should have reached P, and the crank have reached G, and to reach that result admission must begin when the crank is at A, thus making the length of arc A *v* equal to *v* G, in order that the point of cut-off may not occur until the crank reaches G.

It will be seen that to equalize the points of cut-off, then, it has been found necessary to give lead at the port nearest to the cylinder, that being the port which receives steam when the crank is at B, and that there is no valve lead at the other port which corresponds to the end D of the piston travel. Had we commenced at the other stroke and drawn G A first, letting it meet the point B, then by drawing line D H parallel to G A, D H would cut the circle below D, showing negative lead, or, in other words, that port would not have opened until the piston had passed the dead center D and reached the line H D. Or line B G might be drawn to give negative lead at both ends by simply drawing it from point G to a certain amount above the point B, and then by drawing line H D from H to a point as much below D as A was drawn above B, there would be equal negative lead at both ends, either negative lead or else unequal lead being the price that must be paid for equalizing the point of cut-off by employing unequal lap. Obviously, however, the amount of inequality of lead induced by the unequal lap is repre-

sented by the length of arc from A to B, and we may throw this all at one end, as in the figure, or divide it between the two ends as pointed out.

We may now provide for the point of release and the point at which compression is to take place. Let it be required, then, that the compression shall begin when the piston is within, say, 1¾ inches of the termination of the stroke, and we set off, on the line of piston travel, points J and L, distant from points B and D 1¾ inches, according to the scale of piston travel. From these two points, with compasses set to represent the length of the connecting rod, we draw the arcs L *l* and J *j*, locating the corresponding crank pin positions *l* and *j*. From *j* draw line *j k* parallel to H D, and with compasses set to represent the connecting rod length, draw arc *k* K, locating point of exhaust K.

For the point of exhaust on the other stroke, draw from point *l*, line *l m*, parallel to H D, and from *m*, with the compasses set as before, draw arc *m* M, locating the point M where the exhaust is to begin. The reason that the line *j k*, drawn from *j*, or the point where the crank is when compression commences on the crank end of the cylinder, locates the point *k* where the crank is when the exhaust begins at the same end, that is, supposing the crank to be at *j*, and the eccentric at the same point, the compression is commencing with the eccentric lacking distance *j y* of being at mid-throw; hence, as the same edge of the valve that closes the port for compression on one stroke, opens for the exhaust on the other stroke, the eccentric must reach, on its return stroke, the same distance *j y* past the opposite mid-position Z, or, in other words, reach the point K.

Now, since line V *v* represents the line of travel of the valve, Y and Z represent the mid-throw positions of the eccentric, and C the mid-position of the valve, and starting out with a valve having no lap when the eccentric and crank are at H, line Y Z representing mid-throw of eccentric, then the valve would be the distance *d* C from mid-position; hence, exhaust lap, equal in amount to radius *d* C, would be required for the port at the head end in order to close it.

Likewise, when compression is to begin at the crank end, with the eccentric and crank at *j*, or distance *j y* from mid-throw, the exhaust edge of the valve would be the distance *p* C from the exhaust edge of the port, and exhaust lap equal to distance p C would have to be given to the crank end of the valve. Similarly, with the cut-off at G and compression at l, exhaust lap, equal to distance r C, would be required on the head end of the valve. To summarize the data thus arrived at, we have, in Fig. 55, C a as the steam lap at the crank end, and C d that at the head end, while C p and C r are the corresponding exhaust laps, all these measurements being taken on the line V v, the lesser amount of exhaust lap (C r), it will be noticed, belonging to that port of the valve that has the most steam lap.

The angular advance of the eccentric is equal to arc V i or 32°, but, for practical purposes, it is more convenient to express it in terms of the amount the valve is displaced from its mid-travel when the crank is on its dead center. This amount is found by measuring from v to line i C in a parallel direction, and is shown in Fig. 55 to be, in this case, $\frac{2}{3}\frac{1}{2}$ inch.

The amount of steam lap being greater for one port than for the other, it is obvious that, in order that both

letters of reference. It is assumed to be at mid-travel, at which position, alone, will the laps, as obtained from the diagram, show in their proper amounts. To draw, or lay out, such a valve from the diagram, draw the line A B, representing the face of the valve and of the ports, and draw the inside edges C of the cylinder ports equi-distant from the center of the cylinder exhaust port. Next draw the width of ports P P; the valve length, without any steam lap, would then be X added to P P, or distance W. Steam lap C d, corresponding to C d on the diagram, may then be marked for the head end of the valve, and similar lap a C at the crank end. Also exhaust laps C p and r C (corresponding to C p and r C on the diagram) may then be drawn. As the lip H of the valve (corresponding to the width of the port plus the laps $d\,r$ in Fig. 55) is wider than lip L (corresponding to $a\,p$ in Fig. 55), it is important that the valve be placed the proper end foremost, as, were end B placed nearest to the crank, the points of cut-off, compression, etc., instead of being equalized by the un-

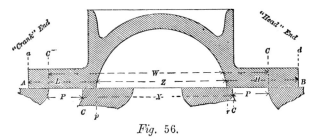

Fig. 56.

ports may open full for the admission of steam, the width of port should equal the greater amount of steam lap, or, in this case, C d, assuming average conditions as to speed, load and pressure, etc. In other words, it is assumed that the area of steam port has been determined as that most desirable for the average conditions under which the engine is to run, and if this area is altered, then the other elements of the diagram must be proportionally altered. Thus, the width of port has been taken as equal to distance $v\,a$; but suppose that this was afterward considered insufficient, and a greater port width be used, then the travel of the valve must be proportionally greater, as also must the laps.

In Fig. 56 we have a valve with the proportions arrived at from the diagram, and with corresponding

equal laps, would be distorted worse than would be the case, if the laps were made equal. This, fact, together with the excessive lead inequality, leads to the consideration of other plottings, in which cut-off equalization is abandoned either wholly or in part.

Thus, in Fig. 57, having located points of cut-off at P and R, perpendiculars P G and R H are drawn, locating the corresponding crank pin positions, supposing the length of the connecting rod to be infinite, or, rather, leaving it out of the question. Line H D is then drawn, and line G B (corresponding to line G A in Fig. 55) parallel to H D, and passing exactly through B (instead of below it, as in Fig. 55), thus showing equal lead, or, rather, no lead at either end. From points J and L, where it has been determined that the compression is to begin, and with the compasses set to represent the length of the connecting rod, strike arcs locating *j* and *l*, and proceed at was explained with reference to Fig. 55, all points being similarly lettered in the two diagrams.

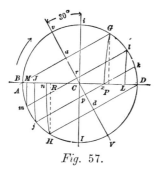

Fig. 57.

Here, then, the lead and compression are equalized for the two strokes while the exhaust is nearly equalized; arcs drawn through H and G, giving points *x* and *n* where the piston will actually be when the cut-off occurs, thus giving an inequality of R *n* at one end, and P *x* at the other, or both together equal to P *x* of Fig. 55. A valve mechanism thus proportioned is suitable for engines in which the cut-off is effected by a separate valve which is capable of independent equalization, in which case the inequality in the points of closure by the main valve is of no consequence, since it may be corrected by the cut-off valve. In this case, also, the lips *d r* and *a p* are of unequal width, but that nearest to the crank (C *a*, Fig. 56) is the widest. This arises from the fact that the exhaust lap is added at C *p* to equalize the compression, while no excess of lead

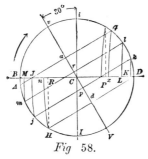

Fig 58.

has been given at A to equalize the points of cut-off; hence, this valve also requires to be placed the proper

end foremost, and since it is generally preferred to have equal lead at each port, the only way to secure at the same time equal compression is to provide it in the manner here shown in Fig. 57.

If the valve is to have equal lips, so that it may be turned end for end on its seat, and, therefore, cannot be put on wrong end foremost, it may be plotted as shown in Figs. 58 and 59. Fig. 58 is constructed in the same way as Fig. 57, the compression being equalized because J and L are equi-distant from B and D respectively. But since the width of lip of any valve is equal to the width of the steam port plus the amount of steam and exhaust lap, therefore the lip of valve, designed as in Fig. 57, would, at end furthest from crank, be width of steam port plus exhaust lap C *r* and steam lap C *d ;* and at crank end would be width of port plus exhaust lap C *r* and steam lap C *a*. Now, suppose enough to be cut off C *a* to make both lips of equal width, and the construction is shown in Fig. 59, in which *a* (corresponding to *a*, Fig. 58) has been made

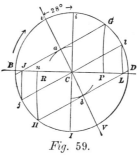

Fig. 59.

the same distance from *p* that *d* is from *r* in the same figure, thus making *a p* equal to *r d* (or lip· H equal to lip L in Fig. 58). A G is parallel to H D as before, but is drawn through *a*, bringing point G further around, thus changing the motion of the piston at the point of cut-off from *x*, Fig. 57, to *x*, Fig. 58. It will be observed that the distance from B to *n*, Fig. 58, is very nearly the same as that from D to *x*, showing that the cut-off is nearly equalized, that the compression is equalized, that the exhausts are good and the lead inequality is only about half that shown in Fig. 56 at B A. The valve is, therefore, symmetrical, and effects a compromise between the excessive lead inequality of Fig. 56 and the great cut-off inequality of Figs. 57 and 58, and, but for the objection that the unequal lead would be too apt to be taken as a defect, it would constitute the best arrangement, as a whole, that could be made.

DR. ZEUNER'S VALVE DIAGRAM.

The principles upon which the foregoing diagrams are constructed, are derived from the form of diagram invented by Dr. Zeuner, and, in order to further explain the base upon which such diagrams are plotted, it it may be as well to trace out the path of the crank pin in connection with the valve positions at the time the

From these dimensions it is required to find the positions of the crank and eccentric for the various events during the stroke, and this may be done as follows:

Let the dotted circle in Fig. 60 be 4 inches in diameter, representing the stroke of the valve drawn full size and the piston stroke one-fifth full size, and its circumference will represent the path of the center of the eccentric drawn full size and the path of the crank pin drawn one-fifth full size. Now suppose the crank to be on the dead center B and to require to move in the direction denoted by the arrow. Suppose the valve was

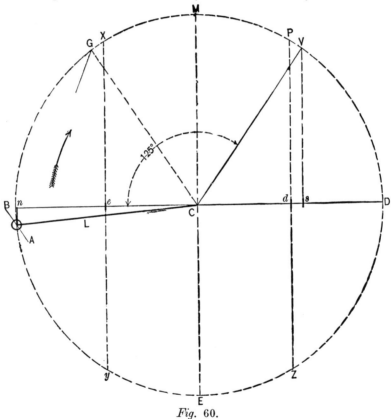

Fig. 60.

events of cut-off, etc., occur. Suppose then, that the engine has the following proportions:

Length of piston stroke	–	–	20	in.
Width of steam ports	–	–	–	1	"
Steam lap	–	–	–	–	1	"
Lead at head end	–	–	–	$\frac{3}{8}$	"
" " crank end	–	–	–	$\frac{1}{8}$	"
Travel of valve	–	–	–	4	"

placed in its mid-position, as in Fig. M, and the throw-line of its eccentric would stand on the line C M in Fig. 60. As the port a, Fig. M, is the one that must be open to the amount of the lead when the crank is at B, Fig. 60, and as the amount of the lead is given for this port as being $\frac{1}{8}$ in., the valve must be moved to the right to the amount of the lap and the lead, or $1\frac{1}{8}$ in.

To find the position of the eccentric when the valve is thus moved to be open for the lead, we mark a point *s,* Fig 60, distant from C to the amount of 1⅛ inches (lap 1 inch, lead at this end ⅛) and erect a perpendicular

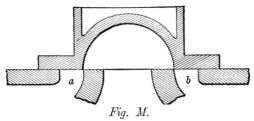

Fig. M.

line from *s* to V. We then draw a line from V to C, and this will represent the throw line of the eccentric with the valve open to the amount of the lead and with the eccentric in the position it must occupy when the

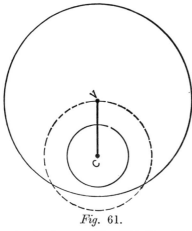

Fig. 61.

crank is on its dead center B. This line V C is shown drawn on the eccentric in Fig. 61, V representing the center of the eccentric and C the center of the crank

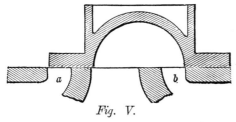

Fig. V.

shaft. The position of the valve when the crank is at B, Fig. 60, and the eccentric at C V, is shown in Fig. V,

6

in which the valve and ports are shown one-quarter full size, the port *a* being open to the amount of the lead. We may now trace the events throughout the stroke as follows:

Suppose the eccentric to have moved from V to D, Fig. 60, and it will have arrived at the end of its stroke, leaving the port *a* full open, as in Fig. D. To find the corresponding position of crank, we may set a pair of compasses to the radius B V, Fig. 60, and resting one point at D (where the eccentric is) mark an arc G,

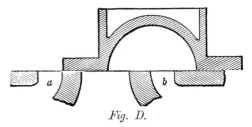

Fig. D.

and a line from G to C will represent the crank position. The eccentric, after leaving position D, begins to move the valve back, and when it has arrived at Z, Fig. 60, will have moved it a distance equal to the amount of the steam lap, thus effecting the cut-off, the position of the valve at this time being shown in Fig. Z.

The point Z is obtained by the following reasoning: Suppose the valve is at the end of its stroke, as in Fig. D (the eccentric being at D, Fig. 60), and it is obvious

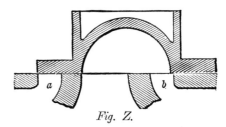

Fig. Z.

that it must move to the left to an amount equal to the width of the port, or in this case an inch, before it closes the port *a* and effects the cut-off, we, therefore, mark point *d* an inch from D, and then draw the vertical line from *d* to Z, showing at Z the eccentric position when the valve is in the position shown in Fig. Z.

The next event that will occur is for the port *a* to open for the exhaust, and it is obvious that this will take place when the valve has moved from its position in Fig. Z to an amount equal to the amount of the steam lap which is, in this case, an inch; hence, we measure off an inch to the left of *d*, Fig. 60, arriving at C, and a vertical line drawn from C to E gives us the position of the eccentric which is then in mid-position, the valve, also, being in mid-position, as in Fig. E, the exhaust being about to open for port *a*.

The next event will be that port *a* will be open full as an exhaust port, and to do this it must move an amount equal to the width of the port (an inch), we therefore mark *e* an inch from C, and draw the perpen-

close *a* as an exhaust port, and to do that it must move back to an amount equal to the width of the port (an inch), passing from the position in Fig. B to the position shown in Fig. X. To find the corresponding eccentric position we mark, on Fig. 60, an inch to the right of B, arriving at *e*, and a line *e* X gives the eccentric position when the valve is in the position in Fig. X and port *a* is about to begin to close to the exhaust. The next event that will occur is for the port *a* to be closed as an exhaust port, and in order to effect this the valve must move from its position, in Fig. X, to its mid-posi-

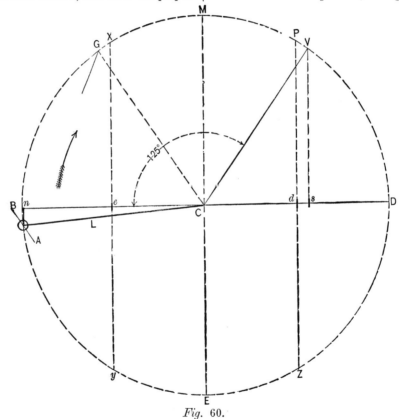

Fig. 60.

dicular line *e* Y, which is the position of the eccentric when the port *a* is open full for the exhaust, as shown in Fig. Y. Continuing the motion and considering the port *a* only, the next event is for the eccentric to move to the end B of its stroke in Fig. 60, and the corresponding valve position is shown in Fig. B. The next event is for the valve to move back until it begins to

tion, as in Fig. M, the distance moved obviously equaling the amount of the steam lap, as will be seen on inspecting the two figures.

As the amount of steam lap is an inch, and C is an inch from *e*, we mark a vertical line C M, giving at M the position of the eccentric when the valve is in the position shown in Fig. M, both valve and eccentric being in mid-position. The port *a* being closed, the compression has begun, and we have found the eccentric position for every event during this stroke, except the crank position at the time the valve is about to open for the lead, and this we may find from the following reasoning :

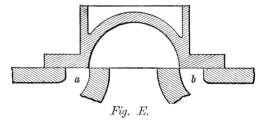

Fig. E.

When the eccentric is at M, Fig. 60, and the valve in mid-position, as in Fig. M, it will require to move to the amount of the steam lap in order to bring the edge of the valve coincident with the edge of the port, as in

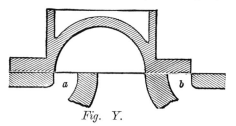

Fig. Y.

Fig. Z, ready to open for the lead, and the amount of steam lap being an inch we measure, in Fig. 60, an inch from the line M C and arrive at *d*. From *d* we draw the line *d* P, giving at P the position of the eccentric

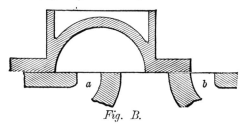

Fig. B.

when the valve is in the position in Fig. P, and the valve is about to open for the lead. To find the crank position, we set the compasses to the radius B V (the lines B C and V C, representing the angle of the eccentric to the crank, or, what is the same thing, the angle of the crank to the eccentric), and from P as a center

mark an arc A, and from the point where this arc cuts the dotted circle we draw a line to C, which will give the position of the crank when the eccentric is at P, Fig. 60, the valve in the position shown in Fig. P, and the port *a* about to open for the lead.

We might find the crank pin position for any other

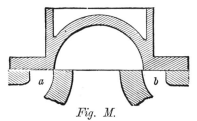

Fig. M.

eccentric position by similar means, because as soon as the line V C is obtained, we have found the position of the eccentric with relation to the crank, and as both are fast on the same shaft their positions with regard to

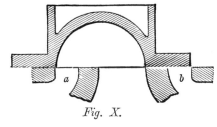

Fig. X.

each other will always be the same wherever either of them may be. Thus, in Fig. 60, we found the position of the eccentric, when the crank was at B, to be at V C, which is 125° ahead of the crank, hence having found

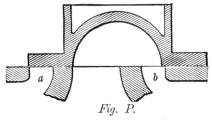

Fig. P.

any eccentric position the corresponding crank position will be 125° behind it.

We may now consider the return stroke, the crank being at D, and we may suppose the lead at the port B, Fig. 62 (which is the one that must now act as a steam port), to require to be ⅜ inch. To find the position of the eccentric when the crank is at D, we mark *s* distant from C to the amount of the lap and the lead (or 1⅜ inches, the lap being 1 inch and the lead ⅜ inch). From *s* we draw the line *s v*, and a line from *v* to C is the required eccentric position, being 135° ahead of the crank as marked, and the valve will be in the position shown in Fig. V, port *b* being open to the amount of the lead. When the eccentric has moved to the end B of its stroke, the valve will obviously be at the end of its stroke, as in Fig. B.

The next event will be for the valve to move back to

valve must move an amount equal to the width of the port or distance C *d* before the port *b* will be full open for the exhaust, hence from *d* we obtain eccentric position Z, Fig. 62 and valve position Fig. Z.

From position Z the eccentric moves to the end D of its stroke, the valve moving to the position in Fig. D. When the eccentric arrives at Z, the valve will have moved back to the position Fig. Z and port *b* will begin to close the exhaust. Then, moving the width of the port or distance from *d* to C, Fig. 62, the eccentric will be at M′ or mid-position, the valve standing in its mid-

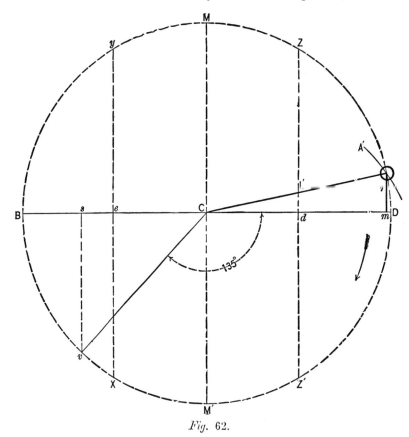

Fig. 62.

an amount equal to the width of the port or distance B *e*, Fig. 62, giving us eccentric position *y* and the valve position shown in Fig. Y. The next event is the opening of port *b* to the exhaust, the valve moving from the position in Fig. Y to that in Fig. M, and the eccentric standing in mid-position M, Fig. 62. The

position, Fig. M, and closing port *b* for the compression to begin. We have thus found the eccentric position for all the events of the stroke, except the crank position at the time the valve is ready to open for the lead, and this we find as follows :

With the valve in mid-position, as in Fig. M, it

must move to the amount of its steam lap before it can open to the lead, hence from eccentric mid-position M' C, in Fig. 62, measure off, on the left of C, point

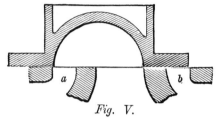

Fig. V.

e, distant from C to the amount of the steam lap. From e we draw the vertical line e X, giving at X the eccentric position when the valve is in the position Fig. Y.

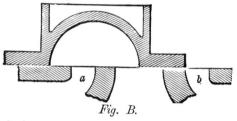

Fig. B.

To find the corresponding crank position, we set the compasses to the radius V D, and from X mark an arc A' and draw from its intersection with the dotted circle

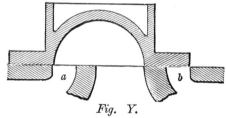

Fig. Y.

line L', which is the crank position when the eccentric is at X, Fig. 62, and the valve in the position shown in

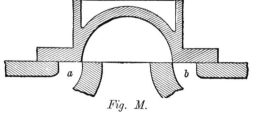

Fig. M.

Fig. Y, ready to open for the lead when the crank moves towards D.

The construction of Zeuner's diagram is as follows: From a center C, in Fig. 63, a circle B I D, called the

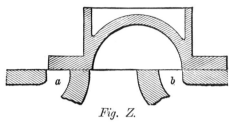

Fig. Z.

travel circle is struck, its circumference representing the path of the center of the crank pin, and its diame-

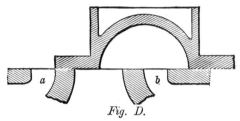

Fig. D.

ter, on the line B D, representing the path or travel of the piston. Now, supposing the crank pin to be at D

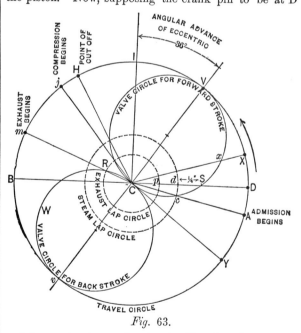

Fig. 63.

and to require to revolve in the direction denoted by the arrow, we may proceed to find the position in which to mark the throw-line of the eccentric. If the valve just covers the ports and has neither steam lap nor lead, the eccentric throw-line would be at a right angle, or

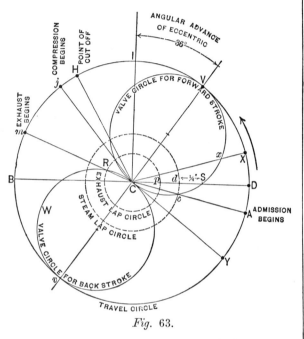

Fig. 63.

angle of 90°, to the line from C to D, which represents the throw-line of the crank when its pin is at D. In proportion, however, as the valve is given steam lap and lead the throw-line of the eccentric must be moved forward in the direction in which the engine is to run, and the line I C is merely used to measure the amount to which in any given case the eccentric is thus moved, or, in other words, to measure the *angular advance*, as it is called, of the eccentric. In the figure, the valve is

supposed to have such an amount of lap and lead as would require the eccentric to have an angular advance of 36° as marked, but the eccentric throw-line V C (which is always referred to as representing the position of the eccentric) is shown in the figure to stand behind the line I C instead of ahead of it, as it would require to do if the crank pin was at D and ran in the direction denoted by the arrow.

This brings us to a feature of the Zeuner diagram that renders it very difficult for the student to understand, and which must be mastered before he can have a thoroughly intelligent conception of its principles, viz., that it is essential to the construction of the diagram that either the engine be imagined to run in the opposite direction to what it would actually run in, or else that the angular advance of the eccentric be given in the wrong direction. In Fig. 64, for example, the crank pin is at D and the piston at the head end of the cylinder, and the crank-revolution being in the direction denoted by the arrow, the throw-line of the eccentric would be at W (the eccentric being marked by a full line), at this point the valve would be in the position shown in the figure, its edge H having opened the port for the head end of the cylinder to the amount of the valve lead. When, therefore, we are considering the piston and valve motions for the crank motion from D to B, we are considering the action of the valve upon the port for the head end of the cylinder (except as regards the compression which will be explained hereafter). The construction of the diagram, however, requires that the throw-line of the eccentric, instead of being marked in at its proper position in advance of the line I C, shall be marked in at the same angle from I, but on the left of I C, as shown in the figure by the

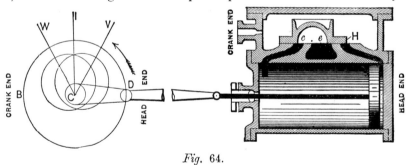

Fig. 64.

dotted line V C. This, be it observed, is the position it would occupy if the crank pin were at B and the piston, therefore, at the other end of the cylinder, the direction of crank revolution being reversed.

To efface from the mind the remembrance that the throw-line V C of the eccentric is on the wrong side of the line I C, and therefore in an assumed or false position, the crank is assumed to be on the opposite dead center to what it actually is, and the engine is assumed to run in the contrary direction to what it actually does. This leads to further complication in the mind because it necessitates that it be assumed that when the crank is at B it is the port at the head end that is acting as a steam port, whereas it is the port at the crank end that is actually doing so. But it is much easier to associate the mind with the idea that the throw-line of the eccentric may be at V C, or behind the crank instead of at W C, and ahead of it, because, in this case, we may think of the crank as running in the right direction and the proper port to be acting as a steam port. Thus, in the figure, it is assumed that the crank is at D, running in the direction of the arrow, while the valve is in its proper position with relation to the crank. During the crank motion from D to B, the valve edge H is, therefore, the one that will act to first open the port and admit the steam and then close it and effect the cut-off, while the edge e is the one that will reopen the port and cause the exhaust, and edge C will effect the compression at the other end of the cylinder.

Resuming consideration of the construction of the diagram, in Fig. 63, the position of the eccentric is, for the reasons stated, assumed to be behind the eccentric to the same amount (36° in the example), as it ought to be ahead of it.

Circle C S V is called the valve circle, and its center is always on the line V C. Its diameter represents one-half the amount of the valve travel. Its circumference always passes through the center C of the travel circle B I D. The valve circle for the other piston stroke has its center upon the line C v, which is a prolongation of the line V C.

On line C D we mark a point d distant from C to the amount of the steam lap of the valve, and from C as a center, with C d as a radius we draw the outer dotted circle passing through the point d. This is called the steam lap circle. On line C D, distance C p is equal to the amount of exhaust lap of the valve, and point s is distant from d to the amount of the valve lead, marked in this case ¼ inch.

Point A represents the position of the crank pin, at the time when the valve first opened the port to admit steam, and as A C represents the throw line of the crank at this time ,therefore, angle A C D is the lead angle of the crank. Line H C represents the center line of the crank at the time when the steam is cut off. Point j represents the position of the crank pin when compression begins at the other end of the cylinder, and m represents the position of the crank pin when the exhaust commences.

The return stroke is similar, in fact the diagram will apply equally well for both strokes if simply turned upside down, if, as we have thus far assumed, the length of the connecting rod be assumed to be infinite. It will be observed that the line from A to C, showing the position of the crank when the port begins to open, is drawn through the intersection of the lap circle with the valve circle at point c, and also that the point of cut-off H is obtained from a line drawn from the center C, and passing through the intersection of the valve circle with the steam lap circle.

The point where the exhaust commences is found by drawing a line from C passing through the intersection of the valve circle with the exhaust lap circle. Now since admission commenced when the crank was at A, and the amount of steam lap is equal to distance C d, therefore the valve must, when the crank pin is at A, have moved from its central position to the amount of the distance from C to c, measured along the line C A (the point c being at the intersection of the valve circle with the steam lap circle), this distance being equal to C d, measured along the line C D. When the crank has moved from A to D, the valve has moved from its central position and opened the steam port to the amount of the lead, or the distance d s. Continuing its motion, the crank will arrive at position X, and to find the position of the valve we draw a line from X to C, and from C to where this line cuts the valve circle is the amount the valve has moved from its mid-position, being shown in the figure by the distance, or radius, from C to x. To find the amount the steam port would be open, we must subtract from C x the amount of the steam lap or radius C d, or what is the same thing, the amount of port opening, when the crank pin is at X, may be measured on the line X C from x (where the line X C cuts the valve circle) to the outside lap circle.

When the crank pin has arrived at V, the valve will have opened to its fullest extent, hence the distance between the travel circle and the valve circle, measured of course along the line V C, represents the greatest amount of steam port opening possible, with the amount of valve lap and travel given in this example. When the crank pin reached point Y, and the line from Y to C does not cut the valve circle, the valve is in its mid-position over the ports, and both ports are there-

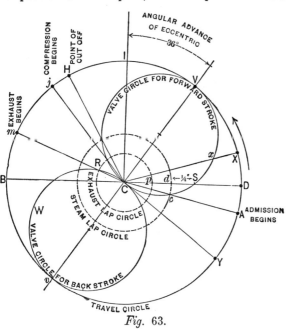

Fig. 63.

fore closed. Thus, in whatever position the crank pin may be, if a line is drawn from that position to the center C, then, from the point where it cuts the valve circle to the center C is the amount the valve will have moved from its mid or central position over the steam ports. Now as the center line A C of the crank, where the steam port first opens, must pass through the intersection of the valve circle with the outside lap circle, as at c in the figure, and as this center line must, at the point H where steam is cut-off, also pass through

the intersection of the valve and the outside lap circle, it follows that if the diameter of the lap circle was increased (giving the valve more lap), or diminished (giving the valve less lap), while the valve circle and travel of valve remained unaltered, their points of intersection would change, hence lines C A and C H (representing respectively the crank positions at the points of admission and of cut-off) would assume different positions or angles, thus locating crank points A and H in different positions and changing the lead angle and also the crank angle at the point or moment of cut-off. Thus, in Fig. 65, the valve travel and the angular advance of the eccentric remains the same as in Fig. 64, but suppose the lap circle C d, Fig. 63, be increased to C

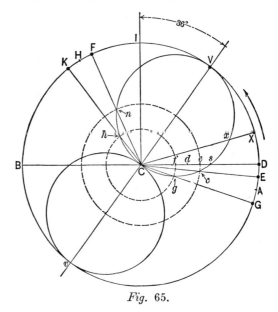

Fig. 65.

e, Fig. 65, and to locate the position of the crank pin when the valve begins to open the port for the admission of steam, we draw a line from the center C through the intersection of the valve circle with the lap circle at c and locating at E the crank pin position at the point of admission the amount of lead being the distance e S measured along the line C D. Similarly, to find the point at which the cut-off would occur, we draw a line from C, passing through the intersection of the valve circle with the lap circle as at n in the figure, and we find point F where the crank pin will be when the steam supply will be cut off by the valve. Similarly, if instead of increasing the amount of steam lap we diminish it, as denoted by the inner dotted circle *f*, the line G, cutting the valve circle and lap circle *f* at *g*, gives at G the crank pin position when the port is first opened for the admission, the amount of valve lead becoming the distance *f s*. The line C K passing through the intersection of the valve circle, with the lap circle at *h*, gives at K the crank pin position at the point of cut-off. It is seen, therefore, that certain elements, as the valve travel, angular advance of eccentric, and the steam lap being given, the amount of lead, greatest amount of port opening and the point of cut-off may be found on the diagram. Or, conversely, the point where the cut-off is required to occur, the amount of

The port we have been considering is port *b* in Fig. 66, the events of port opening and cut-off having been governed by the edge H of the valve while the exhaust has been governed by the exhaust edge *e*. But while the piston is performing this stroke, there will have occurred some compression in the other port (*a*) in the figure, it being obvious that, as the valve is traveling in the direction of the arrow *d*, the edge *c* of the valve will, so soon as it has closed the port *a*, shut in the end *g* of the cylinder whatever steam has not been exhausted from the previous piston stroke, and as the piston P is moving in the direction of arrow *f*, the steam thus shut in will be compressed. To find on the diagram, Fig. 63, at what point this compression will begin (or, in other words, to find in what position the

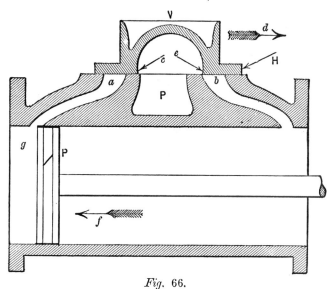

Fig. 66.

lead and the valve travel being given, the necessary amounts of lap and angular advance of eccentric can be found.

Referring again to Fig. 63, the events depending upon the exhaust lap may be similarly located. In the figure, the inner dotted circle *p* represents the exhaust lap circle, and a line drawn from center C and through the intersection of the exhaust lap circle with the *opposite* valve circle W (this point of intersection being at R in Fig. 63) locates the crank pin position when the exhaust begins for the return stroke.

crank pin will be when [the edge *c*, Fig. 66, of the valve closes the port *a* to the exhaust), we draw from *c*, Fig. 63, a line, passing through the point where the valve circle and the exhaust lap circle cross or intersect, and where this line meets the travel circle, which is at *j* in the figure, is the position of the crank when the compression begins. It is obvious that if the valve had no exhaust lap, the exhaust of one port, as *b* in Fig. 67, would open at the same instant that the compression would begin at the other, as *a* in the figure, and that if we add exhaust lap as denoted by the dotted arc in the

7

figure (the valve traveling as denoted by the arrow), we delay the exhaust of port *b* and hasten the compression of port *a*.

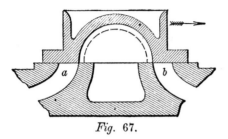

Fig. 67.

Let it be required to find the angular advance of the eccentric, the points of admission, of cut-off and of

travel 4 inches, the width of steam port being 1 inch. From the center C, Fig. 68, we draw the outer or travel circle whose diameter is 4 inches (equal to the valve travel). Then draw the line B D, passing through the center C, and the line C I at a right angle to B D. From C set off on the line C D the point *p*, equal to the given amount (⅜) of exhaust lap, and draw the exhaust lap circle. From C also set off the distance C *d* equal to the given amount of steam lap (¾ inch) and draw the steam lap circle. From *d* set off the amount of valve lead on line C D at *s*. Now, from C and *s* respectively, and with a radius in each case of one-half the length of C D (or one-half the radius of the travel circle) draw the two dotted arcs at *c*. From the center

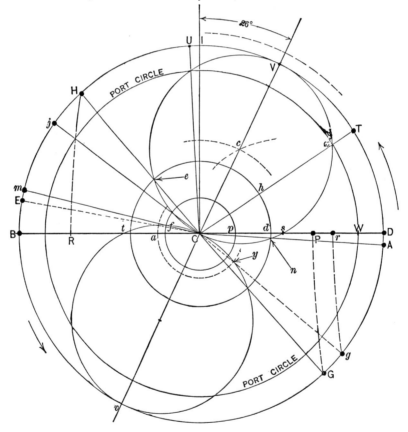

Fig. 68.

exhaust, the valve having ¾ inch of steam lap, ⅜ inch exhaust lap, the valve lead being ⅛ inch and the valve

C draw a line V, passing through the point *c* where the dotted arcs intersect and prolong this line across the travel circle to *v*. This line V C *v* is the throw-line of the eccentric; V C is for the stroke when the piston travels from D to B, and C *v* for the stroke when the piston travels from B to D. The angular advance of the eccentric is the angle V C I, or, as marked, 26°. From point *c*, on the line C V, we draw the valve circle having a diameter of C V, or one-half that of the travel circle B D. For the other stroke we draw a similar circle having its center on the line C *v*.

Now, to find the position of the crank when the port opens for the admission of steam, we draw a line from C through the intersection at *n* of the steam lap and the valve circles, and this line prolonged gives at A the position of the crank pin when the port begins to open for steam admission (it being borne in mind. that the travel circle represents the path of revolution of the crank pin).

To find the position of the crank when steam is cut off, draw a line from C passing through the point *e* of intersection of the valve and steam lap circles, this line giving at H the crank position at the time the valve closes the port and cuts off the steam supply.

Similarly with regard to the exhaust, a line from C and through the intersection of the valve and exhaust lap circles, gives at *j* the position of the crank pin when the exhaust closes *at the other end of the cylinder*, and hence the point at which compression begins, at port *a* in Fig. 68. To find the crank pin position when the exhaust begins, we draw, from C, a line passing through the point of intersection *f* of the exhaust lap circle and the other valve circle (C *v*) giving at *m* the required crank, or crank pin, position. Now suppose that a distance equal to the width of port, in this case 1 inch, is set off from the steam lap circle on the line C D, giving the point W (distant 1 inch from *d*) then a circle drawn through W, and from C as a center, is called the port circle. The distance between this port circle and the steam lap circle is the distance the valve must travel after the admission of steam before it leaves the port full open. Hence when the crank has reached T, where the center line C T of the crank passes through the intersection of the port circle with the valve circle, the valve has moved the distance C *x* from its mid-position over the ports, and if from this we deduct the amount of the valve steam lap, or radius

C *d*, we obtain the amount the port is open when the crank is at T, this amount being the radius *d* W on the line C D, or what is the same thing, the distance *h x* on the line C T.

It is obvious, then, that since the steam port is full open when the crank is at T, the edge of the valve travels, in this case, over the bridge between the steam and exhaust port while the valve is moving from T to V, the valve being at the end of its travel when the crank is at V. The amount of the over-travel of the valve is obviously represented by the radius *x* T, or the distance between the port circle and the travel circle, being, in this case, ¼ inch. It will be noted that in the case we are investigating, the steam lap is ¾ inch and the port 1 inch, and the valve travel being 4 inches is more than twice the width of port and the amount of steam lap, and hence the over-travel. If the amount of valve travel had been made twice the amount obtained by adding the width of the steam port to the amount of steam lap, the travel circle would also have

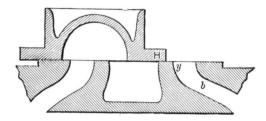

Fig. 69.

served for the port circle. But suppose, on the other hand, that the width of steam port had been 1⅜ inches instead of 1 inch, and then the dotted arc outside or beyond *v* would represent the port circle, and when the valve was at the end of its travel at V, the edge of the valve would be distant from the edge of the port to the amount the dotted arc is distant from V.

Fig. 69 represents the case in which the port circle is at W—the port width being an inch—and the edge H of the valve travels past the edge *y* of port *b*, while Fig. 70 represents the case in which the port circle is at the dotted arc beyond V and the valve edge H does not leave port *b* full open.

In Fig. 63 and 68, we have taken the valve and crank positions for one piston stroke only, and if the length of the connecting rod be left out of consideration, we may invert the diagram and it will serve for the return piston stroke, the various events occurring at the same points in the crank path, although not for the

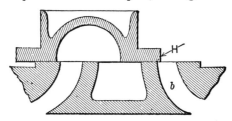

Fig. 70.

points of the piston movement. If instead of a connecting rod a slotted crosshead c, Fig. 71, be employed, the crank pin having journal bearing in a sliding block fitting into the slot of the crosshead and obviously passing once up and once down the slot at each revolution, then the crank positions and the piston positions would correspond, and the events of cut-off, exhaust,

throughout the whole piston stroke from D to B, the piston being at its second, third or fourth inch of movement from D, the crank pin will be at the corresponding

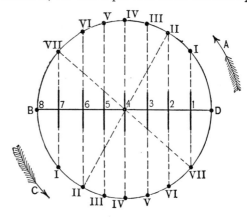

Fig. 72.

ing points I, II, III and IV respectively. When the piston starts from B, and the crank (moving in the direc-

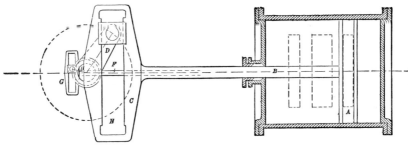

Fig. 71.

compression, etc., will occur at equal points in both piston strokes. Fig. 71 also shows the valve rod F to be operated by a slotted crosshead G, which is done so that the varying angle of the valve rod may not distort the valve motion when considered with relation to the piston motion.

Referring to Fig. 72, for example, the circle represents the path of the crank pin and the line B D a piston stroke of 8 inches. Now suppose the crank pin to start from D and travel in the direction of arrow A, and when the piston had moved its first inch as denoted by the numeral 1, the crank pin will stand at the position I on the upper half of the crank circle similarly

tion of the arrow c) travels over the lower half of the circle, it will be at I when the piston has moved its first inch, at II when the piston has moved its second inch, and so on. Suppose, then, that one steam port is full open when the crank is at position II on the upper half of the circle, and the other steam port will be full open when the crank is at position II on the lower half of the circle, these two crank positions being diametrally opposite. Again, if the steam was cut off when the crank was at position VII on the upper half of the circle, it would, on the return piston stroke, be cut off at point VII on the lower half of the crank circle. The same rule would apply to the exhaust and the compression for the two strokes which would also occur at corresponding points in the piston and crank movements.

The eccentric rod, however, is so long in proportion to the valve travel that its influence on the valve motion is usually too small to be of practical importance, hence, in many diagrams for plotting out valve motions, its length is taken as infinite or as if it were actuated by a slotted crosshead, instead of by an eccentric. The influence of the connecting rod in varying the points of cut-off, exhaust, etc., on one stroke as compared with the

end of the connecting rod when the piston is at N. Then from the point P we mark the arc N J, giving at J the position of the crank when the piston is at N and the cut-off for that stroke takes place. Similarly for the other stroke, we rest the compasses at R and

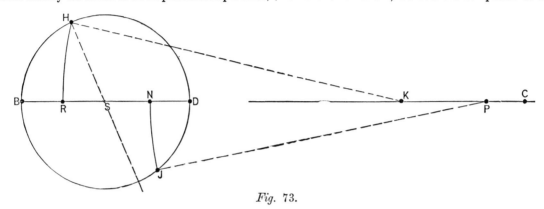

Fig. 73.

mark at K the position of the crosshead and of the connecting rod. Then from K we mark the arc R H, giving at H the position of the crank when the piston is at R. To show the variations in the two crank positions, we may draw the dotted line S, passing through

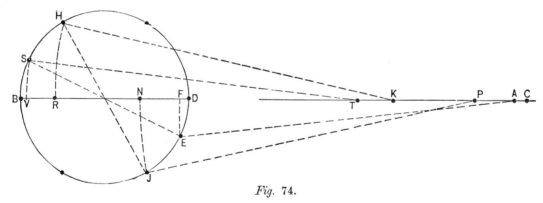

Fig. 74.

other, may be shown as follows: Suppose that in Fig. 73, the valve motion is designed to cut off the steam at N and R, these being equal points in the piston strokes, the length of the connecting rod being twice the length of the piston stroke, or from C to D, and to find the positions of the crank at the points of cut-off, we set a pair of compasses to the radius C D, and resting one point at N, we find at point P the position of the crosshead

the center of the circle, and it is seen that the cut-off is later during the stroke in which the piston is moving towards the cylinder.

Now suppose the valve gear is designed to cut off and exhaust the steam at equal points in the crank path, and we may find the corresponding piston positions by the construction shown in Fig. 74, the length of the connecting rod being twice that of the piston stroke.

Continuing the construction, we set the compasses to twice the radius B D, which will represent the length of the connecting rod, and (H being the position of crank pin at the point of cut-off), we set the compasses at H and mark at K the position of that end of the connecting rod. Then from K we mark the arc H R, giving at R the piston position. Similarly from J we get the piston position N. The difference in the two piston positions may be found by measuring from B to R and from D to N. For the piston positions at

the exhaust of the return stroke we get, from crank position E, the other end of the connecting rod at A, and from A we get arc E F, giving at F the piston position when the exhaust begins. The difference in the points of exhaust being seen by comparing the distance B V with F D.

Referring again to Fig. 68, and to the crank position H, the length of the connecting rod being taken as two and one half times that of the piston stroke, the corresponding piston position is shown at R. On the return

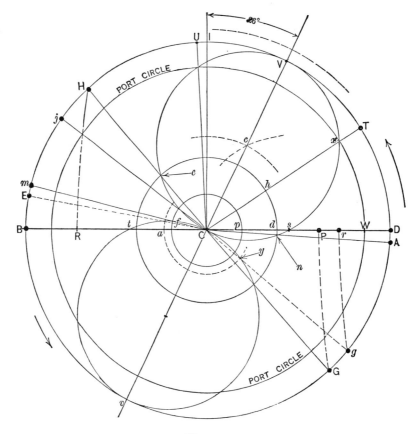

Fig. 68.

the respective points of cut-off, we rest the compasses at S, the crank position when the exhaust is to begin, and mark at T the position of the crosshead end of the connecting rod when the crank pin is at S. From T we mark with the compasses the arc S V, giving at V the piston position at the time the exhaust opens. For

stroke, the crank being at G, diametrally opposite to H, the piston would be at P, or distance P D from the end of the cylinder. To harmonize the points of cut-off with the piston motion, mark r, distant from D to the same amount that R is distant from B, and by means of the arc r g, obtained from the length of the connecting rod, we find the necessary crank position at g. In order to enable the valve to cut off steam with the crank pin at g and the piston at r, corresponding with the piston position R, the valve must be given less steam lap for the port nearest to the crank—this being the port admitting steam while the crank is traveling through the half-revolution B G D—and in order to determine the proper amount of steam lap, all we have to do is to draw a new steam lap circle a that will cut the center line of the crank at the point y where it is intersected by the valve circle. Here, then, we have equalized the points of cut-off by means of varying the steam laps of the valve, the lap for crank position H and its corresponding piston position R being the radius C d, and that for crank position g and the corresponding piston position r being the radius C a. It is obvious that in varying the steam lap to equalize the cut-off, we have varied the point of admission, and to find this point we draw the dotted line E C, passing from C through the point of intersection of the steam lap circle a, and the lower valve circle and giving at E the crank position when the port opens for admission. When the crank reached B the amount of valve lead will be the radius t a, whereas, on the other stroke it is but d s. This shows that the equalization of the points of cut-off can only be done at the expense of unequal lead. Similarly, the compression and exhaust might be equalized by giving different amounts of exhaust lap to the two lips of the valve but there is one condition inevitable, which is, that if the compressions are equalized, the exhausts must be unequal, or vice versa. To summarize, then, it has been shown that with a valve having a given amount of travel, laps and lead, the points of cut-off, of admission, of release, of compression, and the necessary amount of angular advance of eccentric, may be found, but the form in which the problem generally presents itself is as follows: The points of cut-off and of release, the width of port and amount of valve lead for a given engine having been determined upon, it is required to find the necessary steam lap, the amount of valve travel, the angular advance of the eccentric, the amount of exhaust lap, the point at which compression will begin, the position of the crank at admission, and the lead angle.

Suppose, for example, an engine has a piston stroke of 24 inches, and it is required to cut off at 20 inches, release at $23\frac{1}{2}$ inches, the lead to be $\frac{1}{4}$ inch, the steam ports being $1\frac{1}{4}$ inches wide and the connecting rod 6 feet long.

Draw the horizontal line B D, Fig. 75, representing the piston travel, and from its center strike a circle, B H D, to represent the path of the crank pin. Supposing the engine to run in the direction of the arrow, we set off, on line D B, the point R, the position of the piston when steam is to be cut off on the stroke as the piston travels from D to B, and as the length of D B is 3 inches (or 24 one-eighths of an inch), R will be 20 eighths from D. We then set off M, the point in the piston stroke at which the release or exhaust is to begin. Now as the length of D B is on a scale of one-eighth of the piston stroke, we set a pair of compasses to a radius of one-eighth the length of the connecting rod, which being 72 inches gives a radius of 9 inches (72 ÷ 8 = 9 . With the compasses thus set, mark from R and M (in the manner already described with reference to Fig. 73) the arcs R and m, giving at R the position of the crank at the point of cut off, and at m the crank position when the release or exhaust begins.

From H and m we draw lines to the center C, representing at C H the center line of the crank when the cut-off occurs. Draw a line C T bisecting the angle H C D, and from D erect a perpendicular line, meeting line C T at X. Then carefully measure the distance C D, and also the distance C X, and divide C X into C D, thus obtaining a fraction, because C X is greater than C D. Substract this fraction from 1, and divide the width of the steam port, minus one-half the lead, by the remainder, which will give the half travel of the valve and, therefore, the radius of the travel circle.

Thus distance C D $= 1\frac{1}{2}$ inches and C X is 3.4 inches. Dividing $1\frac{1}{2}$ (or 1.5) by 3.4 gives .44 which substracted from 1 leaves .56.

The width of port is $1\frac{1}{4}$ and the valve lead is $\frac{1}{4}$, half of the latter is $\frac{1}{8}$ and $1\frac{1}{4}$ less $\frac{1}{8}$ is $1\frac{1}{8}$ or 1.125. Dividing 1.125 by .56 gives 2, and 2 inches is, therefore, the required half valve travel. Hence a circle, E I F, struck from center C, with a radius of 2 inches, is the travel circle, and E F or 4 inches is the full travel of the valve.

Now, from the point T of intersection of this travel

circle with the line C X, draw a perpendicular line T *t*, and set-off on C D, and on each side of *t*, a distance equal to one-half the lead or, in this case, ⅛ inch, thus getting points *d* and *s*.

line of the eccentric as in our former examples. The angle C I V is the angular advance of the eccentric, being, in this case, about 30° as marked. The eccentric throw-line V C *v* and the travel circle E F being

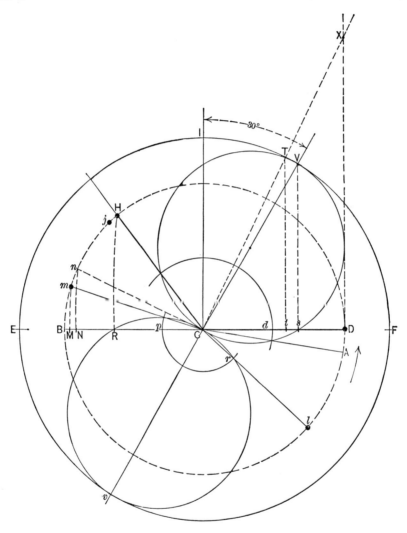

Fig. 75.

Then C *d* is the steam lap and a circle struck from C as a center, and passing through *d*, is the steam lap circle. Now from *s* erect the line *s* V at a right angle to line B D, and from V, where this line cuts the travel circle, draw the line V C *v*, which represents the throw-

obtained, we may now draw in the valve circles and then, as has been explained with reference to former figures, the line A C may be drawn, showing at A the crank pin position at the time of admission.

To find the necessary amount of exhaust lap we draw

from point m (the crank position when the exhaust is to begin) a line to C, and where this line cuts the travel circle, as at p, is the radius of the exhaust lap circle, which may therefore be drawn, from C as a center, with radius C p. The travel, angular advance, and the valve circles have been found for one stroke and they must be the same for both strokes. The steam lap and exhaust lap of the valve for one stroke have also been determined, and these laps may be made the same for the other end of the valve or they may be varied. If it is determined to vary them, it must be remembered that the compression for one stroke depends upon the point at which exhaust begins on the other stroke (as was shown in Fig. 66 and 67), hence the point at which compression begins, cannot be found until the point of release is known. If the laps are to be equal at each end of the valve, then C l, drawn through the intersection of the exhaust lap circle and the valve circle at r, would give l as the point at which compression begins on the return stroke, and a point exactly opposite to l, as point j, would be the point of compression (or closure of exhaust) at the end B of the cylinder. The arc n N is introduced to illustrate the fact that if it had been determined to let the point of release be at the 23rd inch of piston stroke, instead of at $23\frac{1}{2}$ inches, then the crank would, at the time the release occurred, be at n instead of at m, and it would be almost impossible to determine with certainty just where the line n c intersects the valve circle, this point of intersection being necessary in order to find the required amount of exhaust lap.

This defect in the Zeuner diagram, however, may be remedied by drawing it to a scale a certain number of times greater than the given dimensions, and then correspondingly reducing the results obtained from the diagram.

Let it now be required to construct a diagram to show, graphically, the actual and relative openings of the steam and exhaust ports, and also the travel of the exhaust edge (of that end of the valve that is admitting live steam) over the bridge and the exhaust port, the dimensions being as in the subjoined table.

The travel, steam lap, exhaust lap and lead being given, a diagram, similar to that given in Fig. 68, may be constructed from the given data, showing the travel

8

circle and the valve travel, and containing the steam and exhaust lap circles, the valve circles, the angular advance of eccentric and the lead, these lines being repeated in Fig. 76, in which B D represents the valve travel (5 inches). B I D is the travel circle, c p the exhaust lap circle, C D the steam lap circle, d s the lead ($\frac{1}{8}$ inch), V C v the eccentric throw-line, and V h C and C K v the valve circles.

Width of steam port	–	–	$1\frac{3}{16}$	in.
" " exhaust port	–	–	$2\frac{1}{2}$	"
" " bridges	–	–	$1\frac{1}{4}$	"
" " steam lap	–	–	$\frac{3}{4}$	"
Exhaust lap	–	–	–	$\frac{1}{8}$	"
Travel of valve	–	–	–	5	"
Piston stroke	–	–	–	24	"
Length of connecting rod	–	–	84	"

These dimensions being taken from a "Rogers" Locomotive.

From center C, draw above B D a semi-circle as W W, a distance from the steam lap circle equal to the width of port ($1\frac{3}{16}$ inches), and this may be called the steam port circle. Then from center C, and below B D, draw another semi-circle E E, distant from the exhaust lap circle also equal to the width of port; this may be called the exhaust port circle. From C as a center, also draw a semi-circle F F, having a radius equal to the width of the bridge less the amount of exhaust lap; this may be called the bridge circle.

Considering B D to represent the piston stroke, it must be divided into 24 equal parts as I, II, III, IV, etc. (the stroke being 24 inches), each of which will represent an inch of piston travel; and arcs, I 1, II 2, III 3, etc. (obtained with the compasses set to represent the length of the connecting rod, on the same scale as B D represents the length of the piston stroke), will give the corresponding crank positions, the direction of crank-revolution obviously being as denoted by the arrow.

Having found the crank pin positions corresponding to the piston positions, we may draw lines 1 C, 2 C, 3 C, etc., representing the throw-lines of the crank for the respective positions. Now, suppose the crank is, say, at Z, traveling to the dead center D, and beginning with the point at which the lead opens, we may trace out, on the upper half of the diagram, the movement of the steam edge of the valve (H, Fig. 64) and the corresponding steam port openings, and simultaneously upon the lower half of the diagram, the movement of the exhaust edge [*c*, Fig. 64] of the valve. Next we may trace out the movement of the exhaust

are covered by the valve, hence the ports are closed to both the steam and exhaust. But when the crank has reached the point Z, this point being found by a line from C, intersecting the valve and exhaust lap circles,

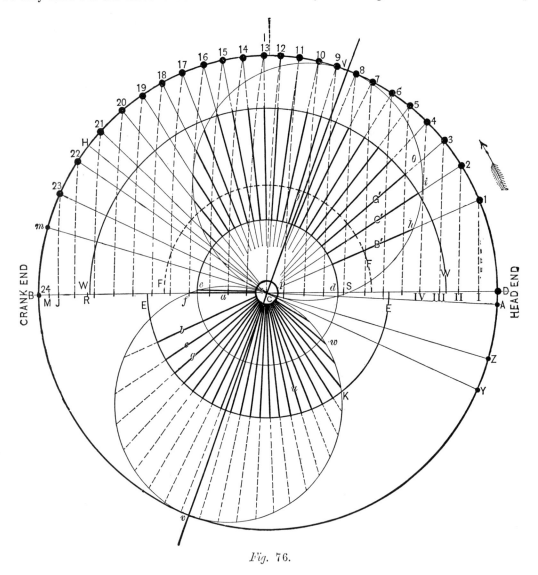

Fig. 76.

edge [*e*, Fig. 64] over the bridge and over the exhaust port.

When the crank stands at Y, or at a right angle to C V, the valve is in mid-travel and both steam ports

the valve has moved from its central position to an amount equal to the exhaust lap, thus opening the exhaust at the end B of the cylinder. When the crank has moved to A, the radial line from crank pin A passes through the intersection of the valve circle with the steam lap circle, and the valve has traveled from its central position to an amount equal to the steam lap C d, and the steam edge of the valve coincides with the edge of the port, hence the admission of steam begins. The exhaust edge at the other end of the valve (c, Fig. 64) has also traveled an equal distance, as shown at C e, and the exhaust is open the distance C e, as denoted by the heavy line a. When the crank has reached the dead center D, the steam edge (H, Fig. 64) of the valve is distant, on line C D, from the steam lap circle to the intersection of the valve circle, or distant d S away from the edge of the steam port, giving an opening d S, or ⅛ inch as lead. The exhaust opening, when the crank is at D, is from the exhaust lap circle to the valve circle at f, or the length of the heavy line from C to f.

When the crank pin is at the position marked 1 on the travel circle, and the piston has moved to 1 on the line D B, the steam port is open the length of the heavy line B′ which extends from the lap circle to the valve circle, while the exhaust at the other end (measured on the line 1 C prolonged beyond C) would be from the exhaust lap circle to the circle E, or an amount equal to the length of the heavy line b. Proceeding to the second inch of piston movement, marked II, and the corresponding crank position marked 2, the amount the valve has moved from its mid-position over the ports is the radius from the lap circle to the valve circle, but the width of the port is only that from the lap circle to the port circle W, hence the actual amount of steam port opening is the length of thickened line C′, which extends from the steam lap circle to circle W, and it appears that the steam edge of the valve has traveled beyond the port edge and over upon the bridge. The amount of exhaust opening at this time is thickened line c, running from the exhaust lap circle to circle E, hence the port is still full open, it being noted that the width of the port, now acting as a steam port, is from the lap circle W, while the width of the other port, now acting as an exhaust port, is from the exhaust lap circle to circle Ш. Similarly, at the third inch of piston travel and the corresponding third crank position, the steam port is opened full, the thickened line G′ extending from the steam lap circle to the port circle W, and lying wholly within the valve circle.

The amount the steam edge of the valve has traveled over the bridge, is shown by the dotted line o, extending from the port circle W to the valve circle. The exhaust opening is still full, because the line g (a prolongation of G′ C), passes from the exhaust lap circle to the circle E without passing outside of the lower valve circle. The steam port, it will be seen, keeps full open for all the succeeding crank pin positions up to the 16th, while the exhaust port remains full open up to the 19th crank position, which corresponds, of course, to the 19th inch of piston stroke. The over-travel of the valve increases until the crank is between the 8th and 9th positions, and then gradually diminishes, finally ceasing just after the 16th piston and crank positions.

In Fig. 77, the valve is shown at the extremity of its

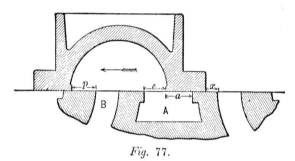

Fig. 77.

travel, or in the position it would occupy when the crank pin stood at V, and the over-travel is shown at X. The exhaust edge of the other end of the valve would be the distance C v, Fig. 76, from its central position, and the port opening being from C to circle E, the exhaust edge of the valve travels past the port the distance from v to the exhaust circle E, this distance being represented, in Fig. 77, by p. Now we have seen that as the crank proceeded from D towards V, the steam edge of the valve traveled across the steam port, opening it full, and finally passed upon the bridge, and at the same time the exhaust edge of the same end of the valve passed over the exhaust port A, as represented at a, Fig. 77. Thus, in Fig. 76, considering piston and crank positions 1, the steam end of the valve was then distance C h from its central position, and hence the distance, on line C 1, between the heavy dotted bridge circle F and the valve circle or distance F h, is the amount (a, Fig. 77) that the exhaust edge projects over the bridge and across the cylinder exhaust cavity. With the crank in posttion 2, this distance [*a*, Fig. 77] would be from *i* to the bridge circle F, measured on the radial line C 2, and so on to position V of the

tance being represented in Fig. 77 by *a*. It is obvious that in all cases in which the width of opening at *e*, Fig. 77, is less than the width of the port B, the width *e* is the effective one and is the one that must be consid-

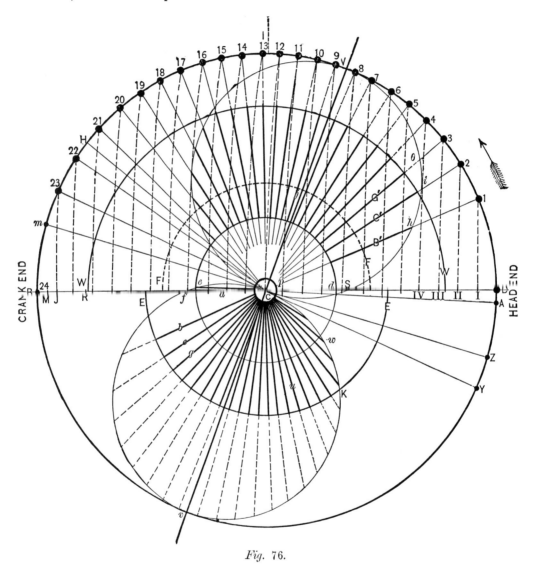

Fig. 76.

crank; at which time this exhaust edge, corresponding to edge *e* in Fig. 64, would project over the exhaust cavity equal to the distance from the travel circle to the bridge circle F, or distance V *j* in the figure, this dis-

ered. Cases are not uncommon where, in increasing the amount of lap on an existing valve, this point is over-looked, and on account of the narrowness of the bridges and of the cylinder exhaust cavity relatively to the amount of valve travel, the exhaust is cramped and partially cut-off. This cannot occur if the cylinder exhaust port or cavity is made equal to the radius V j, Fig. 76, plus the width of the steam port, which would give, in the example, V $j = 1\frac{3}{8}$, plus width of steam port $= 1\frac{3}{16}$, total $2\frac{9}{16}$ inches, as the required width of cylinder exhaust port, and as the dimensions given, in stating the example, were $2\frac{1}{2}$, another $\frac{1}{16}$ has been allowed in drawing the diagram, so as to prevent the exhaust from being cramped at e, Fig. 77. The crank position at the point of cut-off and of compression is, of course, found by the process given in previous examples, and we may now proceed to examine the port openings. Suppose the piston has reached its 14th inch of motion, and the crank stands at position 17, then the amount of steam port opening will be denoted by that part of the radial line C 17 that is thickened, and lies between the steam lap circle d and the port circle W. This port opening decreases until it is closed on the dotted line H C where the cut-off is located.

Similarly referring again to the exhaust of the other port at the 19th inch of piston movement, corresponding to crank position 19, the exhaust port is full open, as shown by the full line from C to the circle E. While, however, the crank is moving from its 19th to its 20th position, the port begins to close the lower valve circle, cutting the circle E at K, so that when the crank has reached its 20th position, the exhaust has begun to close, its amount of opening being represented by the thickened part of the line that runs from 20 through C and extends until it meets the lower valve circle, this thickened part being from the exhaust lap circle to the lower valve circle. At the time that the piston has arrived at the point of cut-off H, the exhaust port is still open an amount equal to the thick line from W to the lap circle, this line being, of course, a continuation of line H C. When crank position 23 is reached, where the radius from 23 to C passes through the intersection of the valve circle with the exhaust lap circle, the exhaust on the crank end of the cylinder closes and compression begins in that end of the cylinder. When the crank reaches m—line m C passing through the intersection of the exhaust lap and the lower valve circle—the exhaust port at the head end of the cylinder opens. Meanwhile compression continues from crank position 23 until the crank arrives at a distance from B equal to A D, when steam is admitted at the crank end port, and the same events are repeated during the next piston stroke.

If we attempt to give to the valve sufficient lap to cut-off the steam at a point earlier than at about five-eighths of the piston stroke, and at the same time keep the lead equal, the points of cut-off, of release and of cushion vary so much on one stroke, as compared to the other, as to render it necessary to adopt a different form of valve mechanism in which a separate cut-off valve is employed.

CHAPTER III.

LINK MOTIONS AND REVERSING GEARS.

With the valve mechanism thus far described, the engine is capable of running in one direction only, and to enable it to run in either direction, the link motion is employed.

The link motion also enables the travel of the valve to be diminished, and this causes the steam supply to the cylinder to be cut off earlier in the piston stroke, thus using the steam more expansively.

STEPHENSON'S OPEN ROD LINK MOTION.

Fig. 78 represents an open rod Stephenson's link motion, of the class used in direct acting locomotives, or locomotives in which no rocking shaft is used. Two eccentrics are employed, one for the forward and one for the backward motion of the engine, and as there is no rocking shaft, the forward eccentric leads the crank—or in other words, is ahead of the crank when considered with relation to the direction of crank revolution—while, with the engine running in this direction, the backward eccentric follows the crank, whereas, when the link motion is placed in backward gear, and the direction of crank-revolution is reversed, then the backward eccentric leads and the forward eccentric follows the crank, as will be seen more clearly presently.

When the eccentrics are so arranged that with the crank on the dead center, shown in the figure, the rods are connected as shown, the link motion is one with

64

open rods, as distinguished from one with crossed rods, but, nevertheless, the rods will be crossed when the crank is on the other dead center, as will be seen presently.

The eccentric rods are pivoted to the link, as shown, and the link is pivoted to a pin upon the saddle s, which is fast upon the link. The link hanger is pivoted to the saddle pin and to the end of one arm of the lifting shaft. On another arm of this shaft is pivoted the reach rod, which is also pivoted to the reversing lever. The sector is (in connection with the reversing lever latch) employed to hold the reversing lever in the desired position. The valve spindle moves in a straight line by reason of passing through a fixed guide. The link motion is said to be in full gear when the link block occupies its furthest position from the saddle-pin, because it is in this position that the link block gives the most or full travel to the valve. The reversing rod is pivoted at its lower end X, and, by pulling the end d of the latch rod towards handle g, the latch is lifted out of the notch in the sector, and the lever may be moved to the left until the latch will engage with one of the other notches, as notch 3, 2, 1, or 0, which is the middle notch. On releasing d, the spring acting against the piece e will force the latch into the notch in the sector and—the latter being in a fixed position—thus lock the lever in place.

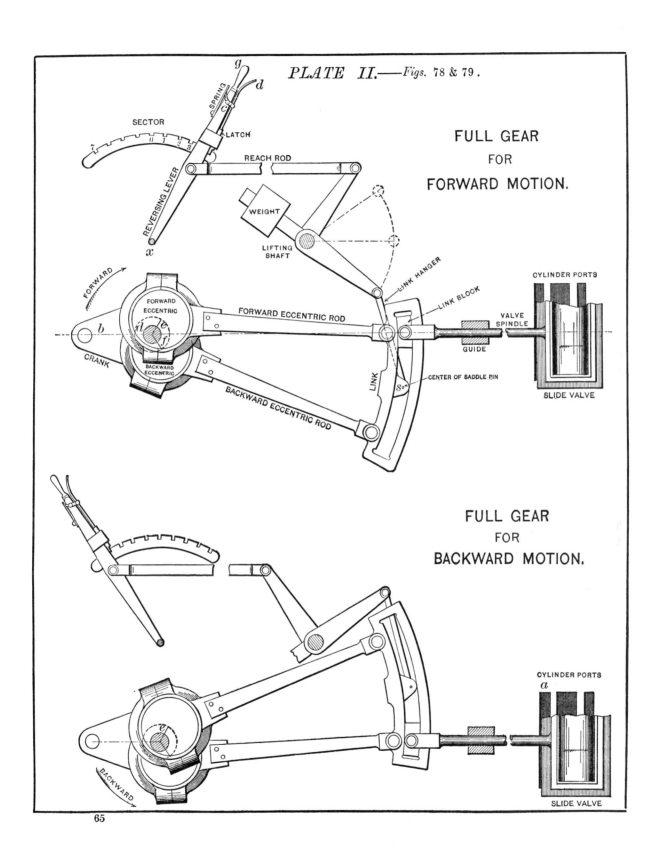

PLATE II.—*Figs. 78 & 79.*

FULL GEAR
FOR
FORWARD MOTION.

FULL GEAR
FOR
BACKWARD MOTION.

65

Now, suppose we hook up the lever one notch, or, in other words, move it so that the latch falls into the notch marked 3, the lifting shaft and link hanger will have moved, lifting up the link and bringing the center of the saddle-pin nearer to the link block, and as the link is pivoted upon the pin, and vibrates upon it, the amount of motion, imparted to the link block, will be reduced as will the valve travel also.

If we move the lever so as to engage its latch with the notch *o* of the sector, the saddle-pin will be brought level with the link block and the valve will have the least amount of travel and the earliest point of cut-off that the link motion can give; this position is called

gear for forward to full gear for backward motion, we have brought the backward eccentric rod in line with the link block, and, as a result, the engine will run backward, or in the direction denoted by the arrow in Fig. 79.

We may find the positions of the parts of a link motion with sufficient accuracy for all practical purposes by the following construction:

In Fig. 82 *n* represents the path of the eccentric, the line of engine centers, *p p'* the center line of the link, *k k'* the eccentric-rod eyes, B the crank on the dead center, *w* the center of the cylinder ports, and *x* the center line of the valve.

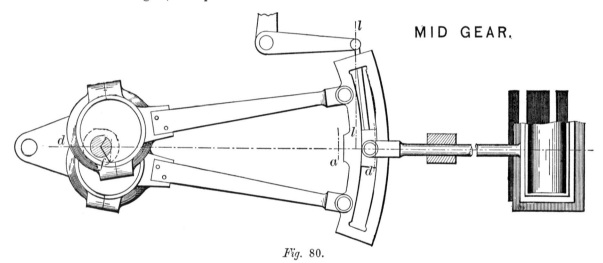

MID GEAR.

Fig. 80.

that of mid-gear, the positions of the parts being shown in Fig. 80. Between these two positions, on this half of the sector, the lever, and therefore the link, may be hooked up in as many positions for the forward gear as there are notches in the sector, each position giving a different amount of valve travel, and, therefore, a different point of cut-off and degree of expansion.

Now, suppose we move the reversing lever so that its latch engages notch 7 of the sector, the parts will then be brought to the positions shown in Fig. 79, which is full gear for the backward motion. It will be observed that the crank and the eccentrics stand in the same position in all three of the figures, and that in moving the link we have moved the eccentric straps around upon the eccentrics, and in moving from full

With the port *a* open for the lead, the valve must have moved from its mid-position to the amount of the lap and lead, or distance *w x*, hence we take this distance and from the center C of the crank-shaft, mark an arc *c*, and then with the length of the eccentric-rod as a radius, and from a point on the line *l*, mark an arc *r r'*, and lines *e* and *f* drawn from C to the intersections of arc *r* with circle *n*, gives the positions of the two eccentrics. From the center of eccentric *f*, we draw the center-line *p p'* of the link-slot, and the arcs *k k'* for the eccentric-rod eyes. This is practically correct but not absolutely so, because we have assumed the distance from F to *p* to be the same as the distance from *e* to K, added to that from K to *p*, whereas the latter is the greatest, and eccentric *e* would, therefore,

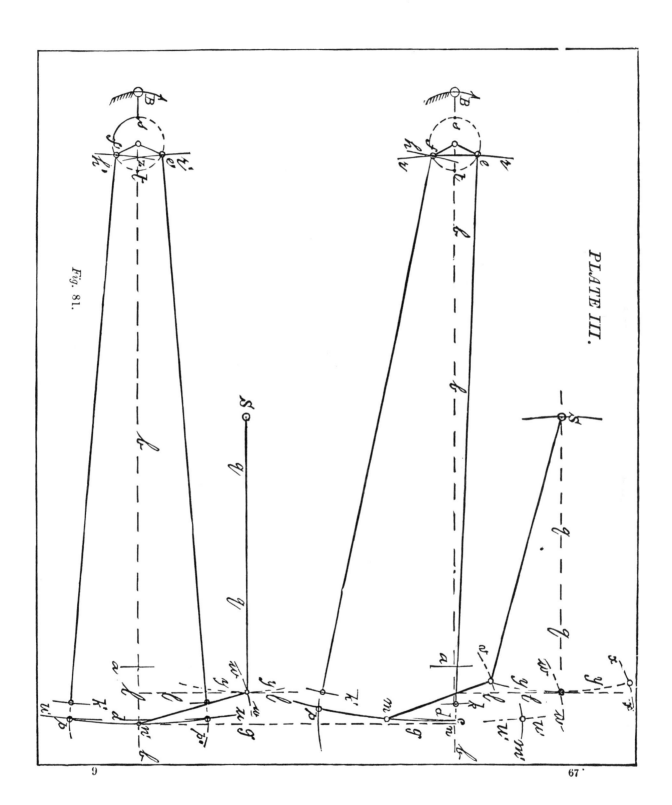

Fig. 81.

PLATE III.

9

67·

actually stand a minute distance further ahead of the crank. The amount of the variation will, with eccentric-rods of ordinary length, be too small to be important.

In the figure the length of the rod is made but about three times the valve travel, whereas in practice, it would be about ten times, hence, the construction here given is practically correct.

In considering the proportions of the parts of a link motion, in which a reversing lever is to be moved by hand power, it is evident that the amount the link must be raised, in moving it from full forward to full backward gear, must be such as will enable the operator to move the reversing lever from one to the other of the end notches in the sector, the leverage of the reversing lever being sufficient to enable him to exert power enough to raise the link. The longer the link is, however, the more perfect the motion.

As the upper end of the link hanger swings upon a fixed pin or pivot, it is obvious that the link swings in an arc of a circle, of which this pin is a center, and, in consequence, the eccentric-rod eye partakes of this motion instead of moving along the line of engine centers, which would give the most perfect action; and

gives very nearly the same port openings as the simple valve gear, the difference being so slight as to have no practical importance, as may be seen by comparing the full line diagram Fig. 32, with the diagram Fig. 128, both being for a valve motion having the same dimensions, but the former is with the eccentric attached direct to the slide spindle, and the latter from the same valve, etc., with the link gear added.

The dimensions below are those used throughout the following examples, unless otherwise stated:

Piston stroke .... 24 in.
Length of connecting-rod .... 72 "
Travel of valve in full gear .... 4½ "
Lead of valve at full travel .... ¼ ".
Length of eccentric-rod .... 17 "
Length of link from center to center
    of link-block when at i's extreme
    positions in the link .... 13 "
Length of link hanger .... 14 "
Width of steam ports .... 1¼ "
Lap of valve .... ⅞ "

In Fig. 81, the parts (represented by their center lines) are shown in full gear for the forward motion in the upper, and in mid-gear in the lower part of the figure.

For the full forward gear we may find the position

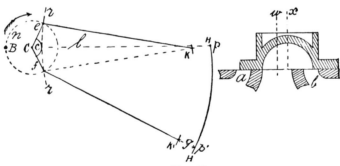

Fig. 82.

it follows that the longer the link hanger is, the less curvature its arc of action will have, and its line of motion will be more nearly parallel with the line of centers, it is usually made, however, but little longer than the link.

The lifting-shaft arm would operate most perfectly if it were as long as the eccentric-rods, but considerations of space and handiness usually limit it to about one-half of that length.

When it is placed in full gear, a link gear, or motion,

of the upper end of the link, as follows:

We take the length of the eccentric-rod (which is measured from the center c of the eccentric strap to the center k of the eye to the other end of the eccentric-rod), and add to it the distance from the center of the eccentric-rod eye, to the center of the link-slot (or distance g, in Fig. 82), and with the radius so obtained, rest one point of the compasses at e, and mark on the line of centers b, an arc, which will give at e the position of one end of the link. For the position of the other

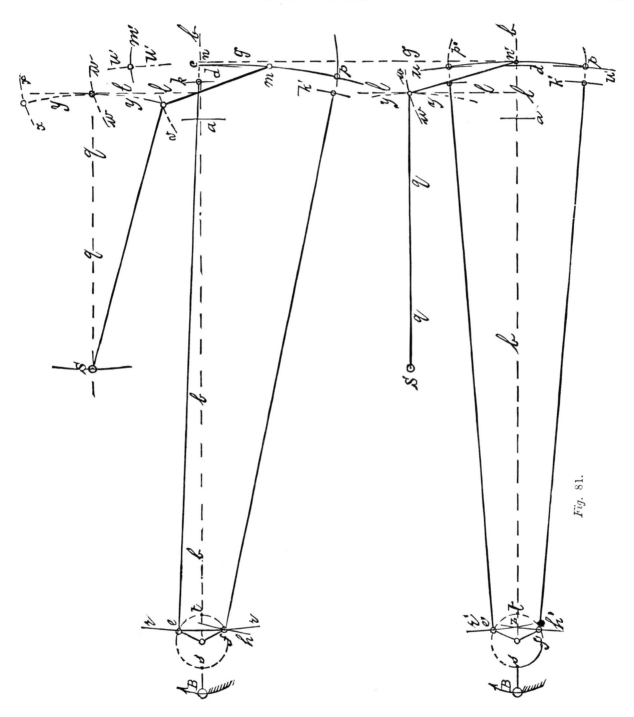

Fig. 81.

end we mark, with the same radius, and from f as a center, the arc from p to c, somewhere on which we know the lower end of the link will come. We then take the length of the link (the radius from p to p' Fig. 82, being taken as the length of the link) and from c as a center, mark an arc p thus getting the position of both ends of the link. To find the center from which the center line of the link slot may be marked, we take the same radius (the length of the eccentric rod, plus the distance g, Fig. 82) and from c as a center, mark an arc $r\ r$, and from p as a center mark arc h, and as arcs $r\ r$ and h intersect at f, the center of the backward eccentric, it shows that from f as a center we may draw in the center of the link curve. To find the center of the link, we divide the length $c\ p$, of the link in equal parts, and find its center at m.

With the length of the link hanger as a radius, and from m as a center, we mark an arc v, somewhere on which we know the upper end of the link hanger will stand, and to find the exact position we proceed as follows: Take the length of the eccentric-rod (that is radius $e\ k$) and from t, mark an arc which will fall coincident with c on the line of centers (this arc being seen at d, in the lower half of the engraving), and then from s, and with the same radius, we mark an arc a. Midway between a and c, we mark a line l at a right angle to the line of centers.

We must now turn to the lower half of the engraving, representing the link motion in mid-gear, and find the position of the link as follows:

With the radius $f\ p$, (or in other words the length of the eccentric-rod, plus radius g, Fig. 82), and from e' as a center, we mark an arc u, somewhere upon which we know the upper end of the link will be, and from f, we mark an arc u', somewhere upon which the lower end of the link will be. Now, as the middle of the link (or, in other words, the center of the saddle-pin) will be on the line of centers, we mark, from the line of centers, arcs p and p', giving the length of the link.

Then (with the length of the eccentric-rod plus the distance g, Fig. 82), we mark from the upper end of the link an arc r', and from the lower end an arc h', and where these two arcs intersect on the line of centers (at Z) is the center from which the center line of the link may be drawn; the center of the link, or of the saddle-

pin, being on the line of centers at n'. From n' we erect a perpendicular line, $g\ g$, to the line of centers of the link in full gear in the upper half of the engraving, thus getting at n, a position corresponding to n', when the link is in mid-gear. We then take the length of the link hanger, and from n as a center, mark an arc w, and, from where this arc cuts the line l, we draw a horizontal line $q\ q$, which will be the plane on which the center S of the lifting-shaft will be; hence we take the length of the lifting-shaft arm, and mark an arc, giving the location of the center of the lifting-shaft at S. From S as a center, with the radius from S to line l, we mark an arc $y\ y$, and where this cuts arc v is the location for the upper end of the link hanger when the link is in full forward gear. When the link is in mid-gear the link hanger will stand, one end on the line of centers at n, and the other at the intersection of arc w with line q, the parts occupying the positions shown in Fig. 79.

It may now be shown that the amount of lifting-shaft motion necessary to lift the link from mid-position to full gear for the backward motion, is less than that required for moving the link from mid-gear to full forward gear. Thus, when the link is in mid-gear, its center will be at n', as has already been shown; hence from m with a radius of half the length of the link, we mark an arc m' somewhere on which the center of the link will be when in full gear for the backward motion. Then from e as a center, and with radius $e\ c$, we mark an arc u', and where u' cuts m' is the position for the center of the link when placed in full gear for the backward motion. From this point, therefore, and with the length of the link as a radius, we mark an arc $x\ x$, and where this cuts arc y is the position for the upper end of the link hanger when in full gear for the backward motion, and if we measure on arc y from x to w, we find it less than from w to v, and this obviously affects the positions of the notches in the sector. We have thus found the positions of all the parts, when the crank is on the dead center B, and the link is in full gear, and also when it is in mid-gear, and it may now be pointed out that in moving the link from full gear to mid-gear, we have increased the lead of the valve. Thus when it is in full gear, the upper center of the link and, therefore, the center of the link-block, is on

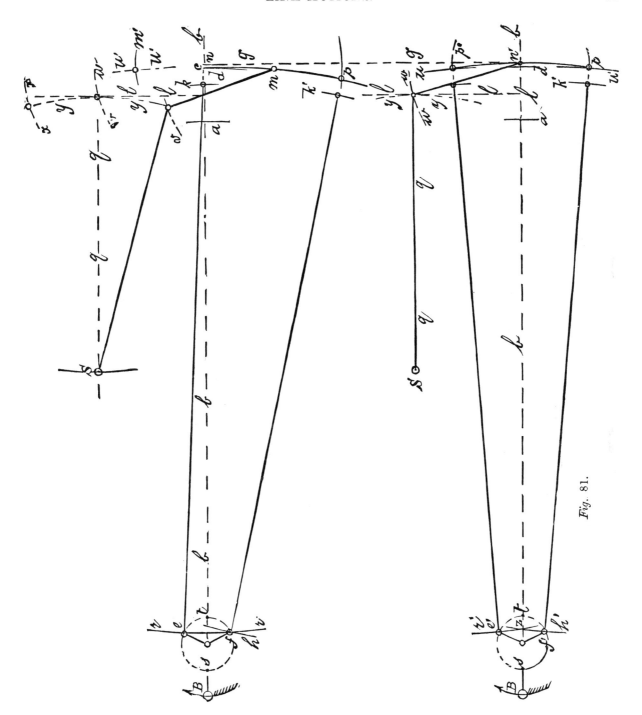

Fig. 81.

PLATE IV.

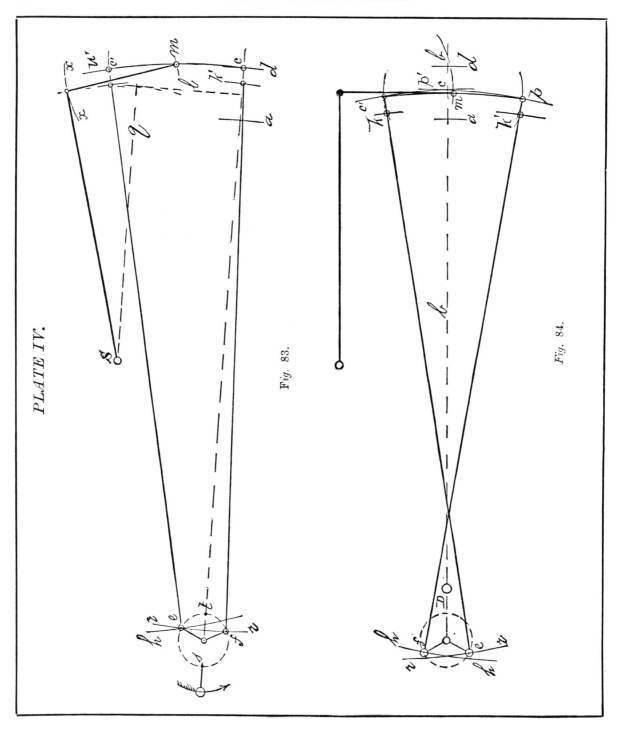

Fig. 83.

Fig. 84.

the line of centers at *c*, and when the link is in mid-gear its center is at *n'*. By means of the vertical line *g g'*, we have transferred position *n'* on the lower half of the engraving, to the upper half at *n*, and it is clear that, as the center of the link-block is always coincident with the ·center line of the link slot, and always on the line of centers *b b* of the engine, therefore it will, in moving the link from full to mid-gear, be moved from *c* to *n*, and the lead will be correspondingly increased.

In Fig. 83, the parts are shown in position · for full backward gear, the construction corresponding to that already described and being as follows; radius *f* K′ is the length of the eccentric rod, to which we add distance *g*, in Fig. 82, and get radius *f d*, and from *f* as a center, mark an arc *d*. With the same radius we mark from *e* as a center, arc *u'*. With the length of the link and from *c* on the line of engine centers, we mark in at *c'*, the upper end of the link. From *c*, with radius *c f*, we mark an arc *r r*, and from *c*, with the same radius, an arc *h*, and as these two arcs intersect at *e*, it shows that *e* is the center from which we may draw in the center line of the link slot. Then by dividing the link arc, we get the center *m*, of the link and of the saddle-pin. Thus it is shown that when the link is in full gear for either the forward or for the back ward motion, the center of the arc of the link slot, is the center of the eccentric that follows the crank, or is behind the crank, in the path of crank revolution.

In Fig. 84, we have the parts in the positions they occupy when the link is in mid-gear, and the crank on the dead center D, and it is seen that the eccentric *f* whose rod is connected to the lower end of the link now leads the crank, and by the same construction, as in the previous examples, that the center line of the link, may be drawn from the center of the eccentric (*e* in this case) that follows the crank, which is found to be always the case when the link is in full gear for either the forward, or backward motion, and the crank is on the dead center. The amount the lead of the valve will be increased by moving the link from full to mid-gear, is shown by the distance on the line of centers, between the arc *c' c*, and the center-line of the link slot. By inserting a pencil in the center of the pin connecting the eccentric rod to the link, we are enabled to trace the path of motion of the end of the eccentric rod. Thus in Fig. 85, we have the path of motion of the forward eccentric, when in full forward gear, and for the stroke, when the piston is moving from the crank end, to the head end of the cylinder. The line *l,l*, corresponds to line *l,l*, in Figs. 83, and 84. During the return stroke, the rod obviously moves in the same path, but the events occur in a different part of the figure, as may be seen from Fig. 86, which is for the return stroke. In Fig. 87, we have the path of motion of the pin, that connects the backward eccentric to the link, the latter being in full backward gear. The events for the stroke, when the piston is moving from the crank end, to the head end, are named outside the line and are lettered *a*, *b*, *c*, etc. For the other stroke when the piston is moving from the head end to the crank end, the events are denoted by letter only, thus at D, the crank is on the dead center, at *a'*; the port is fully opened as a steam port, at *b'*; the valve begins to close the port; at *c'*, the cut-off occurs; at *d'*, the port opens for the exhaust; at *e'* the port is full open for the exhaust; at *f'*, the port begins to close to effect the cushion, and at *g'*, the cushioning begins, lasting of course, until the valve opens for the lead. In Fig. 88, we have a diagram of the openings etc. of the ports, the full lines, representing the full backward gear, and it is seen that the events are very nearly equalized, there being but about $\frac{1}{4}$ inch difference in the points of cut-off.

In Fig. 89, we have the path of motion of the forward gear eccentric-rod eye, when the link is hooked up to cut-off, at one half, in the forward gear. For the piston stroke from the crank end, the letters outside the lines are used, thus, at B, the crank was on the dead center at the crank end; at *a*, the port end was fully opened, and immediately began to close. At C, the cut-off occurred; at *d*, the port opened as an exhaust port; at *e*, the port was fully opened to the exhaust; at *f* it began to re-close, and at *g*, the port was closed to the exhaust and the cushioning began. For the other stroke, when the piston moved from the head end to the crank end, and the head end cylinder port, comes under consideration, D, is the point at which the crank was on the dead center; at *a'*, the port was fully opened; at *c'*, cut-off occurred, and so on, the events for this stroke, all being marked inside the lines of the figure, and corresponding to those used for the same events on the other stroke,

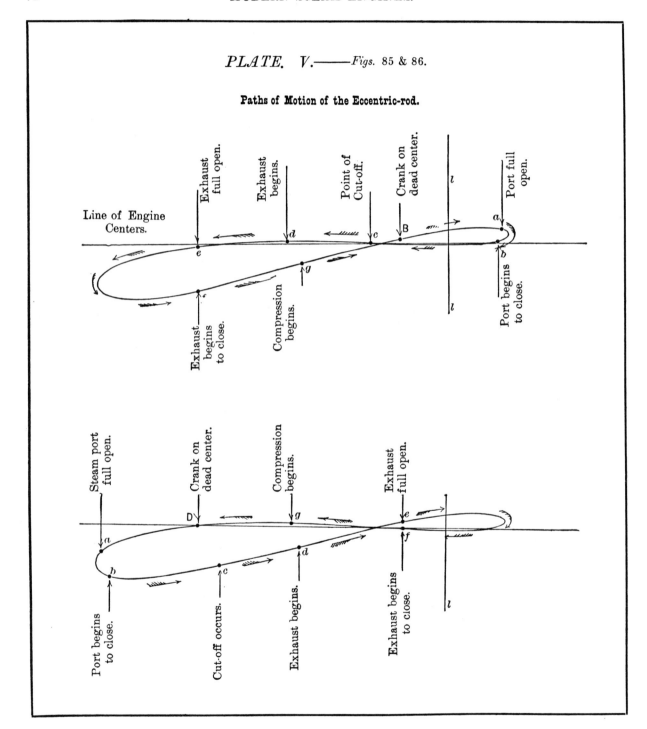

PLATE. V.——*Figs.* 85 & 86.

Paths of Motion of the Eccentric-rod.

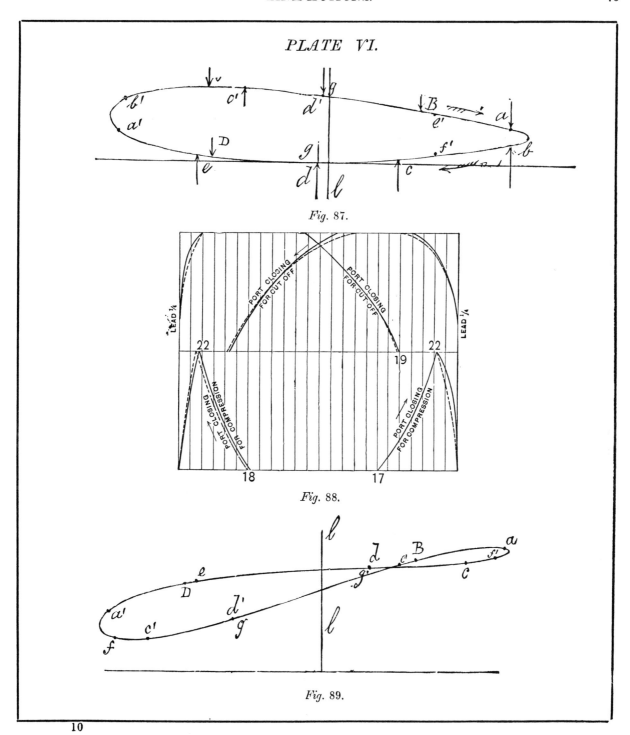

PLATE VI.

Fig. 87.

Fig. 88.

Fig. 89.

except that to each letter a distinguishing dot or mark is added.

The line *l*, in all these motion curves, corresponds to line *l*, in Figs. 81, 82, and 83, representing the link hanger when it stands at a right angle to the line of

represented by the full lines, while the backward gear is marked in dotted lines. It is seen that the points of cut-off are exactly equal, for both gears backward and forward. Thus, considering the port at the crank end B, (which corresponds with B on the motion curves)

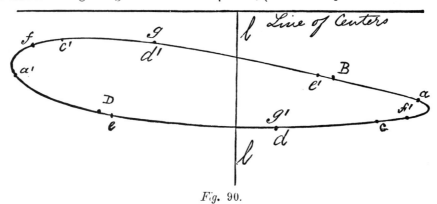

Fig. 90.

centers, and the link is in mid-gear. In Fig. 90, we have the motion curve of the backward eccentric-rod, when the link is hooked up to cut off at half stroke

the cut-off for both gears, occurs at 12 inches of the stroke, while for the stroke from the head end D, of the cylinder, the cut-off occurs at the 14 th. inch of the

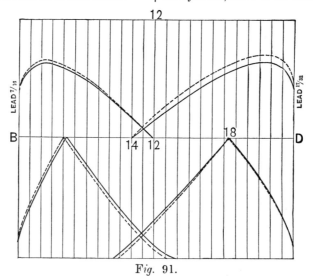

Fig. 91.

In Fig. 91, we have a diagram of the port openings etc., for the cut-off at half stroke, on the stroke from the crank end to the head end of the cylinder, (the designated points of cut-off, refer to this stroke, throughout all the examples, letting the cut-off for the other stroke come where it will). The forward gear is

stroke. The points of release and cushion, are almost identical, and it is seen that the events are as nearly equalized, as possible. The lead, however, has been increased in hooking the link up from the $\frac{1}{4}$ inch, it was with the link in full gear to $\frac{7}{16}$, at end B, and $\frac{17}{32}$ inch at end D, both strokes from end B, and both from D, thus having their amounts of lead equalized.

It will be noted, that the ports open widest on both strokes, when the link is in backward gear, and that the port at the head end, opens wider than that at the crank end, whether the link is in backward or forward gear. This latter point is of advantage, because the piston moves faster when moving from D, to half stroke, than

moving (at the crank pin end), away from the line of centers during one part of the stroke, and towards the line of centers, at another part.

A diagram of the port openings, when the link is hooked up to cut off at one quarter stroke, is given in Fig. 92, the full lines representing the forward, and the dotted lines the backward motion, and it is seen, that

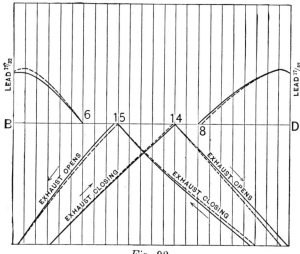

Fig. 92.

it does when moving from B, to half stroke, and therefore requires a wider port opening, in order to admit the steam more readily.

The cause of the faster piston motion, has already

the events are here, also, very nearly equalized, or occur at nearly epual points on both the backward and forward motions, while the port at the head end still gets the greater opening.

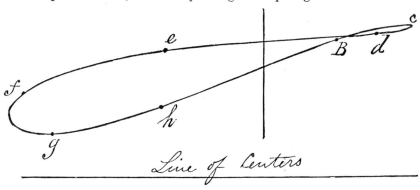

Fig. 93.

been explained, with reference to plotting out valve motions, and is not therefore repeated more than to recall the fact, that it is caused by the connecting-rod

The point of cut.off for the head end D, is still two inches later, on account of the connecting-rod motion, as before explained· When the link is in mid-gear, the motion of the forward eccentric-rod eye will be as in Fig. 93, while that of the backward rod will remain a loop as before. The port openings, are now as in

ports. But this alteration, will act to vary the points of cut-off. Suppose, for example, that instead of locating the line *l l*, Figs. 81, 83, and 84, mid-way between the

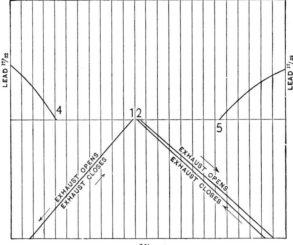

Fig. 94.

the diagram, Fig. 94. The valve lead is here, $\frac{17}{32}$ inch and the ports only open as steam ports to that amount, the valve moving to close the steam port, as soon as the crank pin leaves the dead center on either stroke.

extremes of the eccentric-rod motion, we locate it mid-way between the extremes of the link-block travel (when the link is in full forward gear) the lines *a, d*, Fig. 83, being drawn from the ends of link-block travel, in-

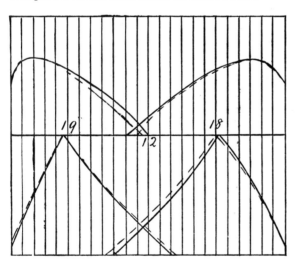

Fig. 95.

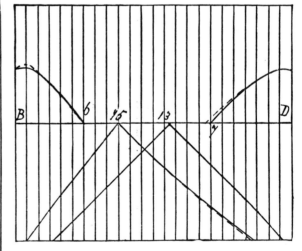

Fig. 96.

We may slightly improve the admission, by putting the point of suspension of the link hanger forward, which will cause the ports to open wider, as steam

stead of from the eccentric-rod travel, and line *l l*, being drawn mid-way between them, thus throwing the center of the lifting-shaft nearer to the link. The effect in full gear, is to cause the point of cut-off to occur (on the stroke from the crank end to head end), ⅝ inch, earlier for the backward, than for the forward gear, and on the other stroke, to delay the point of cut-off, making it a half inch later than in Fig. 88. At half gear the difference is more marked, as may be seen from Fig. 95, which on comparison with Fig. 91, shows the point of cut-off for the backward gear to be nearly equalized, and the amount of port opening nearly equalized, but the points of cut-off for the backward gear, vary from those for the forward, whereas in Fig. 91, they were identical. At quarter cut-off the events are very nearly equalized, as may be seen from Fig. 96. But there is another point to be noted, inasmuch as the cut-off occurs at the 7th. inch, whereas in Fig. 96, it occurs at the 8th. inch, on the stroke from D, to B. In midgear, the events for backward and forward are equalized, but the cut-off is at 4 and 5 inches respectively, as seen in Fig. 97, whereas it was in Fig. 96, at 4 and 5½ inches respectively. Thus, it will be seen that the difference caused by altering the position of the lifting-shaft is not very marked, and possesses both advantages and disadvantages.

SHIFTING THE POSITION OF THE SADDLE-PIN.

If, instead of setting the saddle-pin central to the width of the link-slot as in the foregoing examples, we set it back, bringing it on the chord of the arc of the link-slot, we shall find that the equalization of the events is impaired, as may be seen on a comparison of Fig. 96, with Fig. 98, the latter being a diagram of the port openings, with the saddle-pin thus set back. The point of link hanger suspension is the same for both figures, and the steam supply is seen to still be better for the backward, than it is for the forward gear.

Summarizing these results, we find that for the equalization of the events of cut-off etc., for the forward and backward motions, it is preferable to suspend the link hanger, as in Figs. 81, and 84, while to obtain port openings, as nearly equal as possible for the backward and forward gears, it is preferable to suspend the top of the link hanger, so that the lines *a d*, Figs. 81, and 84, are drawn to represent the ends of travel, of the link-block, when the link is in full forward gear, the line *l l*,

Fig. 81, being drawn mid-way between the lines *a d*, and the lifting-shaft center being set from line *l l*, as described in Figs. 81, 83, and 84.

EQUALIZING THE LEAD.

It has been shown, that in proportion as the link is

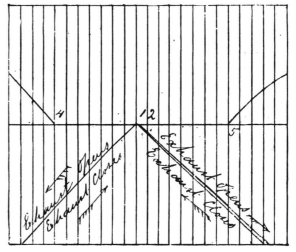

Fig. 97.

moved from mid-gear, to or towards full gear, either

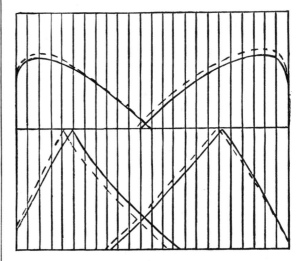

Fig. 98.

backwards or forwards, the valve lead is increased,

and it is a disputed point among engineers, as to whether this is an evil or not. If it is considered desirable we may, however, to a great extent, equalize the lead for the forward gear, by moving the two eccentrics

the forward stroke. The positions of the link, are found by the method given in the previous examples, and it is seen, that putting the eccentrics forward, will not affect the lead in mid-gear, since it merely swings

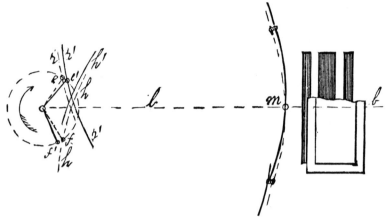

Fig. 99.

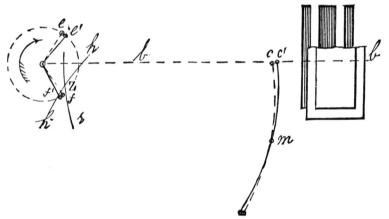

Fig. 100.

forward upon the shaft, and adding steam lap to the valve. In Figs. 99 and 100, for example, we have the center lines of the parts, and the positions of the link, when in full gear and in mid-gear. In both cases, the parts are drawn in dotted lines, and also in full lines.

The dotted lines represent the eccentric and link, drawn in their normal positions, while the full lines represent the eccentrics moved forward, to equalize the lead on

the link on its pivot at m. But in full forward gear, moving the eccentrics forward increases the lead of the valve, making it equal to the lead at mid-gear; the amount of increase of lead, due to moving the eccentrics, being equal to the radius $c\ c'$.

Suppose, now, that as in the case of our previous examples, the eccentrics are set in their normal positions, and the lead at full gear, is $\frac{1}{4}$ inch, and we see in the diagrams, Figs. 88, and 94, that at mid-gear it had increased to $1\frac{7}{32}$ inch. But we may add to the valve sufficient lap to reduce its lead to $\frac{1}{4}$ inch mid-gear, and

This equalization, however, is only obtained at the expense of the backward gear, as may be seen from Fig. 101, in which the link is placed in full gear backward,

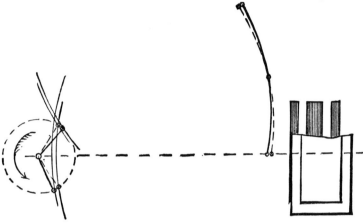

Fig. 101.

then set the eccentrics forward, so as to give $\frac{1}{4}$ lead in full gear, Fig. 99 showing that putting the eccen-

and it is seen that moving the eccentrics forward has reduced the lead. If the amount the eccentrics are

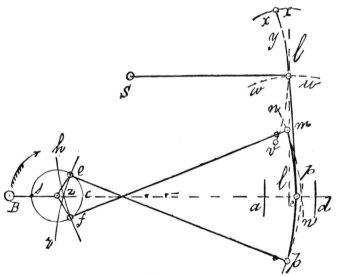

Fig. 102.

trics forward, will not affect the lead at mid-gear, and it being obvious that, for full gear, we must set the eccentrics forward enough to compensate for the $\frac{1}{4}$ inch lead that has been added, and thus equalize the lead for the forward gear.

moved forward is sufficient to about equalize the lead for forward gear, it must have been sufficient to move the valve $\frac{7}{32}$, when in full forward gear, and $\frac{7}{32}$ inch of lap, would require to be added to the lap, and it is evident that if, with the eccentrics in their normal positions, there is ¼ lead at full backward gear; then moving the eccentrics ahead, will have left but $\frac{1}{32}$ inch lead, in the full backward gear, hence we find that equalizing

pond to those used in previous figures, and we find the position of the parts by the same means as in previous examples, except as follows. Having marked from *s*

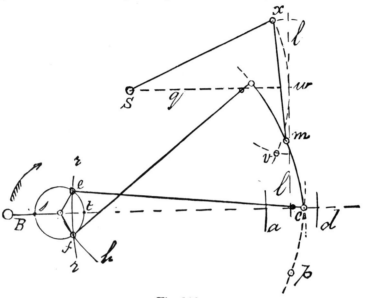

Fig. 103.

the lead for the forward gear, gives ¼ inch lead at mid-gear, and but $\frac{1}{32}$ inch, in the full backward gear.

LINK MOTION WITH CROSSED RODS.

Figs. 102, and 103, represent a link motion with crossed rods, the rods being crossed when the crank is on the dead center B, and open when it is on the opposite dead center.

In this case, the link is brought above the line of centers when in full gear, instead of below it, as in the case of open rods.

This occurs, because the forward eccentric is connected to the lower end of the link, and as a result, the reversing lever requires to be pulled backward, in order to put the link in forward gear, and vice versa, whereas in open rods the reversing lever is moved forward, for forward gear, and backward, for backward gear.

The same eccentric (that which leads the crank), is still, however, the acting one, as may be seen by comparing Fig. 103, with Fig. 81, both being for full forward gear, and eccentric *e*, in both cases operating the valve.

In Fig. 103, the lines and letters of reference, corres-

(and with the radius from the center of the eccentric to the center of the link-slot), as a center the arc *a*, and from *t* as a center the arc *d*, and having drawn *l l* midway between *a* and *d*, we then take the length of the link-hanger, and mark arc *w*, and where this arc cuts line *l*, is the line *q*, on which the lifting-shaft arm will fall for the mid-gear.

We then draw the lifting-shaft, and from its center S, draw the arc *y y'*. Turning now to the full gear with crossed rods (Fig. 103), we mark in the position of the parts by the same means as in previous examples, and with the length of the link-hanger as a radius, and from the center *m* of the link, we draw arc *x*, giving us the point of suspension of the upper end of the link hanger. We may then take the same radius, *m, x*, of the full gear, and mark from *m* the point *v*, which is the point of suspension for full backward gear. It is shown therefore, that in crossed rods, the points *x*, and *v*, are equidistant from *w*, which is not the case with open rods.

The proof is, that we may, by means of a dotted arc, find the position of the saddle-pin when the link is in full backward gear, this point being at *p* in Fig· 103, and it will be found in that figure, that the distances from *p* to *v*, from *c*, to *w*, and from *m* to *x*, are all equal, these positions being those of the link hanger when in full gear forward, full gear backward, and mid-gear respectively.

Similarly we may prove the construction from the mid-gear as follows:

When the link is in mid-gear, the hanger stands from *c* to *w*; at full gear forward, the center of the saddle-pin, and therefore of the link, will be at *m*, and the hanger will stand at *m x*; at full gear backward, the center of the saddle-pin will be at *p*, and the hanger from *p* to *v*; and radii *c*, *w*, *p*, *v*, and *m x* are equal, while *x* and *v*, are equidistant from *w*, which it has been shown is not the case with open rods.

THE VARIATION OF LEAD IN CROSSED RODS.

It may now be pointed out, that in the case of crossed rods the lead is diminished by moving the link from full gear to mid-gear, (instead of being increased as was the case with open rods), and this may be shown as follows:

If, with the radius from the center of the respective eccentrics, to the center of the link-slot, we mark the dotted arcs *p p* and *n n'*, Fig. 102, we get at their intersection on the line of centers, the position of the center of the link-block when the link is in full gear either for backward or forward motion, and the figure shows that the link-block, in mid-gear, stands at *c*, hence the lead has decreased the amount represented by the distance of *c* from the intersection (on the line of centers) of the dotted arcs *p p*, with arc *n n'*.

EQUALIZING THE LEAD WITH CROSSED RODS

We may, however, equalize the lead, by putting the two eccentrics an equal amount backwards upon the shaft, which will have no effect upon the lead at mid-gear, but will decrease it for the full forward gear. Thus in Fig. 104, the parts are shown in mid-gear, the normal position of the eccentrics and link, being shown in dotted lines, and their positions with the eccentrics put back, shown in full lines. By the methods before

11

described, we find the position of the link with the eccentrics at *e* and *f*, and also its position with the eccentrics at *e' f '*, and as the center of the link-slot, and

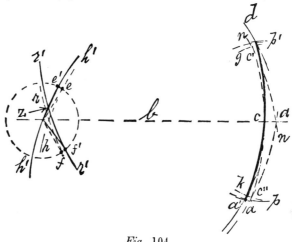

Fig. 104.

therefore of the saddle-pin, falls at *c* on the line of centers in both link positions, it is clear that altering the position of the eccentrics does not move the link center, whereas in full gear it is obvious that moving the eccentric back will decrease the lead, and we may therefore reduce the amount of steam lap so as to give the requisite

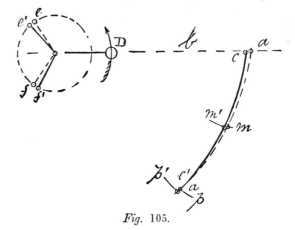

Fig. 105.

amount of lead which will remain perfectly the same at all points in the forward gear, and up to mid-gear. But this will disarrange the lead for the backward gear as may be seen from Fig. 105, in which it is seen that setting the eccentrics back from the dotted to the full lines, brings the link back from the dotted, to the full lines and as the crank is on the dead center D, this increases the lead, whereas in the forward gear it decreases it, and it follows that the equalization of the lead by means of moving the eccentrics, and adjusting the lap to suit, is only permissable when the engine performs its principal duty while running in one direction, because to whatever amount we improve the lead in one full gear, we derange it in the other full gear. The port openings and closures, are practically the same for corresponding points of cut-off, whether the rods are crossed or open.

LINK MOTION WITH THE ALLEN VALVE

We may greatly improve the steam admission, whether

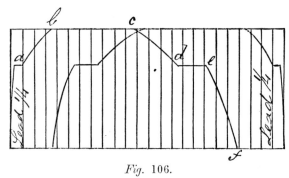

Fig. 106.

the link motion has either open, or crossed rods, by employing the Allen valve in place of the common slide valve.

The construction of the Allen valve, has been shown

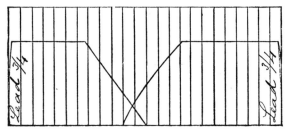

Fig. 107.

in the engravings from Fig. 31, to Fig. 43, but it is for the shorter points of cut-off that it possesses its greatest advantages.

By substituting an Allen valve on the open rod link

motion, shown in Fig. 78, and of the proportions already given, we obtain the port openings given in Figs. 106, and 107.

As the Allen valve is double ported, we may obtain the $\frac{1}{4}$ inch lead at full gear that was given in previous examples, by giving the valve $\frac{1}{8}$ inch lead, the two ports giving an amount of lead opening equal to the $\frac{1}{4}$ inch of a single port, hence, we set the eccentrics back, reducing their angular advance sufficiently to alter the lead from $\frac{1}{4}$, to $\frac{1}{8}$ inch.

The superior admission of the Allen valve is obtained as follows:

In Fig. 108, the port K is shown open to

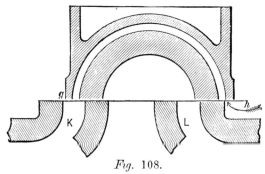

Fig. 108.

the amount of the lead, and as this is $\frac{1}{8}$ inch only, the eccentric is in a better position to open the port quickly, having a less angular advance, and being more nearly at a right angle to the crank, than would be the case if the lead at g were $\frac{1}{4}$ inch. Nevertheless the effective lead is $\frac{1}{4}$ inch, because there is $\frac{1}{8}$ inch opening at h, as well as at g.

When the valve moves, the effective opening remains at double the opening at g, until the valve reaches the position shown in Fig. 109, the opening at g, and at e then being equal, and after this, the amount of effective port opening will remain constant or equal, until the supplementary port is closed as in Fig. 110, because we see in Fig. 109, that to whatever amount e closes (as the valve moves to the right), the opening at g will increase, thus maintaining an equal amount of port opening, while the valve moves from its position in Fig. 109, to its position in Fig. 110. This period of equal port opening is shown on the diagram, Fig. 106 at a, after the valve has reached the position shown in Fig. 110, the port opening is governed by the outer edge only of the valve, the supplementary port being closed, hence, the line of admission from *a*, to full port at *b*, is the same as a common slide valve would give; providing it had ⅛ inch lead.

When the valve has completed its stroke, and returns to close the port, and effect the cut-off, the amount of

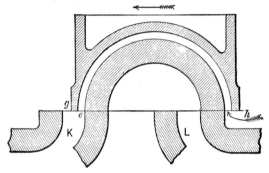

Fig. 109.

port opening is the same as for an ordinary slide valve, until the valve has reached the position in Fig. 111, this period being shown in the diagram Fig. 106, from *c* to *d*. But after this point, the supplementary port will again come into action, maintaining the port opening

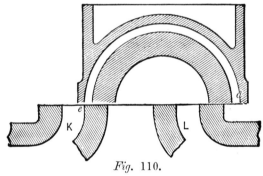

Fig. 110.

equal, until the valve has moved to the position shown in Fig. 110. This equal amount of port opening, is shown in the diagram, Fig. 106, from *d*, to *e*; *d* being for the valve position in Fig. 110, and *e*, for the valve position in Fig. 111. After this point, the amount of opening at *e*, closes as fast as that at *g*, but the effective opening still remains double what it would be, in a common slide valve.

The port openings for the Allen valve, when the link

is hooked up to cut off at half stroke, are shown in Fig. 107, and it is seen that the admission is very much superior to that obtained with a common slide valve, as may be seen by comparing Fig. 107, to Figs. 91, or

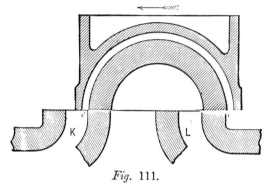

Fig. 111.

95. This superiority is maintained both at quarter cut-off, and at mid-gear, as may be seen on comparing Figs. 111, and 112, with Figs. 92 and 94.

INCREASE OF LEAD WITH THE ALLEN VALVE.

The Allen valve, it may be observed, increases the lead more than the common valve when the link is hooked up, thus at half stroke, the lead is ⅜ for the

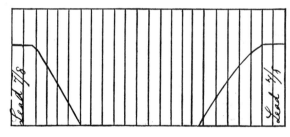

Fig. 112.

edge of the valve, and ⅜ for the supplementary port, or a total of ¾ inch. We may, however, equalize the lead for the forward gear (at the expense of the backward gear), by the construction explained with reference to figures 99 and 100. Similarly, if crossed rods be used for the link motion, we may equalize the lead, by means of moving the eccentrics back, as already explained; but it is to be noted that, as there are two ports, and ⅛ inch of actual valve lead becomes ¼ inch of effectual valve lead, therefore, with ⅛ actual valve lead (the eccentrics not being set back to equalize the lead), the ports would come blind quicker when the link is moved towards mid-gear, than would be the case with the common valve, having the same amount of effectual lead. This corresponds to the action of the Allen valve in increasing the lead to a greater amount when the

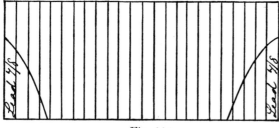

Fig. 113.

link has open rods, and is moved towards mid-gear. But as the eccentrics have less angular advance for a given effective amount of lead in the Allen, than in the common valve, therefore they are in a better position for being moved to equalize the lead, if such be desired.

GOOCH'S LINK MOTION.

In Gooch's link motion, the arc of the link is in the opposite direction to that of Stephenson's link motion, as will be seen in Fig. 114. The hanger is pivoted at its upper end to a fixed pin, and the direction of valve

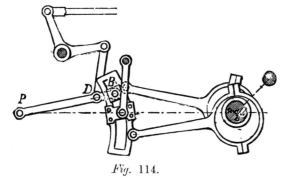

Fig. 114.

motion is reversed by moving the link-block, which is upon an arm that is pivoted at P, to the slide spindle, the lifting-shaft arm being connected at D.

Fig. 115, shows the parts in full gear for the back-

ward motion, the crank being on the dead center B. Shorter points of cut-off are obtained by lifting the arm E, and therefore the link-block, nearer to the line of centers, or to the saddle-pin. The lifting-shaft S, is attached to the arm E, and the curve of the link-slot is an arc of a circle, of which the length of arm E is the radius. As the link stands at a right angle to the line of centers, when the crank is on either dead center, the lead is maintained equal, it being obvious that the link-block can (with the link at a right angle to the line of centers), be moved from end to end of the link, without imparting any motion to the valve spindle A.

The position necessary for the upper end of the link hanger, in order to thus equalize the lead for all points of cut-off, may be found as follows:

In Fig. 115, let *e* and *f*, represent the center lines or throw lines, of the two eccentrics, whose positions for any given amount of lap and lead, may be found as in previous examples, and with a radius equal to the length, from the center of the eccentric to the center of the link-slot, we mark from *e* the arc *n*, and from *f* the arc *m*. From the line of centers *b b*, we mark, with a radius equal to the length of the link (measured from one extreme link-block position to the other), the two arcs *p* and *p'*, and these are the ends of the link. With the radius of the arm E, and from a point *x*, on the line of centers, we mark in the position of the center line of the link, when the crank is on its dead center furthest from the cylinder. Then with the radius *f m*, and from *f'* as a center, we mark arc *m'*, and from *e'* the arc *n'*, giving, at their intersections with the arcs *p p'* respectively, the positions of the ends of the link when the crank is on its dead center D.

From the ends of the link when in this position, and with the length of arm E as a radius, we find, on the line of centers, the point Z, which is the center from which the center line of the link (when the crank is on dead center D), may be drawn.

We have thus found the two link positions for the two crank dead centers, and mid-way between them we draw the line *l*, on which the upper end of the link hanger must stand, and to find the heighth at which it must stand from the line of centers *b b*, we take the length of the link hanger, and from *a*, (the center of the link in one position), on the line of centers mark an arc *a'*. Then from *d*, (the center of the link in the other position), on the line of centers, we mark arc *d'*, and where these arcs cut the line *l*, is the position for the center of the pivot, from which the link hanger on the line of centers, and the link will be vertical, hence the link-block may be moved from end to end of the link without imparting any motion to the valve spindle. To find the positions for the lifting-shaft and

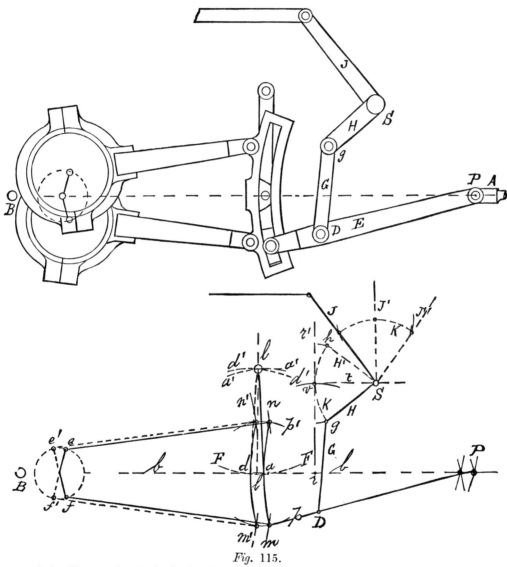

Fig. 115.

must be suspended. The arc, in which the link hanger will swing, is denoted by arc F F, which crosses the line of centers at *a* and *d*, so that when the crank is on either dead center, the center of the saddle-pin will be its arms, we take the radius D P, and from P as a center, mark a point *r* on the line of centers, and from *r*, erect a perpendicular *r r'*; then take the length of arm G, and from *r* mark an arc *v*, and from the intersection of

v with $r\ r'$, draw a line t, then from v, mark, with the length of the arm H, the center S for the lifting-shaft. From S, as a center, the arc K may be drawn, and then with the length of G as a radius, and from D, as a center we mark at g, on the arc K, the position of the upper end of the arm G. When the link-block is in mid-gear, the arm E will be parallel with the line of centers, and the position of G will be from r to v (at a right angle to the line of centers), and we may therefore take the radius $v\ g$, and from v, as a center, mark arc h, the position of arm H, when in full gear for the forward motion; hence the three positions for arm H are: H; on the line t; and at H; and the corresponding positions for the upper arm of the lifting-shaft, are J, J', and J''.

The points of cut-off, with a Gooch link motion thus constructed will, vary to an amount depending upon the proportion the length of the connecting-rod bears to the length of the stroke, as in all previous examples; and if we were to move the positions of the eccentrics (as was done in Figs. 99, 100 and 101) in order to equalize the points of cut-off for either gear, the lead will no longer be equalized because the link will not stand vertical when the crank is on the dead center; hence moving the arm E, Fig. 115, up and down the link will impart motion to valve spindle A and, therefore, to the valve, thus altering the lead. The points of cut-off may, however, be equalized, and the lead maintained equal, by making one steam port wider than the other, as will be explained in connection with a link motion having a rocking-shaft.

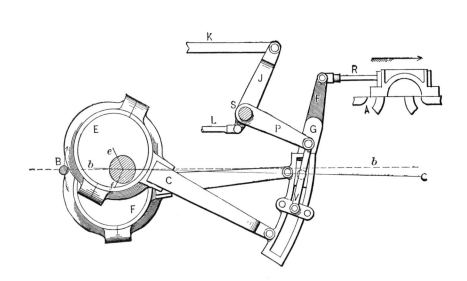

Fig. 116.

CHAPTER IV.

LINK MOTIONS WITH ROCK SHAFT.

Fig. 116 represents a link motion such as is used on American locomotives, the parts being shown in position with the crank on the dead center B, and with the link in full forward gear. The link-block is carried upon the lower arm G of a rocker, rock-shaft, or rocking-shaft, as it is promiscuously termed. The upper arm F' of the rocker drives the valve through the medium of the valve-rod or valve-spindle R. The valve, therefore, moves in the opposite direction to that in which the link-block moves and the eccentrics are therefore at an angle of less than 90° to the crank. The amount to which they are less than 90° is the angular advance.

When the crank is at B, the valve must move in the direction of its arrow in order to open the port A for the admission of the live steam, and as the arm G, which carries the link-block, must move in the opposite direction to the valve, it is clear that it is the eccentric f (which follows the crank) that must be connected to the upper end of the link, and hence the eccentric rods are crossed.

Instead of a weight being used to counterbalance the link, etc., the rod L connects to a volute, or coiled, spring which acts instead of the weight.

THE OFF-SET OF THE ROCKER-ARM.

The center line of the engine is on the line $b\ b$, and it is seen that the center of the link-block is below it on the line $c\ c$, a condition that commonly prevails in American locomotives.

This necessitates that the two arms F G of the rocker be thrown out of line one with the other, in order that the motion of the valve may be equal at all parts of the stroke, to that of the link block.

The upper arm F of the rocker moves the valve in a line parallel to the line of centers, and therefore stands vertical when in its mid-position, while the lower arm G stands, when in mid-position, at an angle to the line of centers $b\ b$, the amount of this angle being called the *off-set*.

To find the amount of off-set, we proceed as in Fig. 117, in which the parts are represented by their centers and center lines and the letters of reference correspond to those in Fig. 116. The line of centers $b\ b$ being drawn, we draw the circle n, its radius equaling the throw of the eccentric, and its diameter the travel of the valve when the link is in full gear. From the center of circle n (and with a radius equal to the distance between the center of the axle and that of the rock-shaft) we draw an arc a, and on this arc measure below line $b\ b$ the distance the link-block is to stand below the line $b\ b$ when the arm G of the rocker is as its lowest point, and through this point draw line $c\ c$. Or, instead of this construction, we may assume a point a and mark it by a dot, then draw above it, and at the required dis-

89

tance, the line of centers *b b*, and on it mark the circle *n*; line C C may then be drawn, passing through the dot *a* and the center of circle *n*. At a right angle to C C, and through the dot at *a* we mark a line *a*, which will be the center line of the lower arm of the rocker when it is in mid-position. From *a*, as a center, and with the length of the lower arm of the rocker, as a radius, we draw an arc *c c*, and where this arc cuts line *a* is the center of the rocker-shaft. From this center, we draw arc *t t*, on which will be the path of motion of the lower end of the rocker-arm, or, what is the same thing, the path of motion of the link-block, and this path of motion will be an equal distance on each side of line *a*.

tremes of motion of the lower rocker arm when the link is in full gear for either the backward or the forward motion. At a right angle to the line of centers, and from the arcs *p p*, we draw lines *u u'*, and where these cut the arc *t' t'*, are the extremes of motion of the eye of the upper arm when the link is in full gear. By this construction, the lower arm G of the rocker vibrates in an arc that is equalized with the line C C, from which it receives its motion, while the upper rocker-arm F vibrates in an arc that is equalized with the line in which it delivers its motion, or, in other words, the line of motion of the slide valve, the valve face being parallel to the line of centers. If the valve face were at an angle to the line of centers, then lines

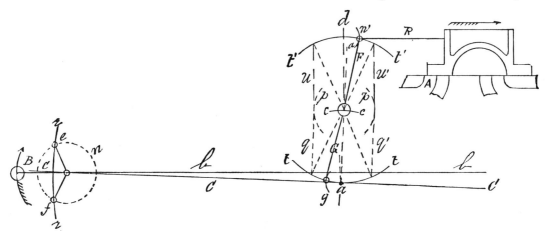

Fig. 117.

From the center of the rocker, and at a right angle to the line of centers *b b*, we draw a line *d*, which will be the center line of the upper arm F of the rocker when it is in mid-position. From the center of the rocker, and with the length of the upper rocker-arm as a radius, draw an arc *t' t'*, on which will be the path of motion of the eye of the upper arm of the rocker, which will vibrate on this arc and an equal distance on each side of line *d*.

To locate the length of arc through which the rocker-arms will vibrate when the link is in full gear, we draw, from the center of the rocker-shaft, two arcs *p p*, their radius equaling that of the circle *n*, and from these arcs we draw lines *q q'* at a right angle to the line *c c*, and where these lines cut the arc *t t* will be the ex-

d, u and *u'* would require to be at a right angle to the valve seat or valve face.

The action of the link motion is not in the least influenced by the off-set in the rocker-arms, providing that, as in this construction, the mid-position of the lower rocker-arm is at a right angle to C C, and that of the upper arm is at a right angle to *b b*, and the action is the same as if the center of the link-block, when at its lowest point, came coincident with line *b b*, and there were no off-set in the rocker-arms.

To find the positions of the eccentrics, we mark from the point *a*, on the line C, and with the steam lap and lead as a radius, an arc *g*, and where *g* cuts arc *t*, is the position of the center of the link-block (or, what is the same thing, the position of the eye in the lower rocker arm) when the crank is on its dead center B. From this center, and with the length of the eccentric-rod as a radius, we mark across the circle *n* an arc *r r*, and where *r r* cuts circle *n*, are the positions for the eccentrics, as marked at *e* and at *f*. In Fig. 118, the construction is repeated for both strokes, the eccentric positions being found by marking arcs *g g'* on the arc *t*, distant from *a* to the amount of the lap and the lead;

gives the eccentric positions when the crank is on the opposite dead center.

To find the position of the rocker-arms.—When the crank is on its dead center B, the valve will obviously have moved from its mid-position over the ports to an amount equal to the lap and the lead, and the rocker-arms will, therefore, also have moved from their mid-positions to the amount of the lap and the lead, hence

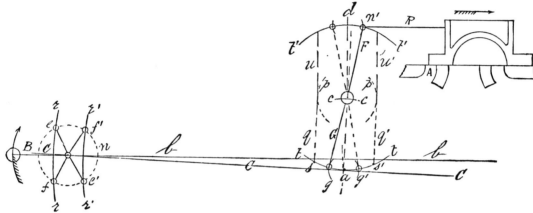

Fig. 118.

arc *r r* (drawn from *g*, with the length of the eccentric-rod as a radius) giving the eccentric positions when the crank is at B, and arc *r' r'*, drawn from *g'* on the arc *t*,

the lower arm will be at *g*, and the upper arm at *n'*, which is distant from *d* to the amount of the lap and the lead.

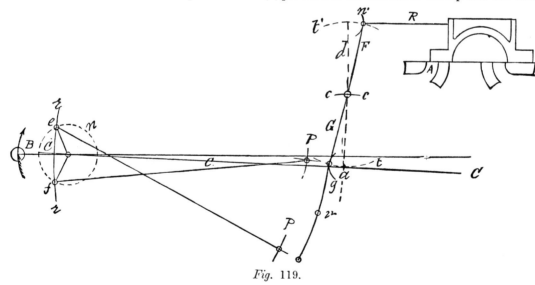

Fig. 119.

To find the position of the link corresponding to crank position B.—In Fig. 119, we have the center lines of the parts (omitting the lines used to find their positions), the link being in full forward gear, and it is found that the center of the link arc is at *e*, or, in other words, the

drawn from the center of the eccentric that leads the crank.

To find the position of the parts when the link is in mid-gear.—From *f*, Fig. 121, as a center, and with a radius from *a*, on the line of centers C C, to the center of the

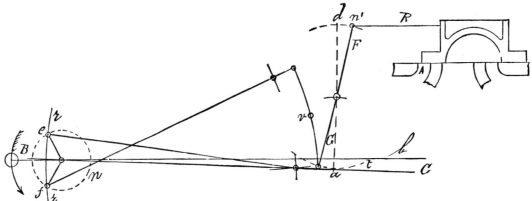

Fig. 120.

center of the eccentric that leads the crank, whereas, in open rods, it is the center of the eccentric that follows the crank. Fig. 120 shows the parts in position for full gear backward, and it is seen that the center of eccen-

crank shaft, we mark an arc E *g*, somewhere on which we know the upper end of the link will be. With the same radius, and from *e* as a center, draw an arc H *g*, and where these arcs cut arc *t* (or at point *g*) is the posi-

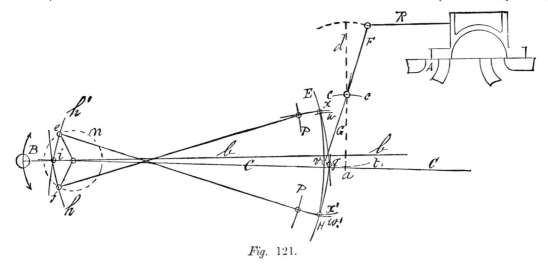

Fig. 121.

tric *f*, that leads the crank, is that from which the link arc may be drawn; hence we have that with the link in full gear, and the crank on the dead center for either the forward or backward motion, the link arc may be

tion of the link-block when the link is in full gear for either the forward or the backward motion.

From *g*, as a center, we mark arcs *w* and *w'* for the ends of the link, and from the points of intersection of these arcs with arcs E and H respectively, and with a radius from *a* (on the line C C) to the center of the crank shaft, we mark arc *h h'*, giving at *i* the center from which the link arc may be drawn.

The position of the center of the link will depend upon the point of suspension of the link hanger. If we suppose this hanger to be suspended at a point coincident with the center of the rocker, then the link-center, as well as the link-block, will move on the arc *t*, and the centers of both will be at *v*; we must therefore mark new arcs *x* and *x'* for the ends of the link.

The increase of lead.—It will now be seen that, in moving the link from full gear to mid-gear, the center of the link-block and of the lower rocker-arm is moved from *g* to *v*, and this moves the valve forward, increas-

action, therefore, so far as the lead is concerned, is the same as a link motion with open rods and no rocker.

The points of suspension of the link hanger.—The points of suspension of the link-hanger may be located with a view to either give the least amount of sliding motion of the link-block in the link slot, and obtain at the same time the widest amount of port opening, or, in the second case, to equalize the points of cut-off for the two strokes. To find the points of suspension for the least amount of sliding motion of the link-block, and for the widest port opening, we proceed as in Fig. 123, in which the link is shown in two positions, viz., full gear forward and mid-gear, and it is obvious that the link suspension would be the most perfect for the full forward gear, if the upper end of the link hanger had

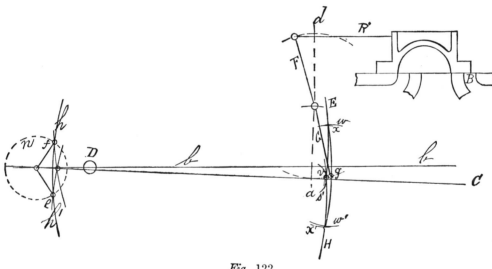

Fig. 122.

ing the lead, the same as in link motions having open rods and no rocker. This only occurs with crossed rods when a rocker is used. The upper end of the rocker-arm obviously stands to the right of the line *d* to the same amount as the lower arm stands to the left of line *a*. In Fig. 122, the link is in mid-gear, with the crank at D, the position of the parts being found by the same means as that employed for Fig. 121, and it is seen that the lead is here also increased by moving the link from full to mid-gear, whereas, in the absence of the rocker (the link having crossed rods), it would diminish. The

its center coincident with the center of the rocker-shaft. so that the link and the link-block would move together in the same arc; but as the lifting shaft arm moves in an arc, therefore, locating the point of suspension for full gear at the center of the rock shaft throws the point of suspension for the mid and backward gear away from the line *a a*, upon which it would require to lie to enable the saddle-pin and the link-block to move as nearly as possible in the same arc, and thus minimize the amount of sliding motion of the link-block in the link slot. Obviously the longer the arm P of the lifting shaft, the more nearly its arc of motion (in lifting the link from full gear forwards to full gear backwards) will coincide with the line *a a*, and the more perfect the suspension will be, or, in other words, the less the sliding motion of the link-block in the link arc. Considerations of room, however, usually limit the length of P to about half that of the eccentric-rods, and the sliding motion of the link-block may be mini-

link-hanger would move in shifting from full gear forward to full gear backward. The points of link-hanger suspension would, in this case, be in the most desirable positions for the full gears, because they are both on the line *a a*, causing the center of the saddle-pin to vibrate on an arc as near as possible parallel to the arc *t t*, in which the link-block moves ; hence there would be a minimum of sliding motion of the link-block in the

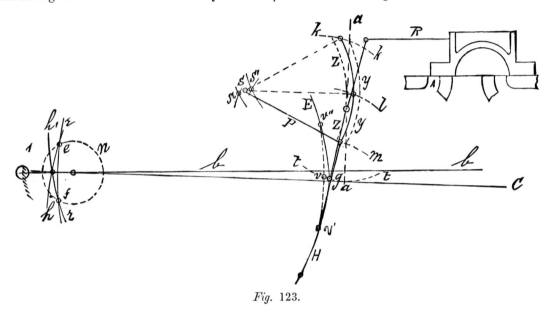

Fig. 123.

mized for the full or mid-gears as follows : For the full gears, we take the length of the link-hanger as a radius, and from *v'*—the center of the link when in full forward gear—draw an arc *m* somewhere on which the upper end of the link-hanger will be when the link is in full forward gear. With the same radius, and from *v* as a center, we mark arc *l* somewhere upon which the upper end of the link-hanger will be when the link is in mid-gear. With the same radius, and from *v''*—the center of the link when in full backward gear—draw an arc *k*, upon which the upper end of the link-hanger will be when the link is in full gear for the backward motion. With the length of the lifting-shaft arm—P, Fig. 116—as a radius—and from the points of intersection of arcs *m* and *k* with the line *a*—find at S' the position for the center of the lifting-shaft, from which may be drawn arc *y y*, on which the upper end of the

link-slot. For the mid-gear, we have so located the hanger suspension that the centers of both the link-block and the saddle-pin arc on the arc *t* at *v*, but as the point of hanger suspension would be to the right of *a a*, the saddle-pin arc would vary from that of the link-block, and sliding motion would ensue.

The reverse of this would be the case if the center of the lifting shaft were located at the point S'', meeting the line *a* at a point coincident with the center of the rocker-shaft, for, in that case, the motion of the hanger, or of the link and of the link-block, would coincide more nearly when the link was in mid-gear than when in full gear, and the amount the block would slide in the link would be lessened at mid and increased at full gear.

By locating the center of the lifting shaft at S, the path of the upper end of the link-hanger, when moving the link from full gear, would be on the full line arc midway between arcs *y y* and *z z*, which passes equally on each side of line *a a*, being as far from that line, when in full gear as it is when in mid-gear, thus equalizing the amount of sliding motion of the link-block in the link-slot.

It may now be pointed out that the employment of a rocker diminishes the amount of sliding motion of the link-block in the link-slot, because the lower arm of the rocker causes the link-block to move in an arc of a circle that more or less coincides with the arc in which the saddle-pin moves, whereas, in the absence of a rocker, the slide spindle is guided to move in a straight line. The link motion, having the same proportions as in previous examples, and as given on page 68, and the length of the arms of the rocker being each 12 inches, the port openings will be as follows :

In Fig. 124, the full lines are forward full gear while

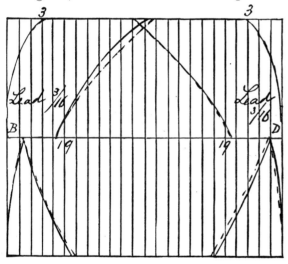

Fig. 124.

the dotted lines represent the full gear backward, the saddle-pin being on the line of the center of the link-arc and the hanger suspended on the arc denoted by the full line in Fig. 123, so as to minimize the amount of sliding of the link-block in the link, and it is seen that the events are as nearly equalized, as is necessary, for the backward and forward gears. Thus when the piston moves from B to D, the cut-off is at 19½ inches for the forward gear and 19⅜ inches for the backward

gear, while when it is moving from D to B, the cut-off is at 19⅝ inches for both gears. Similarly, the points of release and cushion are very nearly equalized.

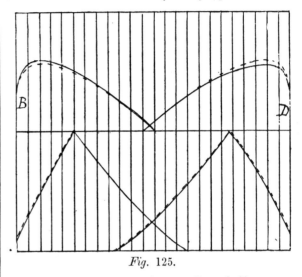

Fig. 125.

The port openings for the cut-off, at half-gear, are shown in Fig. 125, and it is seen that the events are all equalized for the backward and forward gears, but that the point of cut-off is an inch later when the piston is

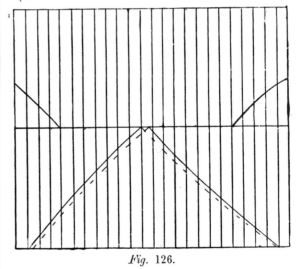

Fig. 126.

moving from D to B than when it is moving from B to D.

Fig. 126 shows the port openings for mid-gear, the

port only opening to the amount of the lead, and beginning to close as soon as the crank leaves the dead center. The lead is here again equalized, and it is seen, from these three figures, that the lead increases as the link is moved from full gear to mid-gear. This, however, may be remedied, for the forward gear, at the expense of the backward gear, by the means described for open rods, and with reference to Figs. 99 and 100, the only difference being that as the rods are in this case crossed, the eccentrics must be set back instead of being set forward, as in the case of open rods.

EQUALIZING THE POINTS OF CUT-OFF.

Referring now to the equalization of the points of cut-off, it may be effected in five ways, first by giving to the valve more steam lap at the head end than at the crank end, as has already been explained in connection with the subject of diagrams for designing valve motions, secured by making the steam ports of unequal widths, as will be explained presently; third by suitably locating the point of suspension of the upper end of the link hanger, or in other words, suitably locating the lifting-shaft; fourth by suitably locating the position of the saddle-pin, or in other words, by setting it back towards the eccentric-rod eye, instead

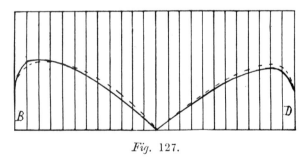

Fig. 127.

of having it on a line with the center-line of the arc of the link-slot, and fifth, by lifting the link to different amounts for corresponding points of cut-off for the two gears, but in proportion as we set either the points of hanger suspension, or the saddle-pin back, we decrease the amount of port opening, especially at the head end of the cylinder.

Fig 127, for example, is a diagram of the port openings, the arc of hanger suspension having been moved back three inches, and corresponding with arc Z Z in Fig. 123, (supposing that arc to be distant 3 inches from the middle or full line arc of hanger suspension shown in that figure.) If we compare Fig. 128 with Fig. 125, we find that the equalization of the points of cut-off has been accomplished at the expense of the port opening, because while we have caused the cut-off to occur an inch earlier on the stroke from D to B, we have reduced the port opening for that stroke. The arc of hanger suspension being moved another three inches back, the port openings will be as in Fig. 128, and it is seen, that although the points of cut-off are still equalized, the port openings are widely distorted, which is a serious defect. It is seen, therefore, that the points of hanger suspension may be located considerably to the left of the line *a a*, without much influence on the points of cut-off.

The cause of the equalization of the points of cut-off and reduction of port opening by means of the points of suspension, may be seen from Figs. 129 and 130, in which the parts are shown in the positions they occupy at the point of cut-off when that point is at half stroke, *b b* is the line of centers of the engine *t t*, the arc in which the link block moves, and P the path of saddle-pin motion, while *x x* is an imaginary line parallel to

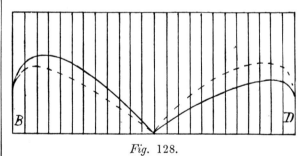

Fig. 128.

line *b b*, and inserted merely to compare the arc P with. It is seen here that the saddle-pin is (on account of moving on the arc P) lifted up towards the line of centers *b b*, and this lifting obviously, from the position of the link in Fig. 129, moves the lower arm of the rocker to the right and the upper arm to the left, thus hastening the point of cut-off; this, however, is obviated by locating the point of suspension lower down on the arc Z, fixing its position so that the cut-off will, on the stroke represented in Fig. 129, occur at half-stroke.

We have now to consider the return stroke, Fig. 130,

ing the point of cut-off on the stroke from D to B, and equalizing the points of cut-off for the two strokes.

THE EFFECT OF GIVING THE VALVE OVERTRAVEL.

In the previous examples, the travel of the valve has

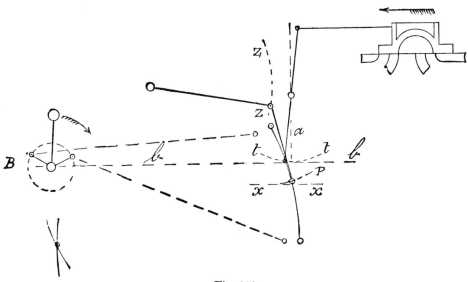

Fig. 129.

and it is here again seen that setting the arc Z back has caused the link to lift on the arc P, and this lifting has, on account of the position the link hangs in, moved the

been $4\frac{1}{2}$ inches, so that the steam edge of the valve traveled but $\frac{1}{8}$ inch more than the amount necessary to fully open the steam port. It is customary in Ameri-

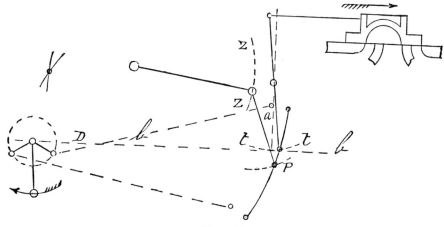

Fig. 130.

lower rocker-arm to the left and therefore the upper rocker-arm and the valve to the right, and so hasten-

can locomotive practice, however, to give to the valve from $\frac{3}{8}$ inch to $\frac{5}{8}$ inch of overtravel. Leaving the ports

1¼ inches and the steam lap at ⅛ inch, as in our former examples, we may, therefore, now increase the travel of the valve to 5⅜ inches, giving ₁⁹₆ inch overtravel.

The port openings in full gear will now be as in Fig. 131, and it will be seen, on a comparison with Fig. 88, that the admission has been hastened, and the points of cut-off delayed, reducing the amount of expansion. Furthermore, the exhaust opening has been reduced by the reclosure from *a* to *b* and from *c* to *d*, which occurs because of the valve partly closing the cylinder exhaust port, as seen in Fig. 132, at *e*, the overtravel of the valve being shown at *x*. This may obviously be remedied by widening the bridges and correspondingly widening the valve.

The overtravel of the valve has no effect upon the port openings at shorter points of cut-off, which will remain the same as if there were no overtravel at full gear. But the overtravel will render it necessary to lift the link more in order to effect the cut-off at a given point, thus bringing the link-block, for any given point of cut-off (less than full gear), nearer to the center of the link, and thus slightly diminish the amount of sliding motion of the block in the link.

EQUALIZING THE POINT OF CUT-OFF BY MAKING THE STEAM PORTS OF DIFFERENT WIDTHS.

It may now be pointed out that the points of cut-off may be equalized by making the steam ports of different widths.

In Fig. 133, for example, we have the parts drawn one-eighth full size, the outer circle representing the path of the crank, and the circle *n* representing the path of the eccentric. The smallest circle has a radius equal to the amount of steam lap, which is equal for both ports.

The connecting-rod has the usual proportion of three times the length of the piston stroke, which is, in this example, 24 inches; hence the connecting-rod is 72 inches. The width of steam port is 1¼ inches, the steam lap ⅛ inches and the valve travel 4¼ inches, the latter being just enough to fully open both ports for the admission.

With these dimensions, it may be found by means of Zeuner's diagram (which has already been fully explained in Chapter II) that the point of cut-off for the

port at the head end will occur at 20 inches of piston stroke, while that for the port at the crank will occur at the nineteenth inch.

Let it be supposed that the points of cut-off are to be equalized at 20 inches, and we proceed as follows :

In Fig. 133 we draw, from a center C, the outer

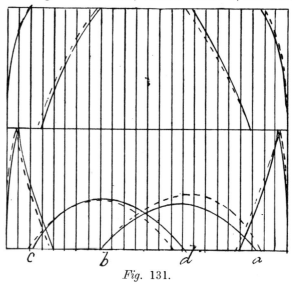

Fig. 131.

circle representing the path of the crank-pin, and a circle *n* representing the path of the eccentric center, the inner circle *d* having a radius equal to the amount of steam lap. From the edge of circle *d*, we draw a

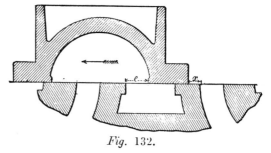

Fig. 132.

vertical line *e*, and a line from C, cutting the intersection of *e* with circle *n*, gives the position of the eccentric at the point of cut-off, the crank-pin being at S.

We may now draw the ports and the valve in position to effect the cut-off. We have now to find how much to cut out the port at the crank end in order to

Fig. 133.

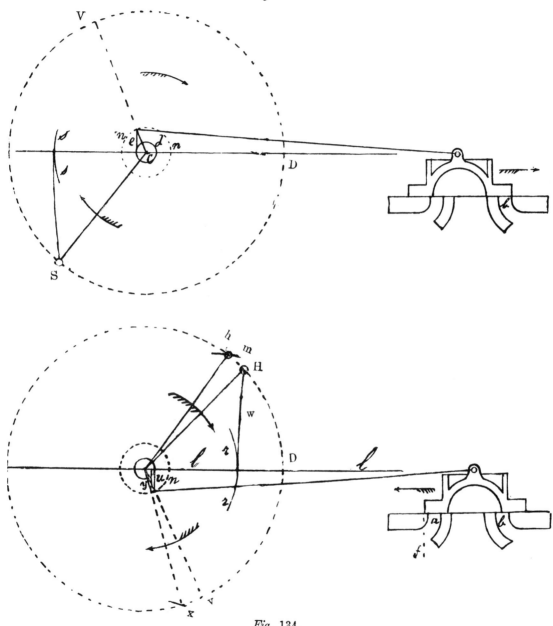

Fig. 134.

cause the cut-off to occur at the twentieth inch instead of at the nineteenth, and this may be done as follows: In Fig. 134, arc *r* represents the position the piston must be in when the cut-off occurs (being the same distance from B that *s* is from D). From the lap circle *d*, we drop a perpendicular line *u*, and a line from C,

13

passing through the point of intersection of u with circle n, gives us the position of the eccentric when the cut-off will occur, if the port a is made the same width as port b. By prolonging the eccentric throw-line to v, we are enabled to find the crank position by taking the radius S V (Fig. 133), and marking from v, in Fig. 134, an arc m. A line drawn from the point of intersection of m with the outer circle to the center represents the throw-line of the crank. We have thus found the positions of the crank and of the eccentric when the cut-off will occur, the ports being of equal width, and we may now find the positions they ought to be in in order to equalize the points of cut-off. To do this, we draw from the arc r, and with the length of the connecting-rod as a radius (this length being represented by three times the diameter of the outer circle), the arc w, giving us at H the position the crank ought to have arrived at when the cut-off occurred, and we find that, as the cut-off occurred when the crank was at h and the eccentric at v, it occurred too early in the stroke, because the crank ought to have arrived at H. Now, suppose it to have reached position H, and we may find the corresponding eccentric position by taking the radius $h\ v$ and marking from H an arc x, giving us the required eccentric position at y.

To find the difference in the positions of the valve when the eccentric is in the positions denoted by v and x respectively, we draw from the point of intersection of line v with the circle n a perpendicular line u, and from u point of intersection of line x with the circle n, a vertical line y, and the distance, measured on the line of centers $l\ l$, between these two vertical lines, is the amount the valve will have traveled past the steam edge of the port.

Now, suppose we cut out the port a to the dotted line f, and it is seen that the crank will be at H, the eccentric at y, and the cut-off just effected, the piston being at r and the points of cut-off equalized by cutting out port a to line f.

Let it be supposed that it is determined to equalize the points of cut-off by altering the width of port b, or in other words, on the stroke when the piston is moving from the head end D to the crank end B of the cylinder, then it is necessary to proceed as follows:

In Fig. 135, point r is located to represent the nine-teenth inch of piston motion from B, and an arc w gives at H the crank position at the time the cut-off occurs. A vertical line e, touching the lap circle d, gives, at its intersection with the eccentric path n, the throw-line V of the eccentric, which is prolonged to v in order to get on the outer circle the position the eccentric will be in when the crank is at H. Having found the positions of the crank and eccentric at the time of cut-off, we may draw in the valve, the cut-off at port a being just effected. For the other stroke we proceed as in Fig. 136, in which the circles correspond to those in Fig. 135, the arc s being the same distance from D that r is from B in Fig. 135, and therefore representing the piston position for equalized points of cut-off.

With a radius equal to three times the diameter of the outer circle E, and from a point on the line of centers $l\ l$, we mark, from s, an arc, giving us at H' the crank position at the time of equalized cut-off.

To find the corresponding eccentric position, we take the angle the eccentric stands at to the crank or radius $h\ v$, Fig. 135, and mark, from H', Fig. 136, an arc g, and from where g cuts the outer circle draw a line x giving us the eccentric position corresponding to crank position H.

But by drawing a vertical line u from the lap circle and a line C c, we find that the cut-off will not occur until the eccentric throw has reached the line C c, which will be too late to give an equalized cut-off, because when the eccentric has arrived at c, the crank will be at H'' instead of at H' as may be proved, because radius H' g equals radius H'' c.

In the figure, the valve is drawn in the position it would occupy when the crank was in its proper position H', the port b being still open, and it becomes clear that, in order to equalize the point of cut-off, we may make the port b narrower, bringing its steam edge on the line p.

The amount to which we must decrease its width may be found by dropping a perpendicular line y from the point of intersection of line x with circle n, and measuring the distance between lines u and y. It is obvious that it is preferable to draw the outer circle to a diameter equal to the full travel of the valve, and let it represent the crank path on a reduced scale, as the lines will be clearer, and correctness may more easily be ob-

Fig. 135.

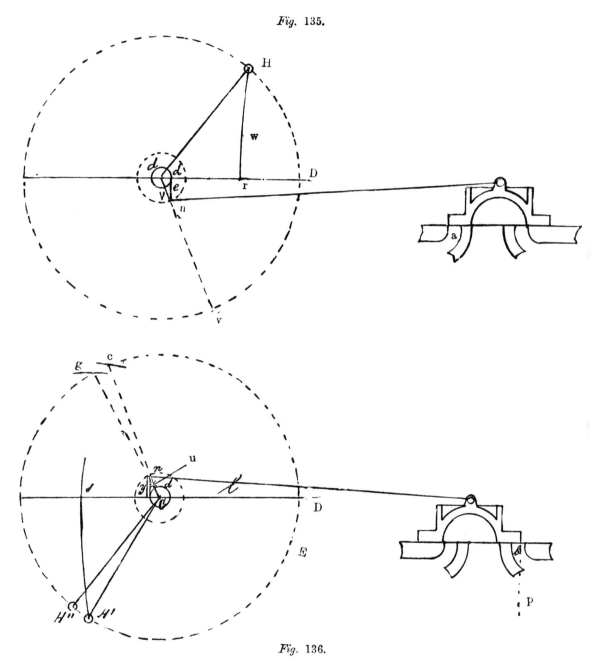

Fig. 136.

tained. Equalizing the points of cut-off, by thus making the ports of different widths, possesses the advantage that the laps of the valve are maintained equal, and the valve may be put on end for end without being put on wrong, which might occur if the laps, instead of the ports, were made of unequal widths.

MODIFIED FORMS OF LINK MOTION REVERSING GEARS.

A link motion, in which but one eccentric is employed, is shown in Fig. 137, which is taken from *The American Machinist.* The saddle-pin is on the line of centers A and remains there, the link-block being moved along the link-slot to vary the point of cut-off, or reverse the direction of motion, as in Gooch's link motion. On the rod E that connects the link-block and slide spindle, there is provided a latch F (on the end of rod D) which, in conjunction with the notches on the concave side of the link, holds the link in its adjusted position. In this arrangement, there is the objection that when the link-block and rod E are at the same end of the link as the eccentric-rod, the weight of both E and the eccentric rod is borne by the eccentric.

In the Stephenson's link motion, shown in Fig. 78, and the Gooch's link motion shown in Fig. 114, the lifting-shaft is shown above the link, but it is often more convenient, from the construction of the engine, to place it below the link, as in Figs. 138 and 139, which are from *Mechanics.* This plan is common upon engines for hoisting purposes.

STEAM REVERSING GEARS.

In cases where the power required to move the link motion is more than can be exerted by hand, a special steam cylinder is employed to assist in moving the reversing shaft, examples of this kind being given in Figs. 140 and 141, which are from *Mechanics.* In Fig. 140, A is a steam cylinder, with ports, slide-valve, steam chest, &c., of the ordinary construction. C is the main reversing lever, the bottom end of which is keyed to the tumbling shaft; its upper end projects a few inches above the quadrant, and is provided with a detent for latching it to the quadrant. The connecting-rod H, of the steam cylinder, is attached to this lever at B. D is the hand lever, having its fulcrum at l on the main lever. Its upper end has a handle and thumb latch, of the ordinary construction, attached by a rod to the detent on the main lever. When the hand lever is in the central position, the pin at its lower end coincides with the axis of the tumbling shaft (if the gear is not at the end of the tumbling shaft, the shaft will require to be cranked to allow for this). On the main lever, at

K, are stops which limit the motion of the hand lever to the travel of the valve. The lower end of the hand

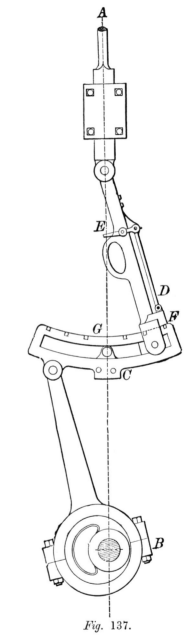

Fig. 137.

lever is connected by links and levers to the valve-stem, as shown. The weight of the links, &c., is balanced by a weight in the usual way. The engineer unlatches the detent, pulls the hand lever in the direction he wishes it to go, when the lever moves until its lower end strikes the stop K. The steam valve is then in the proper position to admit steam to the cylinder A, and its piston assists him to move the lever. When he wishes to

screw can be arranged at E with two nuts, and used for regulating the amount of the travel of the piston, and thus effecting the cut-off in the cylinder by changing the travel of the main valve. This makes a very satisfactory gear for reversing on a reversible rolling-mill engine. For hoisting purposes, where great delicacy is

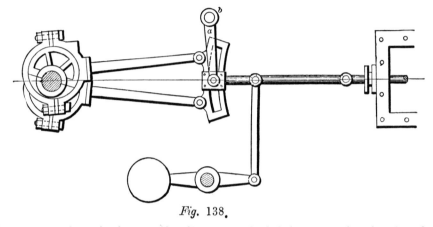

Fig. 138.

reverse, he pushes the hand lever in the opposite direction, which changes the position of the valve and reverses the operation. He can thus move the lever and latch it at any point of the quadrant. In Fig. 141, the hand lever is dispensed with, the steam piston being attached directly to the reversing lever. A is the

required, it is open to the objection that it has nothing positive by which the load may be stopped at a fixed point. This arrangement may be modified by the addition of a cataract as shown in Fig. 145, in which A is the steam cylinder and B a hollow piston, having an arm C, for carrying the rod D, on which is the cataract

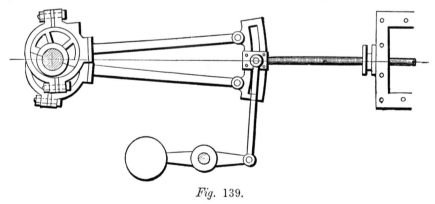

Fig. 139.

reversing cylinder, B the valve chest, C the small lever for moving the valve, and D D the rods connecting to the links. There is a valve on the exhaust pipe of this small cylinder, which is used for regulating the rapidity of the motion of the piston. A right and left hand

piston E; at F, is the rod for the valve that admits steam to A, and at G is the piston rod that attaches to the lifting-shaft arm. H H is a pipe having communication with each end of the cataract cylinder, which is filled, on each side of the piston E, with water or oil.

Now suppose that steam is admitted to cylinder A, and it is clear, that if cock J of the cataract pipe is closed, the piston in A cannot move, because it is connected, through its rod and the arm C, to the cataract piston

until, by passing beneath the auxiliary valve, its ports are again closed, when motion will close, because the steam can neither enter nor exhaust from the cylinder, hence the motion of the piston follows that of the hand

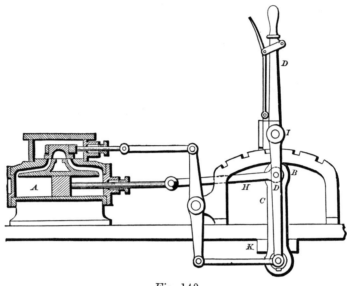

Fig. 140.

E. But if we open the cock J, then the parts will will move, the water, on one side of piston E, passing through pipe H to the other side of E. The speed with which motion will ensue, obviously depends upon how widely cock J is opened, or in other words, upon how fast the water can pass from one to the other side of piston E. By regulating the amount of opening of cock J, therefore, the motion of the steam piston in cylinder A, may be made as slow as desired, enabling the engineer to shut off steam to A, and thus arrest the motion of the reversing gear, at any aequired point with great precision. The cataract cylinder, may obviously be placed at the end of the steam cylinder, as shown in Fig. 143, in which case one piston rod serves for both the steam and the cataract pistons. Fig. 144, (from *Mechanics*), represents a form, in which an auxiliary valve is employed, the main valve connecting to the lifting shaft at B, while the auxiliary valve is operated by a hand lever.

In this arrangement the hand lever may be operated to open the ports in the main valve, which will move

lever, and when the link motion has moved to the re-

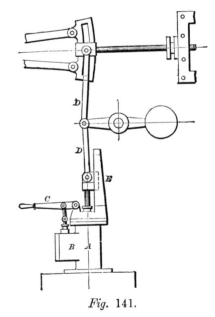

Fig. 141.

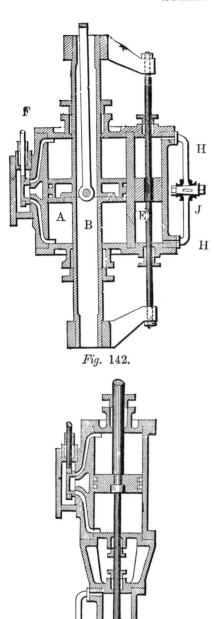

Fig. 142.

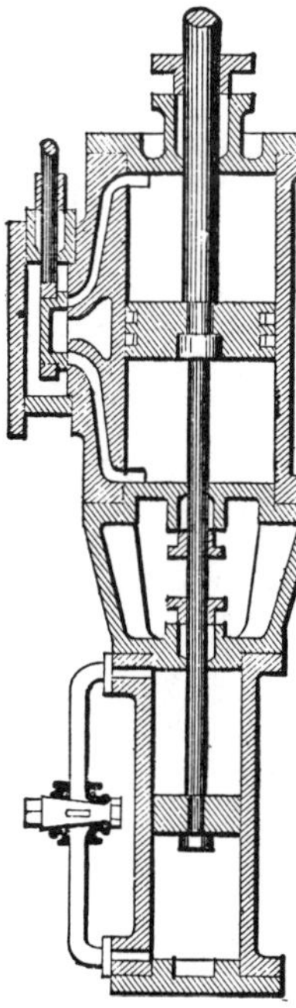

Fig. 143.

quired position, the lever handle may be released and will remain at rest.

Figs. 145 and 146 represent a steam reversing gear, designed by Mr. W. E. Good, and employed on locomotives on the Philadelphia and Reading Railroad. In Fig. 145, the gear is shown in position on the engine, while in Fig. 146, it is shown detached with the cylinder in section. If this gear is simply started, it will continue its movement in the same direction until, by dropping the latch into the quadrant notch at the de-

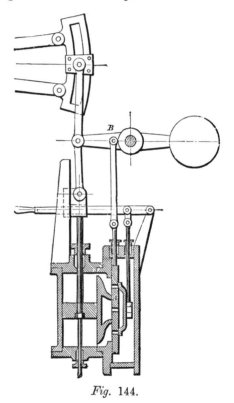

Fig. 144.

sired point of cut-off, further movement is arrested. The link gear of the engine is simultaneously adjusted to the position corresponding with that of the reverse lever. It remains fixed in that position until further change is made by the reverse lever.

From this general statement, it will be seen that not only is the link or other gear of the engine moved by the steam reverse, but also that the reverse lever has communicated to it, by the peculiar arrangement of the reverse gear, the necessary force to move it to any desired direction, without any effort on the part of the engineer, save that of starting it. In Fig. 146, A is a steam cylinder, and B the piston whose rod C is guided by two glands, which are screwed up somewhat tightly. C is attached to the arm D of the reverse shaft. E is the steam valve operated by the rod E², which is pivoted at F¹ to the lever F. The arm D² F² connects D to F, and the rod G connects the handle H to F at the point F³. The valve E covers the steam ports by about

rest. The link is held by friction on the piston rod C. If, for any reason, the piston B should begin to creep, its rod would move the rod D at C², the motion being increased by D² by reason of the increased length of leverage. This increased motion would be conveyed to F, which would operate on its pivot E³, moving F¹ and, therefore, E² to the left. E² would move the valve E so as to admit steam through the passage a' to the side b of the piston. The piston would, therefore, be resisted by the steam pressure, and would be held

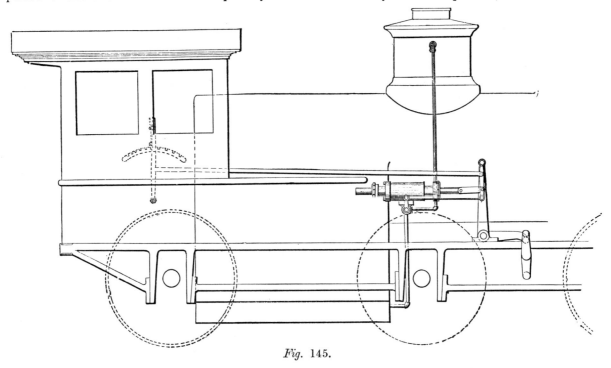

Fig. 145.

$\frac{1}{32}$ of an inch. The gear is shown in the full lines as being in mid-position. If the lever H is moved to the right, it will, through G, move F on F² as a pivot, and give end motion to the rod E², and by this means to the valve E, admitting steam through the passage way a, and moving B in the same direction as the handle H is moved. At the same time, the link will be lowered. When H is brought to a stop, the arm F operates on its pivot F³, and as D continues to move to the right, it moves F² to the right and F¹ to the left, so that the rod E² closes the valve E, and the piston B comes to

stationary by it. Since there is but $\frac{1}{32}$ inch lap on the valve E, it will be seen that the increased motion at D², over that at C², will cause steam to be admitted through a' with a very small amount of piston motion. This amount may be regulated at will by increasing the height of D² above C².

It will be observed, therefore, that this apparatus embodies a very simple and ingenious motion. The action of the parts centering on the fact that moving H forward causes F¹, E², E and therefore B, to move in the same direction. On the other hand, if motion begins at B instead of at H, then F¹, E² and E move | sults are obtained in whatever direction or to whatever

Fig. 146.

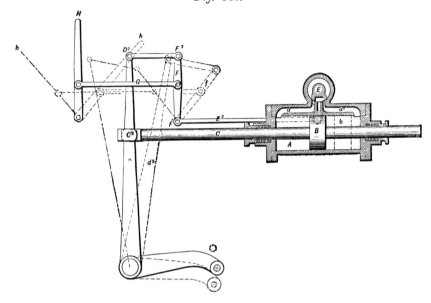

in the opposite direction, causing steam to enter and check the motion of B. It is obvious that the same re- | amount H is moved, and that the checking or detaining action is entirely automatic.

14

CHAPTER V.

Adjustable Cut-off Engines.

When a separate valve is employed to effect the cut-off, and the cut-off valve is set to operate at some fixed point in the piston stroke, this point being varied by altering the position of the valve by hand, the engine is called an adjustable cut-off engine. The construction of such an engine is shown in Fig. 147, which represents a design by the Lane & Bodley Co. The main shaft, driving shaft, or crank shaft, is furnished with two eccentrics, the inner of which (whose rod is shown at A, in the figure) operates the main valve, while the outer (whose rod is shown at B), operates the cut-off valve, c is the guide for the main, and d, that for the cut-off valve spindle. The hand wheel at E, is for operating a screw, which moves the cut-off valve so as to cause it to cut off at the required point in the stroke, there being at F an index, to enable the cut-off to be set at the required point. Representatives of various kinds of cut-off valves are given as follows:

Fig. 148 represents the arrangement of what is called Meyer's cut-off. The main valve rides upon the seat in the same position as the common slide valve, hitherto treated of, the ports in the cylinder remaining the same as for a common slide valve. The main valve, however, is provided with ports or openings, K and L, through which the steam passes to the cylinder ports. Upon

108

the back of the main valve slide two cut-off valves, so called because their sole office is to cut off the steam by closing the ports K and L, leaving the points of admission, the amount of valve lead, the exhaust and the compression to be governed by the main valve, whose action, so far as these events are concerned, is effected precisely the same as it would be by a common slide valve having the same laps and angular position of eccentric. This will be seen from the lower half of Fig. 148, in which a common slide valve is shown in place of the main valve, and it is apparent that the edge h of the cut-off valve will cut off the steam the same as edge h of the common slide valve, and, also, that the edges J of the two valves would cut off alike if their eccentrics occupied the same positions. It is also apparent that when edge h of the main valve has closed the port b, the expansion will begin, independent of any action of the cut-off valve. The longest distance the steam can be allowed to follow the piston is, therefore, governed by the main valve, the action of the cut-off valves being confined to effecting the cut-off at some earlier point in the stroke. This it effects for one stroke by the edge G of the cut-off valve passing over the edge g of port L, and for the other stroke by the edge M, passing over edge N of the port K. The point in the piston stroke, at which these two events will occur, depend

PLATE VII.

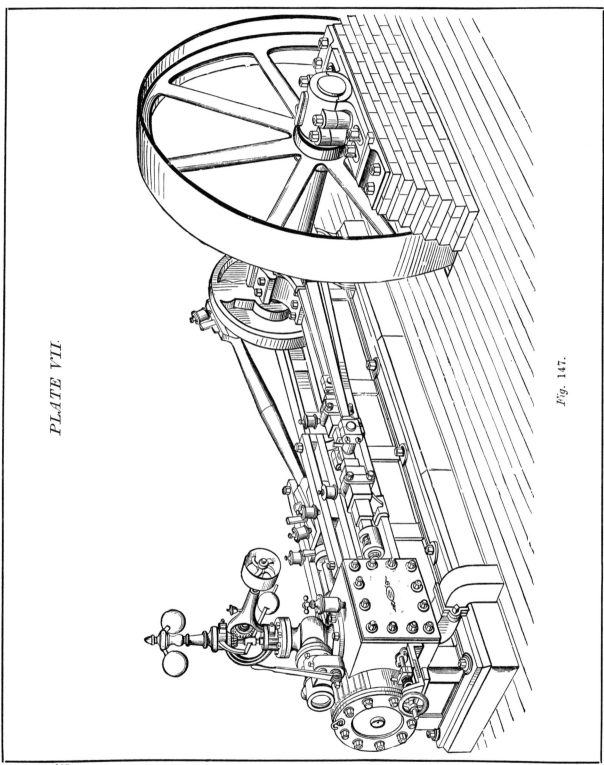

Fig. 147.

upon the distance that G stands from *g*, and M from N, if both the valves were at mid-travel, as shown in the figure. To regulate this distance, so as to effect the cut-off at different points in the piston stroke to suit the amount of power the engine may be required to possess, the cut-off valves are provided with nuts P, which are a sliding fit, in pockets in the backs of the valve, and through these nuts passes a right and left

desirable point, and incapable of moving from that point. But if the amount of work performed by the engine varies, or when the steam pressure varies, the screw affords means of either varying or maintaining the power of the engine by moving the cut-off valves to alter the point at which live steam will be cut-off.

In Fig. 148, the main and cut-off valves are shown in the positions most convenient for explaining their

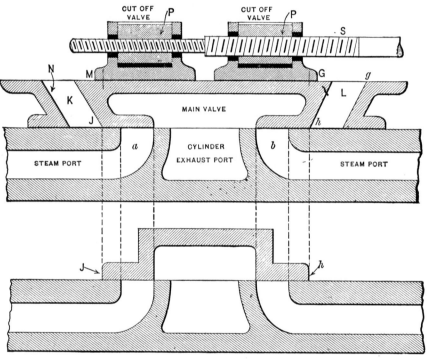

Fig. 148.

hand screw S, which, on being revolved, moves the valves, either lessening the distance G *g* and M N, or increasing it, according to the direction in which S is revolved.

The object of providing the cut-off valves with the nuts P, is to permit them to seat themselves on the main valve (notwithstanding their wear) without bending the screw S.

If the work performed by the engine was constant in amount and the steam pressure was also constant, the cut-off valves might be kept in one position on the screw, being adjusted to cut off the steam at the most

construction, and not in that in which they would stand when properly set upon the engine.

Suppose the cylinder steam ports, and the ports in the main valve, to be an inch wide and the main valve to have an amount of steam lap equal to half the width of the cylinder steam ports, or in other words, $\frac{1}{2}$ inch of steam lap, which will give a cut-off at about $\frac{9}{10}$ piston stroke, and the travel of the main valve being just sufficient to fully open the steam ports the action of the main valve will not be unduly distorted by excessive lap nor over valve travel.

Suppose the outer edges of the cut-off valve, if placed

In mid-position on the main valve, as in Fig. 148, to be distant from the nearest edges of the main valve ports also to an amount equal to half the width of the cylinder steam port, and these being assumed to be average conditions, we may follow the movements of the parts as follows ·

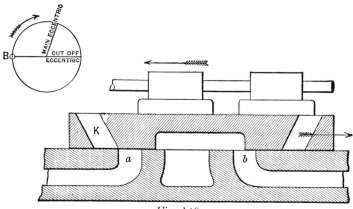

Fig. 149.

In Fig. 149, the crank is on its dead center at the crank end B, the cut-off eccentric being, in this example, set exactly opposite to the crank, and the main valve having no lead. It is here seen that, the valves moving in the direction denoted by their respective arrows, the admission of the steam will occur through ports K and *a*, uninflueuced by the cut-off valve.

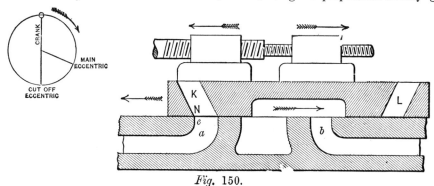

Fig. 150.

Fig. 150 shows the position of the crank, the eccentrics and the valves when the cut-off occurs, the crank being at half-stroke, and it is seen that the cut-off is effected by the cut-off valve independent of the main valve.

Fig. 151 shows the positions of the valves at the point of release or exhaust for the port *a*, thus completing the events for one piston stroke. In moving across the port to effect the cut-off, the cut-off valve reduces the effective width of steam port opening, and it is necessary to take this into account in considering the steam admission. In Fig. 152, for example, it is seen that, although the main valve steam port K is full open to the cylinder port *a*, yet the effective width of port opening for steam admission is the amount the cut-off valve leaves the port K open.

We may clearly perceive the action of a cut-off valve ,having the proportions already given, by means of the diagram, in Fig. 153, in which lines A A and B B are an inch apart, that being, in this case, the width of the steam port, the lengths of these lines is 2¼ inches, representing, on a scale of one-eighth full size, a piston stroke of 20 inches, hence each of the vertical lines indicate a piston stroke of one inch. The piston being moved one inch, the width of steam port opening is measured and found to be (on an engine having a connecting-rod whose length equals three times the length of the piston stroke, or what is the same thing, six times the length of the crank) ⅝ inch. We mark, therefore, on line 1 a dot ⅝ inch distant from line A A.

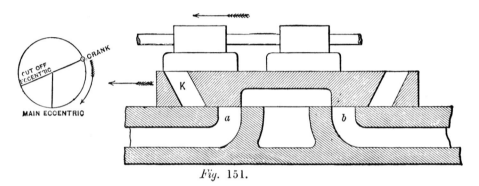

Fig. 151.

The piston may then be moved another inch, the width of port opening measured and marked by a dot on line 2. The piston is then moved to its third inch, the port opening is measured, and its width marked on line 3.

Having continued this process, we may draw through the dots a line which will clearly show the manner in which the steam port was opened and closed; this

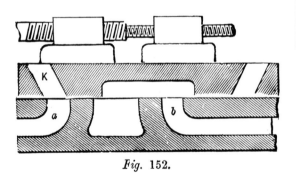

Fig. 152.

being shown in the figure by a full line. Thus we find, from the figure, that the port was opened full when the piston had moved 3½ inches, and that the point of cut-off is at the 9th inch of piston stroke. The manner in which the main valve would have cut off steam is shown on the diagram by dots. Thus we find that if

the cut-off valves were taken off, the main valve would begin to close the steam port at 9½ inches of piston stroke and the cut-off would occur at 17¾ inches of piston stroke. The events for the return stroke are plotted out in the same manner in Fig. 154, and it is seen that the cut-off valve began to act when the piston had moved 4½ inches, and before the main valve had

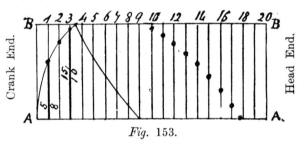

Fig. 153.

cut-off valve has begun to act, is here shown by a broken line. As the compression, lead and exhaust are

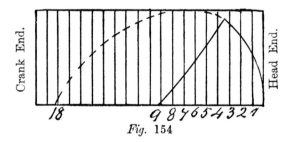

Fig. 154

fully opened the steam-port, which is, therefore, never fully opened. The action of the main valve, after the

entirely uninfluenced by the cut-off valve, it is unneces sary to refer to them in connection with its action.

THE POSITION OF THE CUT-OFF ECCENTRIC.

In an adjustable cut-off engine, the cut-off eccentric is usually fixed upon the crank shaft. When the engine is required to run in either direction, the cut-off eccentric must, in order to enable the engine to run equally well in either direction, be set directly opposite to the

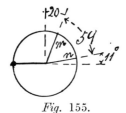

Fig. 155.

crank. But if the engine is required to run in one direction only, there is some latitude in the position in which the cut-off eccentric may be set.

width of the steam port) and we may investigate the limits, within which the position of the cut-off eccentric may be varied as follows: In Fig. 155, the circle represents the path of the crank pin and also the path of the center of the eccentric, the line n represents the throw-line of the cut-off eccentric, and m, that of the main eccentric, the former being set at 169° ahead of the crank and at 59° ahead of the main eccentric, whose angular advance is 20°. The positions of the parts, at the point of cut-off, is shown in Fig. 156, and it is seen that the cut-off eccentric has passed the point x, at which it would move the valve the quickest, it follows therefore, that while the cut-off valve has been moving across the port K to effect the cut-off, the eccentric has been moving through that part of its path in which it moves the valve most rapidly. Continuing the motion of the parts, we have, in Fig. 157, their positions at the

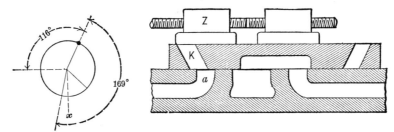

Fig. 156.

Let it be supposed that the cut-off valves are so set that if placed in mid-position (as shown in Fig. 148), on the main valve, the edges G and M would be distant

time that the main valve cuts off, and it is seen, on comparing Fig. 156 with Fig 157, that the main and cut-off valves have moved in the same direction and at nearly

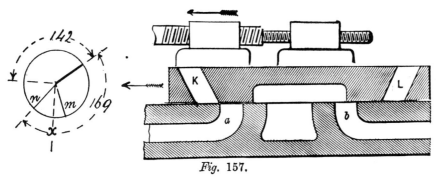

Fig. 157.

from the edges H J (Fig. 148), of the ports in the main valve to an amount equal to the steam lap of the main valve, (which, in this case, is equal to one-half the

the same speed. But at the moment the parts are in the positions shown in Fig. 157, the main valve is moving faster than the cut-off valve, for two reasons, first, because its eccentric throw-line is nearer to its mid-position x, and second, because the eccentric moves the valve faster and further while moving a given amount towards and ending at the line x, than it does while moving an equal distance after passing it.

This is shown in Fig. 158, in which L L represents the line of motion of the valve, m the position of the center of the eccentric when 35° on one side of x, and n its position at the same angle on the other side of x. Setting a pair of compasses to represent the length of the eccentric-rod, rest one point at m, and mark, on

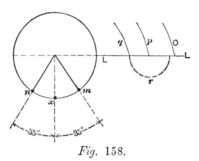

Fig. 158.

the line of centers L L, the arc o. Then rest one point of the compasses at x and mark the arc p, and from o to p, measured along the line L, is the distance the valve would be moved while the eccentric moved from m to x. Rest the compasses at n and mark arc q, and from p to q, on line L, is the distance the valve would be moved while the eccentric moved from x to n. The difference between the two amounts of valve motion being shown by the dotted arc r. Now, suppose that the main eccentric is at m, and it is clear, from the positions of the valves in Fig. 156, that the position of the cut-off eccentric must be such that, from and after the cut-off, it will move its valve at least as fast as the main valve is moved, or else the steam-port K will be reopened, and live steam again admitted until such time as the lap of the main valve itself effects the cut-off.

In Figs. 159 and 160, we have diagrams showing the steam distribution, effected with the eccentrics set as in Figs. 156 and 157, and if we compare these diagrams with those shown in Figs. 153 and 154 (for which the cut-off eccentric was set directly opposite to the crank), we find, in the case of the forward strokes, the cut-off is at 9 inches in one case, and at 13 inches in the other;

for the return strokes, it is at 9 inches in one case and at 16 inches in the other. We also find that from the time the cut-off valve commenced to close the port, until final closure and cut-off took place, there was, for the two forward strokes, 5½ inches of piston stroke in one case and 9 inches in the other, and it appears that by setting the cut-off eccentric back 11°, we have delayed the point of cut-off, and wire drawn the steam. The term *wire drawn* meaning that the pressure of the

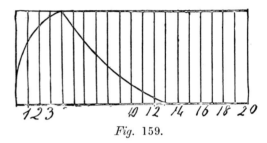

Fig. 159.

live steam has, in passing through the steam ports into the cylinder bore, been reduced by reason of the amount of port opening being diminished on account of the slow movement of the valve.

The reason that the point of cut-off occurs at such unequal points in the two strokes, is that in this case, the cut-off eccentric has been set in such position, and its rod made of such length, as would give the longest point of cut-off in each case without letting the port reopen.

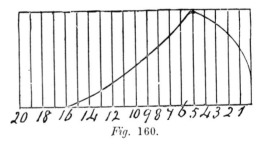

Fig. 160.

To equalize the points of cut-off, we shall require to move the cut-off eccentric, and shorten its rod. In Fig. 161, for example, the positions of the crank, cut-off eccentric, and valves are shown at the time the piston is on the back stroke, and at its 13th inch of motion (this being the point of cut-off for the other stroke), and it is seen that the steam port L, is not yet closed.

If, to close it, we move the cut-off eccentric ahead (increasing its angle with the crank), we hasten the point of cut-off for the other stroke, while if we lengthen the eccentric-rod enough to close L, we shall delay the point of cut-off for the other stroke (and that would, in this case, cause the port to reopen). The course to pursue

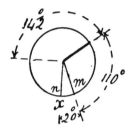

as in Fig. 156 (both valves moving together after the point of cut-off), the cut-off valve will not act on the forward stroke after the 13th inch of piston stroke, and since the main valve cuts off at the 18th inch of piston motion (as is shown in Fig. 153), therefore no cut-off can be effected between these two points.

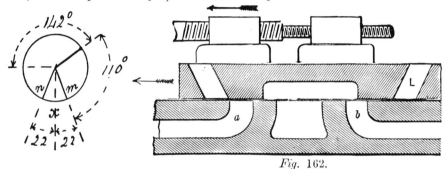

Fig. 162.

Fig. 163.

is to both lengthen the rod and move the cut-off eccen-

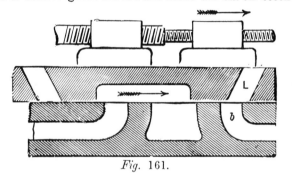

Fig. 161.

tric as well, adjusting the two until the ports just escape

That this cannot be remedied by moving the cut-off eccentric position, may be shown as follows: Referring again to Fig. 156, if the cut-off eccentric was moved further ahead, its angle of 169 degrees being increased, the cut-off would have occured earlier, while if this angle was diminished the cut-off would not be effected by the cut-off valve, because it would not fully cover the port.

In Fig. 162, for example, it is moved nearer to x, and as a result the cut-off is effected by the main valve, Finally, if we moved the cut-off eccentric to position x. Fig. 163, at the time the main eccentric stood at m, the cut-off valve would not effect the cut-off at all, since

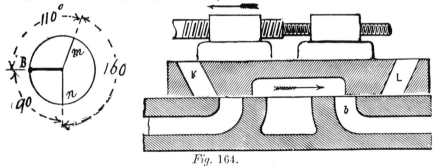

Fig. 164.

reopening and the points of cut-off are equal. In this way, the longest possible equalized points of cut-off are obtained.

It may now be pointed out that with the eccentrics set

15

it would not pass entirely over the main valve port. The cut-off eccentric is thus shown to be in position to cut-off at the latest possible point, without reopening the port when it is set as in Fig. 156.

Having limited the position of cut-off eccentric in one direction, we may now proceed to find its limit in the other, or, in other words, find in what position to

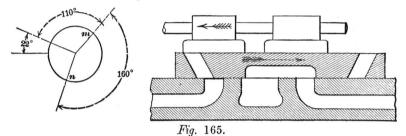

Fig. 165.

set it, in order to cut-off as early as possible in the piston stroke. In our previous examples, we have set the cut-off eccentric either at 180°, or opposite to the

and the crank being on its dead center B, and the main valve having no lead, the port *a* is closed.

In Fig. 165, the parts are shown in the position they would occupy at the point of cut-off, the crank having moved but 22°. On continuing the motion, the parts will arrive at the positions shown in Fig. 166, from

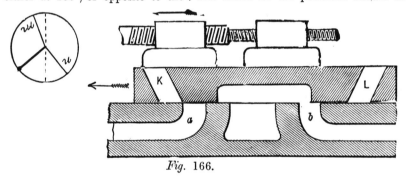

Fig. 166.

crank, or else at some lesser angle ahead of the crank, but we may set it at some angle behind the crank instead of ahead of it (it being understood that ahead

which it will be seen that the cut-off eccentric being at *n* and the main eccentric at *m*, the latter (being nearer to its mid-position than *n* is to its mid-position) will

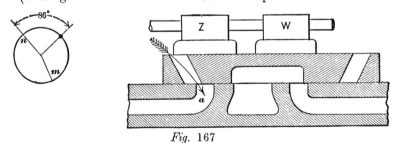

Fig. 167

means less than 180° measured in the direction of crank-revolution, and behind the crank means less than 180° measured in the opposite direction to that of the crank motion). In Fig. 164, for example, it is set at 90° behind the crank, the valve laps, travel, etc., remaining the same as in the previous examples,

move the main valve farther than the cut-off valve will be moved, and it is clear that if the cut-off eccentric were set at an angle of less than 90° behind the crank, the cut-off valve would first effect the cut-off as in Fig. 166, and then lag behind and permit K to reopen. In Fig. 167, for example, we have moved the cut-off eccentric so that it stands at 86° behind the crank, and, as a result, the cut-off valve, after having cut-off the steam laps behind, causing the port to re-open and live steam to re-enter, as denoted by the arrow. The least permissable angle behind the crank for the cut-off eccentric is, therefore, 90°.

The steam-port opening of the valves, in Figs. 164 and 165, is given in Figs. 168 and 169, where it is seen

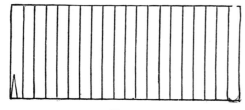

Fig. 168.

that the ports open less than a quarter inch and close when the piston has moved one-half inch on the forward and three-quarters inch on the return stroke.

Fig. 169.

We have thus found the limits within the cut-off eccentric can be moved on the shaft in either direction when the valves have the proportions given. In the following examples the proportions are as follows:

Stroke of Piston   –   –   –   18   inches.
Length of Connecting-rod   54   "
Width of Steam Port   –   –   1   "
Steam Lap   –   –   – $\frac{1}{4}$ "
Lead of Valve   –   –   – $\frac{3}{16}$ "
Travel of Main Valve   –   3   "
Travel of Cut-off Valve   –   3   "

In Figs. 170, 171, 172 and 173, we have the port opening when the eccentric is set to cut-off at $\frac{1}{6}$, $\frac{1}{3}$, $\frac{1}{2}$ and $\frac{2}{3}$ of the stroke, the proportions of the valves and ports remaining unchanged, the eccentric having been moved upon the crank shaft in order to effect the cut-off at the respective points.

Here we find that when the cut-off is to occur at less than half-stroke, the cut-off eccentric is at an angle of less than 180°*behind* the crank, while, when it is to occur later than at half-stroke, the cut-off eccentric is at an angle of less than 180°*ahead* of the crank, the extremes of its position being for the shortest cut-off ($\frac{1}{3}$) 135° behind the crank, and for the longest 169° ahead of the crank. But its position, at the time the cut-off occurs, varies but 6°; thus at the $\frac{1}{6}$ cut-off, it stands, at the point of cut-off, 5° ahead of its mid-position *x*, while for the longest it stands 11° ahead, a difference of 6° only for the whole range of cut-offs.

But if we alter the amount of lap on the cut-off valve, and adjust the length of the cut-off eccentric-rod so as to equalize the points of cut-off for the two strokes, we shall alter the position of the cut-off eccentric. Suppose, for example, that we reduce the cut-off lap to $\frac{1}{4}$ inch, that is to say, let the edges of the cut-off valve each come (when it is in mid-position as in Fig. 148,) within $\frac{1}{4}$ inch (instead of the $\frac{1}{2}$ inch, in our previous examples) of the nearest edges of the ports, in the main valves, and the effect is shown in the following figures: In Fig. 174, the cut-off is at $\frac{1}{6}$ stroke, and we find that the cut-off eccentric stands 5° behind the line *x*, instead of 5° ahead of it, as it was in Fig. 170; also we find that the point at which the port begins to close, is at 1$\frac{3}{4}$ inches, instead of at 1$\frac{1}{4}$ inches as before, thus giving a greater and therefore better port opening.

On the return stroke, we find the widest port opening in both cases at 1$\frac{3}{8}$ inches of piston motion, but in Fig. 174, the port is a trifle wider open at the second inch, which is a slight gain, as the steam is less wire drawn.

In Fig. 175, the cut-off being at $\frac{1}{3}$ stroke, the port openings are about equal to those in Fig. 171, and the cut-off eccentric stands (at the point of cut-off) 11° behind the line *x*, instead of 1° ahead of it, as it did in Fig. 176. In Fig. 179, (the point of cut-off being at $\frac{1}{2}$ stroke) the port openings are about the same as those in Fig. 172, the cut-off eccentric, in this case, standing at the point of cut-off at 8° behind *x*, instead of being ahead of it, as in Fig. 172. In Fig. 177, the cut-off being at $\frac{2}{3}$ stroke, the cut-off eccentric stands at the point of cut-off on the line *x*, instead of 11° ahead of it, as in Fig. 173. Thus we find, that the added lap has made but a very slight difference in the port open-

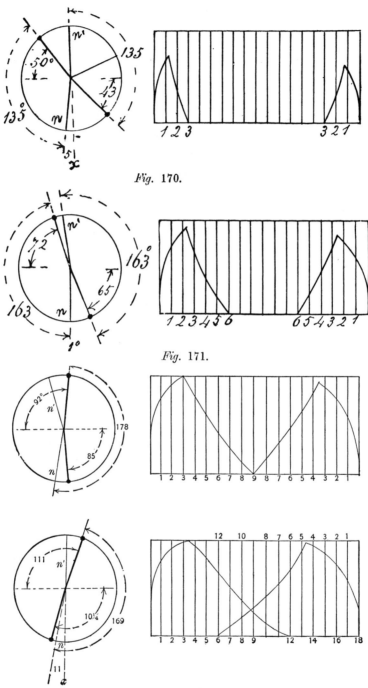

Fig. 170.

Fig. 171.

Figs. 172 & 173.

ings, although it has thrown the cut-off eccentric back | lesser cut-off lap, 12 inches was the latest point at which

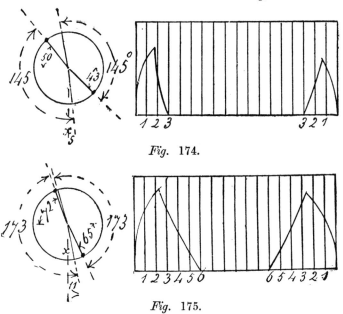

Fig. 174.

Fig. 175.

in each case. But by this means, the valve is enabled | the cut-off could be effected without reopening the port.
to cut-off at a later period in the stroke, without reopen- | It will be noticed that both in *Figs.* 1 77 and 178, the

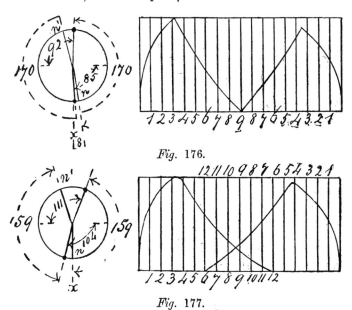

Fig. 176.

Fig. 177.

ing the ports as may be seen from Fig. 178, in which | line is rounded somewhat at the point where the port
the cut-off occurs at the 15th inch, whereas, with the | begins to close, and this occurs because the main valve

begins to close the port before the cut-off valve comes into action, as may be seen from Fig. 179. The rounded corner *e*, Fig. 178, shows the cut-off by the

or directly opposite to, the crank, so that its position, with relation to the crank, will be the same, let the direction of revolution be what it may.

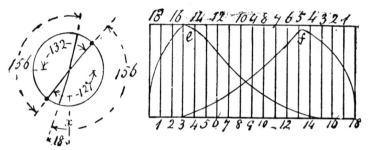

Fig. 178.

main valve. The rounded corner *f*, Fig. 178, is caused by the valves acting in opposition to one another, the

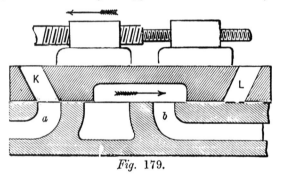

Fig. 179.

main valve moving in the direction to open, and the cut-off in the direction to close, the port. It may now

VARYING THE POINTS OF CUT-OFF BY MOVING THE CUT-OFF VALVES.

We may now assume the cut-off eccentric to be fixed upon the crank-shaft, variations in the points of cut-off being effected by moving the valves by means of a right and left hand screw, which was shown in Fig. 148. Let the engine be required to run in one direction only, and the proportions to be as before:

Let it be required that the cut-off be made adjustable at all points between ¼ and ¾ stroke, and the first question that arises is, at what point in the stroke are the two points of cut-off to be equalized, because the cut-off eccentric must be set in such a position as to accomplish that end at some particular point in the stroke,

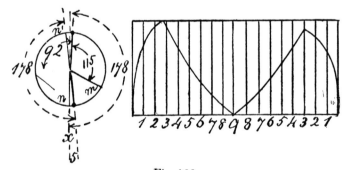

Fig. 180.

be pointed out that if the engine is required to run backwards as well as forwards, as in the case of marine engines, the cut-off eccentric is usually set at 90° from,

letting the variations come as they may at other points of cut-off.

If we were to select the point of equalization of cut-off to be mid-way between the two extremes of ¼ and ¾, or in other words at half stroke, the positions of the parts and the port openings, will be as in Fig. 180, but when we come to move the valves apart with the screw S, Fig. 148, to effect the cut-off at ¾ stroke, we find the port will reopen on the stroke when the piston is moving from the crank end to the head end, and cut-off too early on the stroke when the piston is moving from the head end to the crank end.

a diameter equal to the travel of the main valve, and whose diameter, on the line B C D, also represents the piston stroke on some scale; as the diameter of the circle is, in this case, 3 inches and the piston stroke is 18 inches, the scale is one-sixth full size. From center C draw a circle *d*, whose radius C *d* equals the steam lap of the main valve. Then mark the point *s*, distant from *d* to the amount of lead the main valve is to have (in this case $\frac{3}{16}$ inch), and with half the distance C B

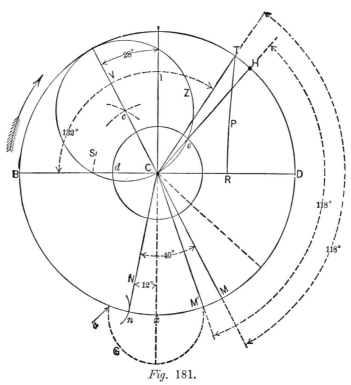

Fig. 181.

It is evident, therefore, that the position of the cut-off eccentric must be adjusted to conform to the requirements of the longest point of cut-off required, which in this case, has been selected at three-quarters of the piston stroke, or at 13½ inches. Furthermore, since it has been shown that the port nearest to the crank is the one in which the port reopening occurs first, this is the one to be dealt with in finding a cut-off eccentric position that will not effect a reopening of the port, which may be done as follows:

From a center C, Fig. 181, draw a circle B D, having

as a radius, mark from *s* and from C the dotted arcs intersecting at *c*, draw the line V, passing from C, through the intersection *c*, and this line will represent the throw-line of the eccentric, set as much behind the line I as it will stand ahead of it when the crank pin is at B, the path of crank-revolution being as denoted by the arrow.

Having found the angular advance of the eccentric, which is the angle V I C (equal to 28° in this case), we may proceed to find the position of the crank at the time the main valve would cut off steam, and from this we can then find the position of the main eccentric at the point it would cut off. From these two, we may next find the position of the cut-off eccentric required to cut off at three-quarter stroke, without reopening the port.

To find the position of the crank at the time the main valve would cut-off, we simply draw, from the center C, a line, passing through e where the steam lap circle crosses the valve circle Z, and this gives us at H the required crank position, being as marked 132° from B. From H we may find the position of the main valve, at the time it cuts off, as follows :

We have found that with the crank at B, the main eccentric would stand 28° *ahead* of line I, and as I is 90° from B, we have 90° plus 28°, or 118° as the num. ber of degrees the main eccentric stands ahead of the crank, hence we mark the line M' at 118° ahead of H, showing the position of the main eccentric when the crank is at H and the main valve cuts off. We now find the positions of the crank and main eccentric when the piston is at three-quarters stroke, because from these we may locate the necessary position of the cut off eccentric. First, then, we mark point R three-quarters of the distance from B to D (this distance representing the piston stroke), and then with compasses set to represent the length of the connecting-rod on the same scale as B D represents the piston stroke (one-sixth full size), we mark, from R, the arc P, giving at T the position of the crank when the piston is at R. Now as the main eccentric is 118° ahead of the crank at T, we may mark M, which is the position of the main eccentric when the cut-off valve is to cut off, the crank being at T, and the piston at R.

Now while the main eccentric is moving from M to M', the cut-off eccentric must move its valve as fast as the main eccentric moves *its* valve, and all we have to do is to find where to place it in order to enable it to do so, which we may accomplish as follows :

If we draw from x, as a center, a semi-circle G, having a radius x M', we shall find at a the position of the cut-off eccentric at the time the main valve has arrived at M' and would cut-off the steam, and from this point a we may obtain the position of the cut-off eccentric at the time its valve is to cut-off by setting the compasses to the distance between M and M', measured on the

circle, and marking from a an arc n, from which we may mark line N, giving the position for the cut-off eccentric at the point of its valve's cut-off, the crank being at T, the piston at R and the main eccentric at K. Now suppose the piston, the crank and the main and cut-off valves to stand in these respective positions, and the cut-off, by cut-off valve, will occur while the main eccentric is at M and the cut-off eccentric is at N, and being nearer to its mid-position x than M is to its mid-position x, therefore N will move its valve fastest and the port cannot reopen.

While the main eccentric moves from M to M', the cut-off eccentric moves from N to d, being, at each point in its movement, nearer to x than M is, and therefore moving its valve faster. On arriving at a, however, N and M will be equi-distant from x, and they will move their valves at equal speeds. At this time, however, the main valve will have closed the port and reopening cannot occur.

Having found the positions for the crank, the main eccentric, and the cut-off eccentric, at the time of cut-off on the piston stroke from B to D, the crank having started from B, and moved to T, we may now find the corresponding positions for the valves. In the upper half of Fig. 182, we have marked the positions of the crank, the cut-off eccentric and the main eccentric, simply transferring them from Fig. 181. To find the corresponding positions of the main valve, we mark from the main eccentric position, the dotted line m p, giving the point p. Now since the line B D represents the stroke of the valve, and line x, the valve's mid-position, therefore the radius from p to the line x, measured on the line B D, is the distance the main valve must have moved from its mid-position, when the main eccentric stands at m.

Now, if the center of the main valve stood on the line X, it would be in mid-position, and all we have to do is to mark from X, a line P distant from X to the same amount that p is distant from the line x x and from this new center P, we draw in the main valve.

The cut-off valve we draw in position, to just close the port K at the crank end. For the return stroke, we draw, on the lower half of the cut, B D, representing the valve travel, and also the piston stroke as before; then mark the piston position R, at the point of cut-off, and from this obtain the crank position, as in previous examples. From the crank position, we mark the main eccentric throw-line *m*, 118° ahead and from this the cut-off eccentric throw-line 40° ahead of the main eccentric. From *m*, we draw the line *r*, showing what distance the main valve has moved from its mid-position, this distance being from *r* to *x*, measured on the line B D, or in other words, from *r* to the center of the circle. We transfer this distance from X, obtaining P

cut-off valves, and from this we can easily ascertain the lap.

Now in moving from the point of cut-off on one stroke to that of the other, the cut-off eccentric has moved from *n* in the upper half to *n* in the lower half of the figure, the difference in its distance from mid-position *x* being shown by the dotted lines E F; all we have to do then, is to draw a dotted line C from the edge of the valve in the upper half of the figure, and

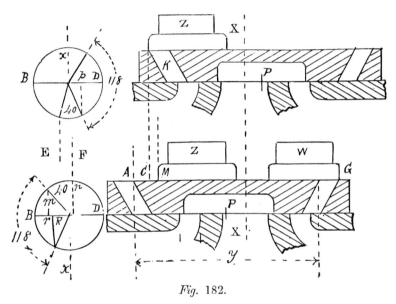

Fig. 182.

as a center wherefrom to draw in the main valve, and we then draw in the cut-off valve W, in position to just close the port; knowing that to be its position at the point of cut-off, and corresponding to the positions of the crank and the eccentrics.

Having found, for both piston strokes, the positions of the main and cut-off valves at the point of cut-off we may find, therefore, the amount of lap the cut-off valves are given by this construction, the method being as follows: We have in the lower half of Fig. 182, the valves in position at the point of cut-off for the return stroke, or while the piston is moving from the head end of the cylinder towards the crank, and we require to find the position of the cut-off valve Z, on the lower half of the diagram, because this will give us the distance between the two outside edges of the

16

draw in the cut-off valve, letting its edge M be put back from dotted line C, to an amount equal to the distance between the lines E F, and the positions of the eccentrics and both valves will be shown on the lower half of the figure.

To find the amount of cut-off lap, subtract the width apart M G, of the cut-off valves from the width apart of the inside edges of the main valve ports, or in other words, from distance *y* in Fig. 182.

If we move the cut-off valves apart, by means of the adjusting screw, sufficient to effect the cut-off at half stroke on the forward stroke, the point of cut-off will be at 10 inches on the return stroke, as is seen on the diagram of the port openings in Fig. 183.

By moving the valves still further apart to effect the cut-off at ¼ stroke, there is a variation of ½ inch in the two points of cut-off, as is seen in the diagrams of the port openings in Fig. 184. This may be obviated, to some extent, by giving to the screw-thread, that moves the cut-off valve W, a coarser pitch than the thread that moves Z, so as to move it more, a plan that is not infrequently resorted to.

main valve and the crank, at the point of cut-off, by the construction in Fig. 185, which is, in this respect, merely a repetition of former examples; B D represents the full travel of the main valve, and the large circle the path of the crank drawn to scale; *d* is the lap circle, *d s* is the valve lead, V I C the angular advance

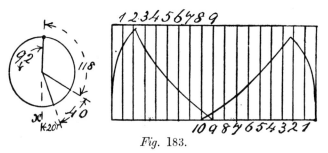

Fig. 183.

It will be noted that, throughout the whole of the diagrams of the cut-off valves, the steam port openings are grealy reduced by the action of the cut-off valve. which, assuming the full area of the steam port to be required for the admission, causes the steam to be wiredrawn, especially at the early points of cut-off. This may, to some extent, be remedied, by giving to the cut-off valve, an increase of travel.

of the main eccentric (placed for reasons already explained, behind instead of ahead of the line I), H is the position of the crank, and *m'* the main eccentric, these two latter standing in the position they would occupy at the time the main valve would effect the cut-off; R is the position of the piston at the time we wish the cut-off valve to effect the cut-off, and T the crank position corresponding to piston position R. We mark

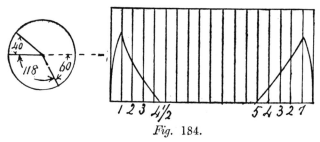

Fig. 184.

It has been shown that, in a simple slide-valve, the port openings may be increased by increasing the amount of valve travel, but when a cut-off valve is used, it is not permissible to give the main valve overtravel, because the main valve would then begin to close the port, instead of permitting the cut-off to be effected entirely by the cut-off valve. The effect of giving to the cut-off valve more travel than the main valve, and the method of finding the positions of the eccentrics for any given amount of cut-off valve overtravel, is shown as follows: We first find the positions of the

m behind *m'* to the same amount that T is behind H, and thus have the positions of the crank and of the main eccentric, at the time the cut-off valve is to effect the cut-off. To find the position for the cut-off eccentric, we mark arc Q Q, distant from the circle B D, to half the amount of overtravel of the cut-off valve, (hence as Q Q, is $\frac{1}{2}$ inch from circle B D, the cut-off valve will have, in this case, an inch more travel than the main valve).

We then set a pair of compasses to the radius *x m*, on circle B D, and transfer it to circle Q Q, getting the radius *x a.* Then transfer the radius *m' m* from *a,* and get arc *n* on Q Q, from which get *n n* or the position for the cut-off eccentric, at the time its valve is to cut-off.

lines the cut-off at one-quarter stroke, and it is seen, on comparing these with previous examples, that the port openings are improved. At the three-quarter stroke, the

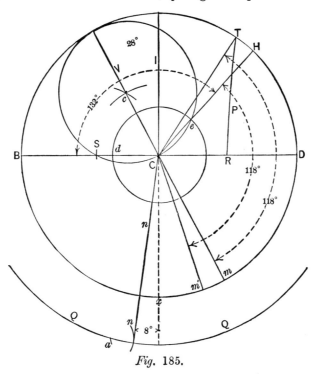

Fig. 185.

The method of finding the required amount of lap on the cut-off valve, is the same as has already been explained with reference to Fig. 182, hence we may now examine the port openings given by the new condition, viz., one

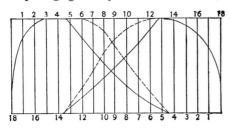

Fig. 186.

inch of overtravel for the cut-off valve. Fig. 186 shows in the full lines the port opening for the cut-off at three-quarter stroke; Fig. 187 shows in the full lines the cut-off at half-stroke, and Fig. 188 shows in the full

port openings are considerably increased, and reopening avoided. At half stroke the port openings are also increased, as may be seen by a comparison with Fig. 183, while at one-quarter the openings are wider at the

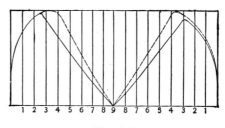

Fig. 187.

2nd and 3rd inch of piston motion on the forward stroke, but about the same, on the back stroke as those in Fig. 184.

But we may further increase the port openings by

giving to the cut-off valve a further increase of travel, making it 5 inches instead of 4 inches, and leaving the main valve at its original stroke of 3 inches. The port openings, with 5 inches of cut-off valve stroke, are marked on the three last diagrams in dotted lines, and it is seen that the increase is considerable in the longer points of cut-off, but not so great in the shorter ones, and, also, that the points of cut-off are more nearly equalized for the two strokes, let the cut-off occur where it may. It is usual, therefore, to give to the cut-off valve more travel than the main valve, and from what has been said, the student may readily plot out diagrams from any given dimensions, and work out the port openings for himself.

To obtain a wider steam port opening, and therefore a fuller supply of steam, previous to the point of cut-off (an object that is of great importance when the

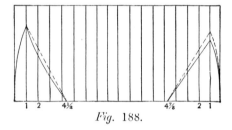

Fig. 188.

cut-off is to occur at early points in the stroke), double or treble ported valves may be employed, such valves being called *griddle* valves. Thus the form of valve in the foregoing figures may be double ported, doubling the port openings, or the same amount of port opening may be had with less valve travel.

In Fig. 189 is shown an example of griddle, or multi-ported, valves, the main valve having the two ports, K and L, on its face, and the four, *g h p q*, on its back. The cut-off valve has four corresponding ports (*v u t*, etc.), and, as only one cylinder port can admit steam at one time, a full supply of live steam is obtained so long as the combined openings of the ports, *g h p q*, equal the port opening of the cylinder steam port. Thus, if the cylinder ports are an inch wide, and the ports, *g h p q*, are each one-quarter open, the total opening will be an inch, because the four quarters equal an inch, and if, at this point in the piston

motion, the cylinder port is open an inch, the steam supply is not decreased by the cut-off ports.

The ports in the cut-off valve are made wider than *g h*, etc., in the back of the main valve, because by this means the cut-off eccentric is brought into a better position to effect a quick cut-off and avoid wire-drawing, as will be understood from Fig. 190, in which the full lines represent the ports in the cut-off valve made twice as wide as those in the back of the main valve, and the dotted lines show the cut-off ports if made of equal width with the main valve ports *g h*, etc. The main eccentric is at *m*, and with the cut-off ports made twice as wide as *g h*, the cut-off eccentric will be at *n*, when the valves are in the positions shown, and when the cut-off takes place the cut-off eccentric will have arrived at *n′*. Thus it will be seen that through widening the

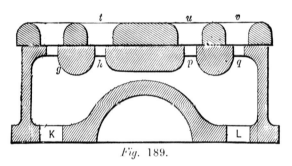

Fig. 189.

ports in the cut-off valve, the cut-off has been effected with the cut-off eccentric in the most desirable position for cutting off quickly.

Now suppose that the ports in the cut-off valve were as denoted by the dotted lines, or in other words, were of equal width with those in the back of the main valve, and, in that case, the cut-off eccentric will be at *n* when the cut-off is effected, the port edges *a b c d* of the cut-off valve having closed ports *g h*, etc., and the cut-off valve will not have been moved so quickly, hence the steam would be more wire-drawn. Another and important consideration is that in order to effect the cut-off at late points in the stroke, the cut-off eccentric would require to be placed nearer to the main eccentric, and the two would travel at a more nearly equal speed, causing a very slow cut-off port closure, and therefore, great wire-drawing.

This may be seen as follows: Taking the dotted

lines to represent the cut-off valve ports, and the cut-off will be effected with the crank at T, the main eccentric at *m*, and the cut-off eccentric at *n*. To enable the cut-off to take place at a later point in the crank path, we must put the cut-off valve back, (or to the right of the figure), and to do this, we must put the cut-off eccentric *n* back, or nearer to the main eccentric *m*, causing both valves to travel at a nearly equal speed and, as before stated, effecting a very slow cut-off port closure.

The extra width of the cut-off valve ports must be situated equally on each side of the ports in the main

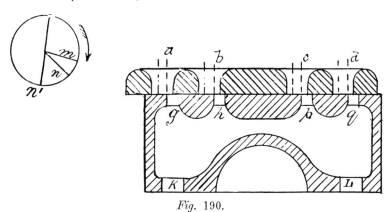

Fig. 190.

valve when the two valves are placed in mid-position, as in Fig. 198, so that the ports may all close simultaneously, as in Fig. 191.

To investigate the action of the valves, and locate the positions of the eccentrics for any point of cut-off, we first find the positions of the various parts, at the point of cut-off by the construction shown in previous examples, and repeated in Fig. 192, H being the position of the crank, and *m* that of the main eccentric

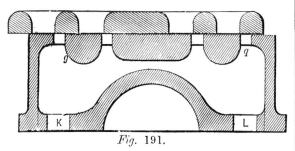

Fig. 191.

when at the point of cut-off. It being assumed that the cut-off valve is to be set to cut-off at its longest point at the same time as the main would cut-off, we may locate the corresponding position of the cut-off eccentric from that of the main eccentric as follows: In Fig. 193, we have the positions of the valves corresponding to crank-pin position H, and main valve position *m* in Fig. 192, and in Fig. 194, we have the ports *t u v*, etc., in the cut-off valve twice as wide as those in the back of the main valve, and it will be found that the cut-off eccentric must be set in advance of the main eccentric, sufficently to move the cut-off valve the distance *e*, or that between the dotted lines W and Z, the former being fair with the outside edge of the main valve port, and the latter fair with the opposite edge of the cut-off port. Supposing the ports *g h*, etc., to measure $\frac{1}{2}$ inch, and ports *t u v*, etc., to measure an inch, and the distance *e* (the two valves being placed in their mid-positions) will measure $\frac{3}{4}$ inch. Continuing the construction of the diagram, Fig. 192, we draw the line G, passing from *m* to line B D, and at a right angle to it. Then on line B D, measure off from G the distance *e*, Fig. 194, and draw line *n'*, giving, at its intersection with the outer circle, the location of the cut-off eccentric as shown by the line *n' n*.

The construction is simple enough when we consider that the line B D represents the valve travel full size, and Fig. 194, the valves one-quarter full size, and that the main eccentric being at *m* (Fig. 192), we have by lines G *n'*, (distant apart equal to four times *e*, or $\frac{3}{4}$ inch in this example) found a position for the cut-off eccentric *n′ n′*, that would move the cut-off valve ¾ inch as required. Suppose, now, that it were required to find the position of the cut-off eccentric, when the

center C, the arcs *c*. A line V drawn through C *c*, gives the angular advance of the main eccentric. From *c*, on V, we draw the circle X, and from C,

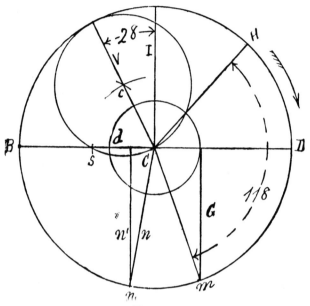

Fig. 192.

piston is to have its steam cut off at half-stroke, and we draw a circle B D, Fig. 195, equal in its diameter to the stroke of the valve, and a circle *d* whose radius

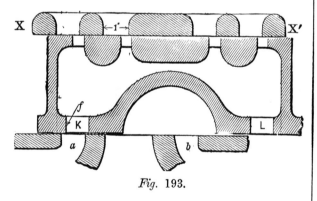

Fig. 193.

equals the amount of lap on the main valve. We mark point S distant from *d* to the amount of lead the main valve has. With a radius, equal to one-half the radius of circle B D, we mark from S, and from the

through *e*, where circles *d* and X intersect, we get at H the position of the crank at the time the main would cut off.

To find the corresponding position of the main valve, we add to 90° the angle V I, whatever it may be (in this case it is 28° as marked) and get at *m′* (118° from H), the position of the main valve at the time it would cut off the steam, independently of the cut-off valve. We have next to find the position of the cut-off eccentric, when set to cut off at the same point as the main valve, and to do this, we draw from *m′* line G. We then take the full distance the edge of the cut-off valve stands from the edge of the main valve (when both valves are in mid-position as denoted, by *e* in Fig. 195), and mark line *n′* distant from G to the amount of *e*, Fig. 195, and this gives us the point *n″* on the outer circle, and from this we draw the line *n″ n″*, which is the position of the cut-off valve, when the crank is at H, and the main valve at *m′*. Then we mark, at T, the position of the crank when the piston is at half-stroke, and it is clear that, having moved the crank back from

H to T, we must move the main and cut-off eccentrics back an equal distance, hence, with compasses set to radius H T, we mark from *m'* point *m*, which gives the position of the main eccentric when the piston is at

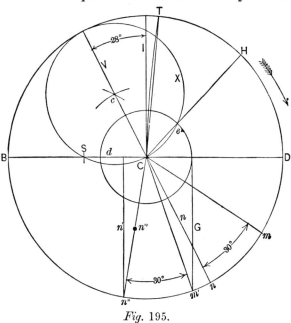

Fig. 195.

half-stroke, and with the same radius we mark from *n''* point *n*, which is the position of the cut-off eccentric when the piston is at half-stroke. The proof of the construction is, that the two eccentrics stand at 30°

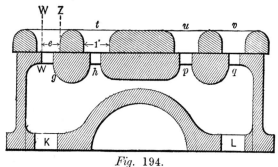

Fig. 194.

apart in either case, and the main eccentric is 118° ahead of the crank, hence we have merely first found the position of the parts when the main valve cuts off and then moved them all back together to get their position at another point of cut-off.

Now suppose the positions of the crank and eccentrics to have been determined by this method, the cut-off valve ports being 1 inch wide, those in the back of the main valve $\frac{1}{2}$ inch wide, the cylinder ports being 1 inch, the main valve having $\frac{1}{2}$ inch steam lap and $\frac{3}{16}$ lead, the travel of both valves being obtained by adding the width of the port to the amount of the steam lap and then multiplying by 2 (this amount of travel being sufficient to let the main valve fully open the steam ports); suppose also the main and cut-off eccentrics to be so set on the shaft that with the cut-off eccentrics moved as close together as possible, both will cut-off the steam at the same point in the stroke, and the crank being at H, Fig. 195, the main eccentric will be at *m'* and the cut-off at *n''*, and the port openings will be as in Fig. 196 (these openings having been obtained, as in previous examples, by moving an engine piston and measuring the port opening at each inch of piston motion).

It is seen from this diagram that up to the eleventh inch of piston motion, the admission is as free as if no cut-off valve were employed, since the amount of port opening is equal to that given by the main valve. The admission is, therefore, more free and the cut-off more sharp than in the cases in which the main valve had but one port. We have now to consider the means by which the valves are to be enabled to cut off at other points in the stroke, and this can, in the case of a single cut-off valve, as in the examples now under considera-

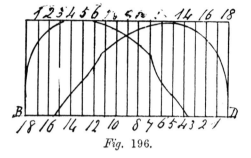

Fig. 196.

tion, be done by increasing the stroke of the main valve, whereas if two cut-off valves were used on the back of the main valve, the earlier points of cut-off must be effected by means of moving them apart by a screw, such as in Fig. 148.

Now suppose that the eccentrics having being set by the construction given in Fig. 195 (the longest point of cut-off effected by the cut-off valve equalling the point at which the main valve cuts off), and suppose that we increase the main valve travel sufficiently to effect the

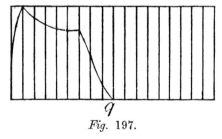

Fig. 197.

cut-off at half-stroke, and we shall find that the admission will occur as in Fig. 197, the port closure from the first to the sixth inch of piston motion occurring because the extra travel given to the main valve causes its port edge to partly cover the port K, as in Fig. 198. This may obviously however, be remedied by cutting away that edge as in Fig. 199, which will not affect in any other way the action of the valve.

To find the amount the travel of the main valve must be increased to effect the cut-off, at some earlier point

in the piston stroke, as say for example, at half-stroke, we proceed as in Fig. 200, in which the position of the crank, when the piston is at half-stroke, and the corresponding positions of the main and cut-off eccentrics, are found by the same construction as in previous

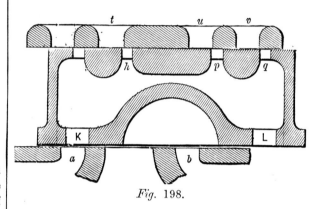

Fig. 198.

examples. Now, it will be found that if we place the two valves in their mid-positions, the amount the *throw* (not the *travel*) of the main eccentric must be increased, in order to change the point of cut-off from what it was in Fig. 196, to half piston stroke, will be equal to the distance *e*, Fig. 201, and as the position of the cut-off eccentric will not be influenced by increasing the

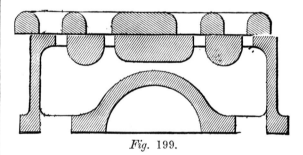

Fig. 199.

main eccentric throw, we draw a line *n'*, on the diagram Fig. 200, and from this, a second line G, distant from *n'* the amount represented by *e*, in Fig. 201, (which being $\frac{3}{16}$ inch, and the illustration being one-quarter size, makes the distance from *n'* to G become $\frac{3}{4}$ inch) we then draw an arc J J, passing through the point where line G and the cut-off eccentric throw-line intersect, and the distance between the circle and the arc

J J, or radius *a*, is the amount the throw of the eccentric must be increased, in order to effect the cut-off at half piston stroke. By increasing the throw of the eccentric, we have merely put the main valve further

increased its lead, making it ¼ inch instead of 3/16 as before. The port openings, with the cut-off at half stroke, are shown in Fig. 203, and it is seen that in-

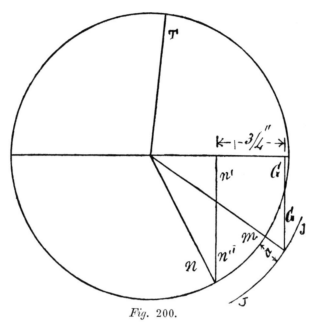

Fig. 200.

eccentric, we have merely put the main valve further back, as will be understood from Fig. 202, in which with both eccentrics having the same throw, the valves would occupy the positions they occupy in the figure; but by increasing the throw from the circle to arc J,

creasing the stroke of the main valve has made the steam port open full at 1⅛ inches of piston motion for one stroke, and at 1½ inches of piston motion for the other, which is a great advantage, while the ports have remained wide open to 5½ inches of piston motion on one stroke, and 6⅛ inches on the other. The points of

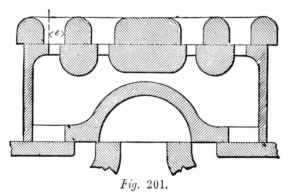

Fig. 201.

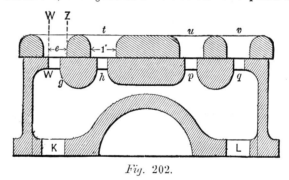

Fig. 202.

(Fig. 200), we have pushed the main valve back, so that the cut-off will be effected.

We have, also, by increasing the main valve travel,

cut-off also, are very nearly equalized, and it is clear that the steam distribution has been greatly improved, as may be seen on a comparison with previous figures.

17

To find the throw of the main eccentric necessary, in order to cut off the steam at one-quarter stroke, we find the crank position with the piston at quarter stroke,

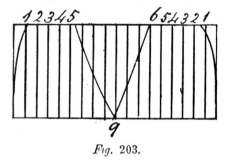

Fig. 203.

as in Fig. 204, and then, by the construction explained with reference to Fig. 195, find the corresponding positions of the main and cut-off eccentrics. We then draw the line *n'*, and, from this, mark point G distant from *n'* the amount represented by *e* in Fig. 201, or in this case, ¾ inch. From G we mark the arc J, and the

Least or normal travel of main valve, – – 3 in.
Travel of main valve to cut off at half stroke, 3½ "
    "   "   "   "   " " " " one-quar. " 4⅛ "

Increasing the main valve travel to 4⅛ has, however, again increased the lead, making it ⅜ inch full.

In Fig. 205, we have the port openings for the cut-

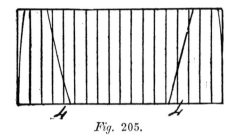

Fig. 205.

off at one-quarter stroke, and it is seen that they are much greater than in Fig. 184, where a single-ported cut-off valve was used.

If two separate main, and two separate cut-off, valves

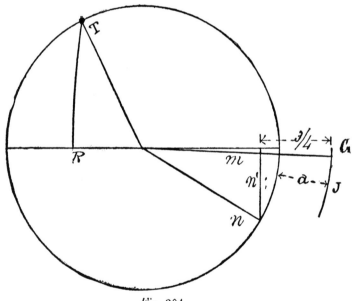

Fig. 204.

radius *a* is the amount of increased travel, above that required for the longest cut-off, necessary to cut off at one-quarter of the piston stroke.

As the radius *a*, measures in this case $\frac{9}{16}$ inch, the increase in the valve *travel* is ⅝, and we have:

are employed, being connected by arms A and B, Fig. 206, the construction is the same except that the valve does not need the outer ribs X X' in Fig. 193, which are necessary, in that case, to close the end ports in that figure; it being obvious that, in the absence of X', the left hand port in the main valve would be left open, while the others are closed. In Fig. 206, however, the

one and a half times the width of the steam port, and the valve exhaust port (or, more properly, exhaust *cavity*)

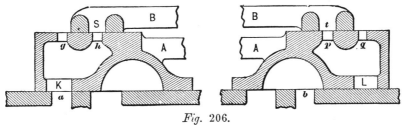

Fig. 206.

cut-off ports for one cylinder port being separated from those of the other (by reason of separate valves being

need only equal the width of the cylinder exhaust port added to twice the width of the bridge. It may also

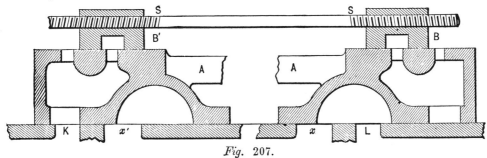

Fig. 207.

used), the cut-off valves need not bridge the end ports *g q*. In all other respects the valves are essentially

be noted that were it not that the main valve stroke is increased in order to effect the earlier points of cut-off, the cylinder exhaust port would only require to be of

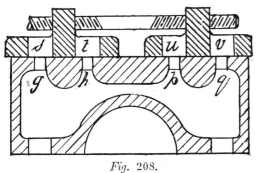

Fig. 208.

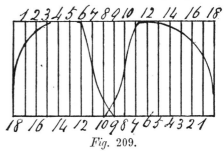

Fig. 209.

alike, and so also is their action, which may be investigated in the same manner, and (under equal conditions) with like results, by means of the diagrams already explained. It may be pointed out, however, that the exhaust ports, in the main valves and in the cylinder, need not be so wide where two main and two cut-off valves are used, because they act for one steam port only, hence the cylinder exhaust port need only equal

the same width as the cylinder steam ports. The ports K L, in the main valve, are made wider than the cylinder ports *a* and *b*, so that they may not unduly close them when the main valve stroke is increased to effect the earlier points of cut-off, a matter that was explained with reference to Fig. 198.

It is obvious that, instead of increasing the throw of the main eccentric in order to effect the cut-off at different points of the piston stroke, we may employ separate cut-off blocks and move them apart by a right and left hand screw S S, Fig. 207, giving an example of the arrangement. The distance the cut-off blocks must be moved, in this case, in order to vary the point of cut-off to any given amount, may be found by the con. struction in Figs. 181 and 182, the amount the cut-off valves must be moved exactly equalling the amount of increased travel the main valve must have to cut-off at any point earlier than the longest point of cut-off.

It is better, however, to effect the earlier points of cut-off by increasing the main valve travel than it is to move the cut-off valves, because a more free admission of steam is given. Suppose, for example, that the valve, in Fig. 198, had cut-off valves adjustable by a screw, as in Fig. 208, and that the cut-off eccentric being set to cut-off at the same time as the main valve, we move the cut-off valve $\frac{1}{2}$ inch (by means of the screw) wider apart so as to effect the cut-off at half-stroke, and the port openings will be as in Fig. 209, which, on comparison with Fig. 203 for the cut-off at half-stroke by means of increasing the main valve

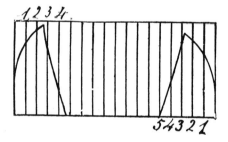

Fig. 210.

travel (all the other elements being alike for the two cases), shows the steam admission to be more tardy and the cut-off less sharp. Fig. 210 gives the port openings with the cut-off valves adjusted to cut off at quarter-stroke, and compares very unfavorably with Fig. 205, in which the cut-off was effected by increasing the main valve travel, the cases being identical in all other respects, that is to say the dimensions of valve ports, etc., are alike in the two cases.

Fig. 211 represents a cut-off valve, which operates on a fixed seat in a steam chest divided into two compartments, *a* and *b*. It is obvious that, in this class of valve, the steam that surrounds the main valve in

the lower compartment *b* of the chest, is not affected by the cut-off and acts to maintain the pressure of the steam in the cylinder after the point of cut-off, hence

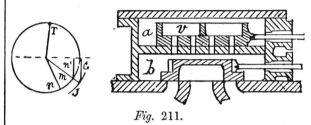

Fig. 211.

it is desirable to keep the size of the lower compartment as small as possible, and thus limit its cubical contents.

In this design of valve gear, the longest point of cut-off is obtained by the shortest amount of valve

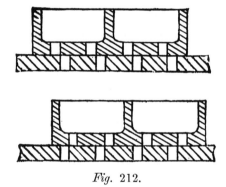

Fig. 212.

travel, and if the ports in the cut-off valve are of the same width as those in its seat, the least amount of travel that will effect the cut-off, is that equal to twice the width of the ports. The two extreme positions of the valve are shown in Fig. 212, these being the positions at their respective points of cut-off for the two piston strokes, and it is seen that the valve, moving from its position in the upper to that in the lower half of the figure, has traveled a distance equal to twice the width of the port.

The useful range of this form of valve, is limited by reason of the extreme amount of wire drawing of the steam that occurs if it is attempted to cut-off steam at late points in the piston stroke, and the design is therefore, useful for the earlier points of cut-off only, as from quarter to half-stroke.

TO FIND THE LIMITS OF THE RANGE OF CUT-OFF.

The longest or latest point of cut-off being given to find the earliest point at which the valve can effect the cut-off, the amount the valve stroke must be increased, and the position for the cut-off eccentric, we proceed as in Fig. 213, in which the inner circle B D represents the shortest amount of valve travel (its diameter equalling twice the width of the ports), and T the position of the crank at the latest point of cut-off, S represents a portion of the valve seat containing one port (which is all that is necessary since the openings at all the ports in the seat will be alike), and V represents a portion of

and the shortest valve stroke, from these may be found their positions for the shortest point of cut-off and the longest valve stroke as follows:

The crank being on its dead center B; the cut-off eccentric at n; and the edge of the valve at d when the parts are set in the position necessary for the shortest cut-off, the question is, how much we must lengthen the valve stroke in order to enable the valve to cut-off as early as possible, and to find this, we mark a dotted line f, distant from edge e of the port to the amount the port must be open for the lead when the crank is at B. We then take the distance $d f$, and mark from n the point g, and draw line g g, we then prolong

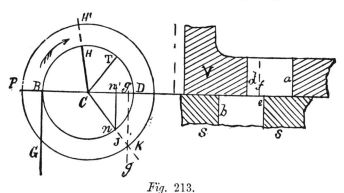

Fig. 213.

the valve containing one port. Now, it is clear that the cut-off eccentric must stand as much behind the dead center B as the crank stands ahead of it when the cut-off occurs, hence we take the radius B T, and from B mark, at n, the position of the cut-off eccentric when the crank is at B, and it follows that while the crank moves from B to T, the cut-off eccentric will move from n to B, effecting the cut-off when it arrives at B, at which time edge a of the valve will be fair with edge b of the port. To find the position of the valve when the crank is at B and the cut-off eccentric at n, we mark line n n' and take the radius C n; with this radius, and from edge b of the port, mark edge d of the valve, it being clear that edge d of the valve must be as far from its mid-position b (when the valve is in mid-position, edges d and b coincide) as n' is distant from C. From d mark off edge a of the valve equal to the width of the port in the seat. We have thus found the positions of the parts for the longest point of cut-off

n by a dotted line J, cutting g g at K, and through K we draw a circle P, representing the path of the cut-off eccentric when set for the shortest point of cut-off. We have thus increased the throw of the eccentric to the amount n K, and, by doing so, moved the edge of the valve from d to f, leaving the port open to the amount $e f$, which is necessary for the lead, which will obviously equal as many times the opening $e f$ as there are ports in the cut off valve.

Having found the path of the eccentric for the shortest cut-off, we find the position of the crank at the shortest point of cut-off by taking the arc, or radius, K G and mark it from P, thus getting at H' the required crank position at the shortest point at which it can cut off and leave the valve open to the amount $e f$ of the lead.

In Fig. 214, we have a diagram of the port openings for the longest, and in Fig. 215 a diagram of the port openings for the shortest, points of cut-off found by the diagram Fig. 213, and it is seen that, in the latest point of cut-off, the steam is extremely wire-drawn,

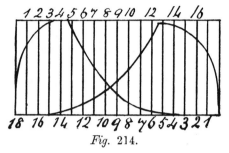

Fig. 214.

which occurs because as the cut-off eccentric approaches

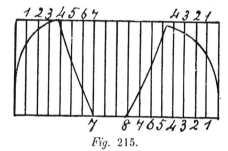

Fig. 215.

the point B, it necessarily moves the valve very slowly,

We may, to a certain extent, improve the admission, and prevent the wire-drawing, by making the cut-off valve ports wider than the port in its seat, and setting the cut-off eccentric to correspond. Thus, in Fig. 216, the cut-off valve ports are 1 inch wide and the seat ports are ½ inch wide. The least amount of cut-off valve stroke is, therefore, 1½ inches, or the width of the port in the valve added to the width of port in the seat.

The circle B D represents the least amount of valve stroke, and T the crank at the longest point of cut-off. We mark *n* (as far behind B as T is ahead of B), and thus get the position of the cut-off eccentric when the crank is at B. To find the corresponding position of the valve, we take the radius from the point D to *n'*, and, with this radius, mark from *e* the edge *d* of the valve, it being obvious that when the cut-off eccentric was at D, edge *d* of the valve was at *e* and at one point of cut-off, hence while the cut-off eccentric moved from D to *n'* (or what is the same thing to *n*, since *n'* shows the amount of linear motion caused by the eccentric in moving D to *n*), the valve edge *d* moved from *e* to its position in the figure. Now it is clear that in lengthening the cut-off eccentric throw,

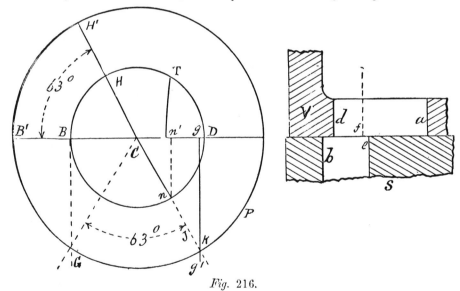

Fig. 216.

scarcely moving it at all during the 2 inches of piston motion previous to the final point of cut-off.

from *n* towards K, we move the valve edge *d* back towards *e*, and the utmost we can do this, and still leave the port open for the lead, is the radius *d f;* hence we mark *f*, distant from *e* to the amount of lead the valve is to have, and taking radius *d f* we mark from *n'* the point *g*, which gives the amount we can push the valve back to lengthen its travel while still leaving it

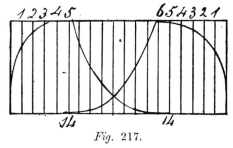

Fig. 217.

the amount *e f* of lead. From *g* we draw the line *g g'*, and then prolong the eccentric throw-line by a dotted line J, and through the intersection, at K, of J and *g'*, we mark circle P which represents the path of the center of the eccentric when its throw is increased for the shortest point of cut-off.

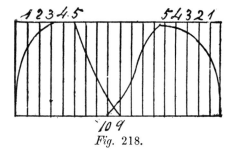

Fig. 218.

To find the crank position at the shortest point of cut-off, mark line B G at a right angle to B D, and take the arc K G (in this case 63°), and from B' mark

H', which is the crank position at the shortest point of cut-off, the crank throw-line being at H' H. For the longest point of cut-off the valve will effect, we have, then, the piston at its dead center B and the cut-off eccentric at *n*, while for the shortest point of cut-off we have the crank on its dead center B', the cut-off eccentric at K, and the edge *d* of the valve at *e* open for the lead. Having set the parts by this construction, Fig. 217 shows the port openings at the longest point of cut-off, and it is seen that, although the wire-drawing is less than in Fig. 214, it is still inadmissably great during 6 inches previous to the point of cut-off. But

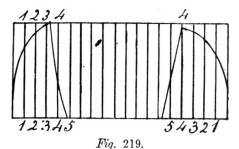

Fig. 219.

we may set the valve by this construction, and not employ it to cut-off at a later point than half-stroke, the steam admission for which is shown in Fig. 218, and it is seen, on comparison with previous diagrams representing the port openings when the cut-off is at half-stroke, that the steam supply is here fuller and the wire-drawing less. Lengthening the valve stroke sufficient to effect the cut-off at quarter-stroke, we get the port openings shown in Fig. 219, and here again it will be seen, on comparison with previous diagrams, that the port opening is unusually wide and the cut-off sharp and, therefore, advantageous in both respects.

CHAPTER VI.

VARYING THE POINT OF CUT-OFF BY SHIFTING THE ECCENTRIC ACROSS THE CRANK-SHAFT.

Instead of employing a separate cut-off valve, the points of cut-off may be varied by employing a single valve and shifting the position (on the crank-shaft) of the eccentric so as to reduce its throw, and therefore the travel of the valve. The line in which the eccentric may be moved, is shown as follows:

In Fig. 220 is shown an eccentric whose bore is slotted so that it may be shifted across the shaft. The circle r represents the path of the center of the eccentric when moved to its position of greatest throw, as shown in the full lines, the center of the eccentric being at e. The circle f represents the path of the center of the eccentric when shifted to its position of least throw, as denoted by the dotted circle H, the center of the eccentric being at x.

The circles r and f, Fig. 221, correspond to circles r and f in Fig. 220, the radius C x representing the lap of the valve. The valve is shown to have no lead, the eccentric-rod R being shortened for convenience of illustration; the crank is supposed to be on its dead center at B. Now suppose the eccentric-rod R to be pivoted on the valve, the other end being disconnected, and if we move it across the outer circle (which represents the path of the eccentric center when at its greatest throw) the center of the eccentric-strap bore will

138

move in a path denoted by the line from e d. During that part of this line that runs from e to x, the throw of the eccentric will be reduced and the cut-off hastened, the engine running in the direction denoted by the arrow, but after the eccentric has passed the line x, the valve would be in position for the engine to run in the opposite direction, its throw increasing as it is moved towards the point d. As the valve and crank have remained motionless while the eccentric-rod was moved across the shaft to mark the line e d, it is clear that so far as this crank position (B) is concerned, the eccentric may be shifted to any position between e and d without altering the valve lead.

Now suppose the engine to make a half-revolution, the crank being on the dead center D, as in Fig. 222, the port b being about to open for the live steam, and supposing the eccentric-rod to be pivoted to the valve as before, and the valve and crank to remain at rest, then the center of the eccentric-strap, in being moved across the shaft, will move in a path denoted by the line e' d' But while the crank moved from B to D, the line e d, Fig. 221, has moved to the position shown in Fig. 222, and it becomes evident that the port b would be given an amount of lead represented by the distance between e' and e. This amount will obviously

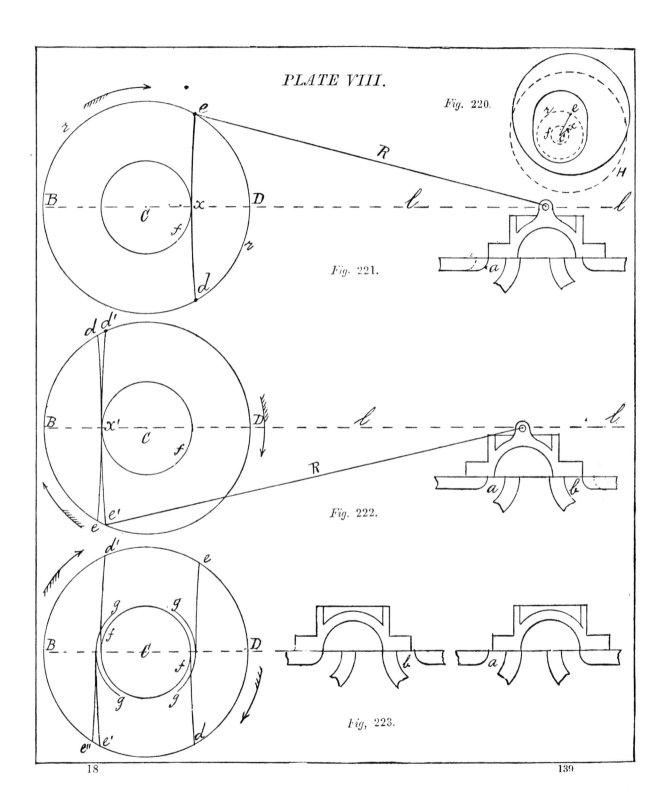

PLATE VIII.

Fig. 220.

Fig. 221.

Fig. 222.

Fig. 223.

diminish as the eccentric is shifted across the shaft towards the position x of least throw until, on arriving at x, there would be no lead for either port, because at x the arcs $e\ d$ and $e'\ d'$ coincide.

It is obvious, however, that the longer the eccentric-rod is, the nearer the arc $e\ d$ will approach to a straight line, and that if it were a straight line the distance $e\ e'$ would be less and the lead variation would also be less. while, if the eccentric-hanger was infinitely long so as to move the eccentric-hanger in a straight line, the lead would be equal for the two ports.

The eccentric is also attached to an arm or hanger that swings on a center, and it is obvious that the longer this hanger is, the nearer the arc $e\ d$ will be to a straight line and the less the distance $e\ e'$ and the lead variation will be.

For the purpose of considering the lead variation, due to swinging the eccentric across the shaft in an arc of a circle, it is sufficient to suppose the eccentric-rod and eccentric-hanger to be of equal lengths.

The amount of lead variation, thus shown to accompany the shifting of the eccentric across the shaft, will obviously increase in proportion as the range of cut-off is increased.

In order to make the lead variation show plainly in the figures, the length of the eccentric-rod has been taken, as equal to but six times the amount of the valve travel, and as the eccentric hanger has been taken as of the same length, the same radius serves for arc $e\ d$ and $e'\ d'$.

In Fig. 223, the valve is supposed to have lead, the arcs g being distant from circle f to the amount of the valve lead, and therefore distant from C to the amount of the lap and the lead. With the crank at B, the eccentric center, when in its position of greatest throw, will be at e, and the cut-off will occur when it has arrived at d. This is clear, because if we placed the valve in position at cut-off, and moved the strap end of the eccentric-rod across the shaft, its center would move in the arc from f to d.

It will, therefore, be perceived that whatever position the center of the eccentric may be shifted to, between the points from e to the line l of centers, the cut-off will occur when the eccentric center passes the line $f\ d$. Similarly when the crank is at D, and the port b is open

to the amount of the lead, the eccentric center ought to be (if the valve is to have equal lead) at e', but it will, for reasons already explained with reference to Fig. 222, be at e'', hence the lead will be greater for port b than for port a. The cut-off, however, will occur when the eccentric center crosses the arc from f to d', this arc being the path that would be described by the center of the eccentric strap if the eccentric-rod was pivoted to the valve, and its strap end moved across the shaft. It follows, therefore, that the cut-off will be later for the stroke when the piston is moving from the head end to the crank end, and it is also seen that as the two arcs e' and e'' coincide on the line of centers, therefore the lead will be diminished in proportion as the eccentric is moved towards the line of centers for earlier cut-offs, and will be all taken up when the eccentric is at its inner position for the shortest cut-off.

Instead, however, of shifting the eccentric from a pivot or pin situated on the line of engine centers, we may do so from a pin on a line $m\ m$, Fig. 224, at an angle to the line of centers which will reduce the valve lead for the longest points of cut-off, and either increase the lead at the early cut-offs for both strokes, or for the head end port only, according to the angle of the line $m\ m$ to the line of centers. Thus, in Fig. 224, the eccentric is supposed to be shifted across the shaft from a point or pivot on the line $m\ m$. The arcs $g\ g$, have a radius from the center C, equal to the amount of lap of the valve. The arc $e\ x$ is drawn with the length of the eccentric-rod as a radius, and from the center of the valve, when the latter is in position ready to open port a for the lead, and, therefore, represents the path in which the eccentric must move across the shaft in order to maintain equal lead for all points of cut-off on this piston stroke. But the path in which the eccentric would actually move (its point of suspension being on the line m) is shown by the arc $e'\ x'$, and it is seen whatever the amount of lead at e', it will increase, as the eccentric is shifted towards x' for the earlier cut-offs.

The amount of difference in the lead will obviously depend upon how far distant the point of eccentric-hanger suspension is located away from the line of engine centers.

Turning now to the stroke when the piston is moving from D to B, and the port b is taking steam, the arc e''

x'' being struck from the line of centers l of the engine, and with a radius equal to the length from the center of the cylinder ports to the center of the crank-shaft is, as in previous examples, a line parallel to which the eccentric must move in order to keep the

drawn, this being the arc on which the eccentric will actually move when shifted across the shaft, as may be seen as follows:

Suppose the crank to be at D, and the point of eccentric suspension to be on the line m to the left of the

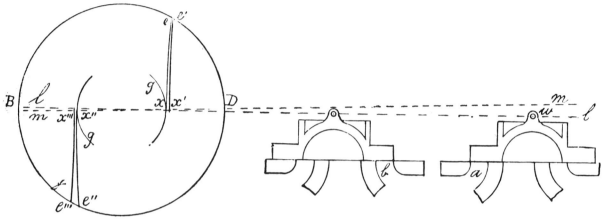

Fig. 224.

lead equal on this stroke for all points of cut-off. To find the line in which it will actually move, the line m m must be prolonged to the left hand of the figure, and

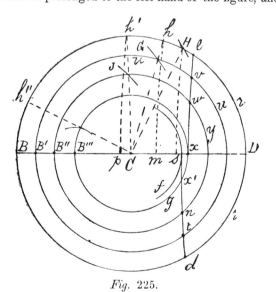

Fig. 225.

from a point on this line, and with the length of the eccentric-hanger as a radius, the arc $e''' x'''$ may be

figure, and if a pencil be inserted in the center of the eccentric and the eccentric-hanger were operated by hand, then the pencil point would mark or describe the arc $e''' x'''$. For this stroke, therefore, the amount to which the lead will vary as the eccentric is moved from its outermost to its innermost position is that represented by the difference of the distance between e'' and e''' and x'' and x'''. The variation in the amount of the lead, for the two ports, is shown, for the longest point of cut-off, by the difference in distance between the points e and e' and that between e'' and e''', while, for the shortest points of cut-off, it is shown by the difference between the points $x\ x'$ and $x''\ x'''$.

TO FIND THE PISTON POSITION FOR A GIVEN ECCENTRIC POSITION.

In Fig. 225, the arc f has a radius equal to the lap of the valve, and circle g a radius equal to the lap and the lead, and it follows, from what has already been explained, that with the crank at B, and the valve open to the amount of the lead, the eccentric, when at its greatest throw, will be at e, and the cut-off will occur when it arrives at d. As the crank will move through the same number of degrees of arc that the eccentric does, we may set the compasses to radius *e d*, and mark, from B, the position H of the crank at the point of cut-off. To find the corresponding piston position, we set a pair of compasses to represent the length of the connecting-rod on the same scale that the outer circle B D represents the path of the crank-pin, and from a point on the line of centers *l l* of the engine, mark from H an arc, giving at *s* the position of the piston at the point of cut-off. Suppose, for example, the engine stroke was 24 inches, and the outer circle *r* being 2½ inches will represent the path of the crank-pin on a scale of ⅛ full size, or ⅛ inch per inch. Supposing the connecting-rod to bear the ordinary proportion of three times the length of the piston stroke, and its length will be 72 inches, hence the radius of the arc H *m* will be 72 eighths of an inch, or 9 inches. Then as the distance from B to *m* is 1¼ inches, we multiply this by 8 and get 15 inches as the piston position at the point of cut-off.

If now the eccentric be moved to position *v*, its path will be on the circle *u* and the cut-off will occur when the eccentric arrives at *t*, because at that time the valve will be moved from its mid-position to an amount equal to the amount of the steam-lap; hence the arc moved through by the eccentric from the beginning of the piston stroke to the point of cut-off, is the arc *v t*. To find the position of the crank at the point of cut-off, we take the radius *v t*, and from B', on circle *u* at the line of centers, mark an arc G. A line from the center C, and cutting the point of intersection of arc G with circle *u*, gives at *h'* the position of the crank at the point of cut-off. An arc whose radius represents the length of the connecting-rod, and whose center is on the line of engine centers, gives at *m* the position of the piston at the point of cut-off.

If the eccentric be moved further across the shaft to the position represented by *u*, the path of its center will be on the circle *y*, and to find the position of the crank when the cut-off occurs, we take the radius *w u*, and from B'', where circle *y* crosses the line of centers, mark an arc J, and a line from C, passing through the intersection of J with circle *y*, gives, at *h'*, the crank position at the time the cut-off occurs, while the arc from *h'* gives, at *p*, the corresponding piston position.

We have now to consider the earliest point at which the cut-off can be effected, hence let it be supposed that the eccentric center is moved to position *x*, and the crank being on its dead center at B, the port will be open to the amount of the lead, and as the cut-off will occur when the eccentric reaches *x'*, we take the radius *x x'* and mark it from B''', obtaining, by the same lines as before, the crank position *h''*. It is obvious, however, that if the valve had no lead, there would be no admission, because when the crank was on its dead center B, the port would be blind, and as soon as the crank moved, the valve would move back further over the port.

TO FIND THE AMOUNT OF STEAM PORT OPENING FOR EACH POINT OF CUT-OFF.

In Fig. 226, the outer circle represents the path of the eccentric center when at its greatest throw, while the circle *f f* has a radius equal to the steam lap of the valve. Now, suppose the path of the eccentric center

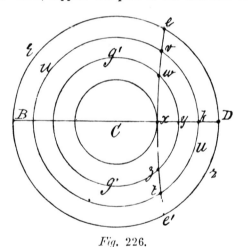

Fig. 226.

to be on the circle *r r*, and when it is at *e*, the valve will have moved from its mid-position to the amount or distance C *x*, the point *x* will, therefore, represent the edge of the steam port and the edge of the valve when the eccentric is at *e*. While the eccentric is moving from *e* to D, it is opening the port, hence the maximum amount of port opening may be measured from *x* to

D. If the valve has no overtravel, the maximum port opening will be the radius from x to D, when the eccentric path is on the circle $r\ r$. But suppose the eccentric path to be on the circle $u\ u$, then the maximum of steam port opening will be equal to the radius $x\ k$; or, if the eccentric-path be on circle g', then the maximum amount of steam port opening will be represented by the radius $x\ y$. The area of port opening will obviously depend upon the dimensions of the ports and upon whether the valve is double or single ported.

THE POINT OF ADMISSION.

If the valve has no lead, the admission occurs when the crank passes the dead center, but if the valve has lead, the point of admission may be found as follows:

In Fig. 227, the outer circle represents the path of the crank-pin and that of the eccentric as before. Circle f has a radius equal to the amount of valve lap, and circle g a radius equal to the lap and the lead. When the crank is at B, and the valve open to the amount of the lead, the eccentric, when set for the latest cut-off, will be at e, and as the cut-off will occur when it reaches point d, therefore the arc it moves through, while the crank moves from the dead center B to the point of cut-off, is the length of the arc $e\ d$, hence we take this distance, and from n (where the eccentric will be when admission occurs) mark arc A, and a line A′, drawn from the intersection of A to the center C, represents the crank-pin position at the point of admission for the latest point of cut-off.

Now suppose the eccentric to be moved to e' for the earliest cut-off, and the length of arc, it will move through from admission to cut-off, is from v to d', and as it will move through half this arc, or from v to e', while opening the valve to the amount of the lead, and the other half while closing it, we take the radius $e'\ d'$ (half $v\ d'$), and from point E, where circle g cuts the line of centers, mark arcs J and F, and then draw lines H C and a C, and line a C will be the crank position

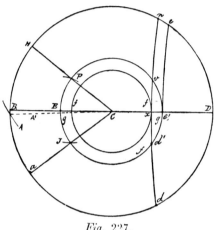

Fig. 227.

at the point of admission, while H C will be its position at the point of cut-off.

This is clear because arc $v\ e'$ equals arc E J, and arc E F equals arc $e'\ d'$.

It is seen, therefore, that if the valve is given lead, then in proportion as the eccentric is moved across the shaft for the earlier points of cut-off, the point of admission is hastened.

CHAPTER. VII.

EXAMPLES FROM PRACTICE.—AUTOMATIC CUT-OFF ENGINES.

Automatic Cut-off Engines may be divided into two classes, as follows: Those in which the valve is released from its operating mechanism when closing to effect the cut-off, which is done to accelerate its movement, and those in which its action is positive throughout.

The Automatic Cut-off Engine possesses the following advantages.

1st. It contains, within itself, means of altering the amount of its power to suit changes in the amount of its lead or duty, while permitting the steam to flow unchecked from the boiler to the steam chest.

2nd. It contains, within itself, means of maintaining a constant proportion between its piston power, and the lead it drives, notwithstanding ordinary fluctuations of boiler pressure, and it does this without checking the flow of steam from the boiler to the steam chest.

3rd. It accomplishes both the above named results by varying the amount of expansion, and, therefore, avoids the loss that accompanies their accomplishment by means of wire-drawing the steam.

4th. It adjusts its power to the load more quickly than is possible by means of varying the steam chest pressure, as is done when a throttling governor is used, and thus maintains the engine speed more uniform.

5th. It possesses all the qualifications of an engine having a fixed point of cut-off, because if the load remains constant, it retains the point of cut-off at a fixed point in the stroke, and only varies the point of

144

cut-off when the conditions of a varying load demand it.

When an engine has a fixed point of cut-off, and the engine speed is regulated by varying the pressure at which the steam from the boiler is admitted into the steam chest, three evils are induced:

First, suppose the engine to be running at its normal speed and a sudden decrease occurs in its load, and the engine speed will increase, causing the speed of the governor to increase and check the flow of steam into the steam chest; but the steam chest will, at this time, be filled with steam admitted to it before the governor checked the inflowing steam, hence the steam in the steam chest forms a reservoir acting in opposition to the governor, and this obviously prolongs the time necessary to bring the engine power down to meet the requirement of a reduced load, and it follows that the engine speed will fluctuate because of the sluggishness of the method of governing it.

Secondly, suppose the engine to be running at its normal or proper speed, and that its load suddenly increases, and its speed will decrease, causing the governor to increase the flow of steam into the steam chest, but the inflowing steam will be partly expanded in raising the pressure of the steam already in the steam chest, and the latter, therefore, acts, to a certain extent, in opposition to the governor, and, therefore, delays its action in governing the engine speed. In either case, therefore, there elapses a certain amount of time between the action of the governor and its effect upon the engine speed.

The third evil is that a throttling engine, with a fixed point of cut-off, must either wire-draw the steam considerably or else be deficient in governing power under the changes that are liable to occur in the boiler pressure. Thus, suppose the boiler pressure to be at its lowest permissible point, and the governor will open the steam pipe to its widest, and if this is sufficient to maintain the engine speed, then when the boiler pressure is at its highest point, the governor must close the steam pipe sufficiently to keep the steam chest pressure down to what it was under the lowest boiler pressure, hence supposing the amount of the engine load to remain constant, then the wire-drawing action of the governor must, at least, equal the greatest amount of fluctuation that may occur between the highest and lowest boiler pressures.

THE PORTER-ALLEN ENGINE.

The Porter-Allen was the pioneer in high piston speed stationary engines, and has attained a piston speed as high as 1,100 feet per minute, without noise or shock.

Its usual piston speed, however, is from about 650 to 850 feet per minute.

The prominent features in the design are as follows :

1st. The valve motion is so designed as to proportion the steam admission and supply to the piston velocity, and thereby act to counterbalance the reciprocating parts of the engine.

2nd. The valves are so constructed and operated as to give a rapid opening and steam supply early in the stroke, and a quick cut-off, while their amount of travel is limited and their speeds diminished during that part of the travel in which lap of the valve is being taken up.

3rd. The variation in the point of cut-off is effected by means of altering the amount of valve travel, which the mechanism of the engine effects automatically. This variation is governed by the amount of resistance of the load or duty of the engine, the point of cut-off occurring later in the piston stroke in proportion as the load or duty is increased.

4th. A single eccentric and link operates the two steam and two exhaust valves, yet the exhaust valves are actuated independently of the steam valves.

5th. The valves are balanced to relieve them of undue pressure to their seats, thus causing a minimum of friction and wear, and may be set up, as occasion may require, to a proper adjustment to their seats.

Fig. 228 is side view of the engine, from the connecting-rod side, and Fig. 229 an outline view of the engine, from the link side, and having a part of the bed broken out to expose to view the arrangement of the valve rods.

The eccentric E operates the link L, which is pivoted to a pin a at the top of the arm A, which is pivoted at its lower end e by a pin in the bracket N.

In the slot s of the link is the usual sliding block, to which is pivoted the rod R, operating the wrist motion at W. This wrist motion consists of a three armed lever, the lower arm receiving motion from rod R, the upper arm operating the rod C for the slide valve spindle F, and the lower one operating rod D for the other slide valve, whose spindle is seen at K. The exhaust valves are driven from the top of the link by the rod X which vibrates arm y, whose shaft passes to the other side of the engine and operates an arm, to which is attached a rod Z, which is seen in the side view, Fig. 229.

The exhaust valves are in the same plane, and are operated by the same rod, but the steam port slide valves are not in the same plane, the crank end valve being higher than the head end valve, as is denoted by the cover H being higher than cover J.

The motion of the link and valve mechanism, will be fully explained hereafter, it being sufficient for the present to point out that, as the top of the link has more motion than its point of suspension on arm A, therefore the valve has most travel when the link-block is at the top of the link slot f, in which position it is shown in the engraving.

To the valve rod R (which carries the link-block), is pivoted the rod G, which is in turn pivoted to the weighted lever P, whose ball weight counterbalances the weight of the rods G and R and of the link-block. The lever P is actuated by the governor as follows:

When the load or duty of the engine is lightened,

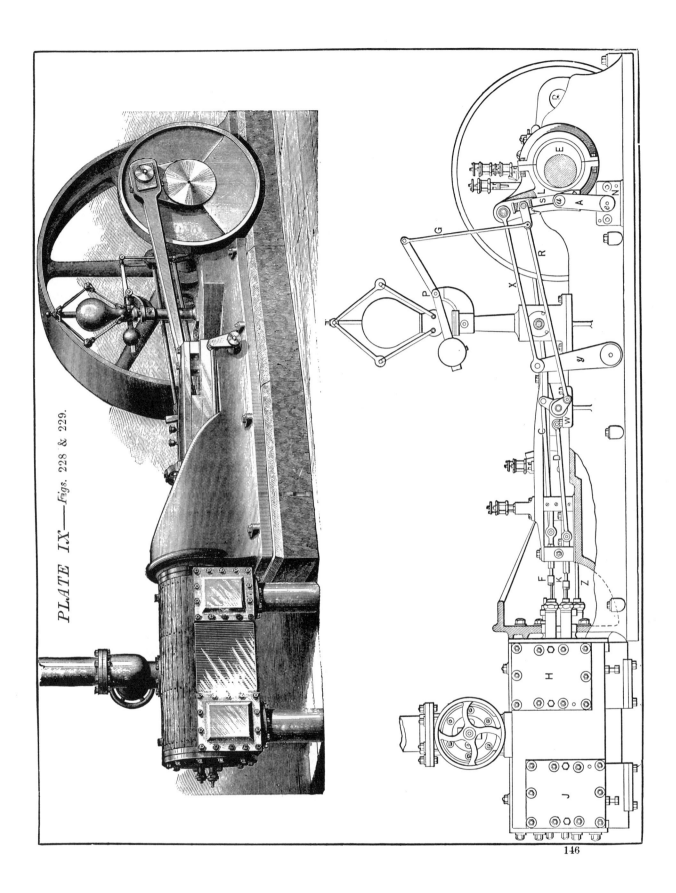

PLATE IX.—Figs. 228 & 229.

146

the governor causes the ball end of lever P to raise, its outer end depressing the rod G, which, therefore,

form the power of the engine to a lightening of the engine load. The point of cut-off is varied, in ordinary

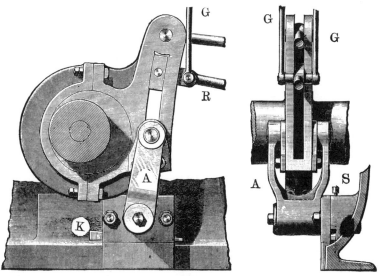

Side Elevation. *Fig.* 230. *End Elevation.*

The Link and its Connections.

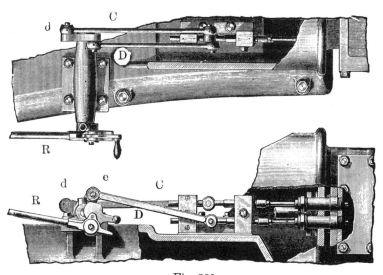

Fig. 231.

The Wrist Motion.

moves the link-block down the link-slot *s*, diminishing the valve travel and hastening the cut-off, so as to con-

Porter-Allen engines, between the limits of from quarter-stroke to half-stroke, but in engines, such as those

19

for iron-rolling mills, where occasion may call for greater power, the limit is extended between ¼ and ⅝ stroke.

THE CYLINDER AND VALVES.

The details of construction are more fully seen in the following figures: Fig. 230 gives a side and end view of the eccentric, the link and the link supporting arm A. The eccentric is formed solid with the main shaft, and the link solid with one-half of the eccentric-strap.

head end of the cylinder. C and D are the rods for the respective steam valve spindles. The letters of reference correspond to those in the figures illustrating the valve mechanism.

In Fig. 232, the cylinder is shown cut in half horizontally and viewed from above. The spindle F operates the valve at the crank end of the cylinder, and through this valve passes the spindle K for operating the valve at the head end. Spindle M is for the exhaust valves, these letters of reference corresponding to those given in the outline view, Fig. 229.

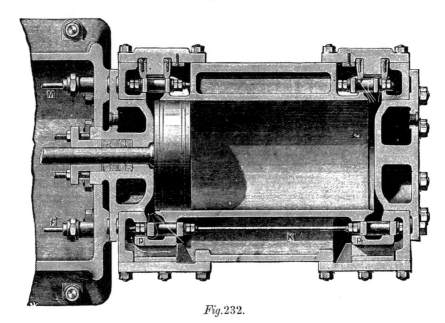

Fig.232.

Horizontal Section through the Cylinder and Valves.

The link is slotted so that the valve rods pass centrally through it, as shown in the end view. The arm A supports the link on both sides, and has provision, by means of key K and set screw S, for adjusting the height of the pins by which the link is pivoted, this being necessary to effect a proper adjustment so as to equalize the lead and the points of cut-off, as will be explained hereafter.

The wrist motion is shown in Fig. 231, R being the rod from the link block, *d* the arm for operating the valve at the crank end, and *e* that for the valve at the

A cross sectional view of the head end of the cylinder is shown in Fig. 233, and referring to these two views, it is seen that when the steam valve opens the port for the admission of steam, the steam passes through four different openings, as denoted by the arrows.

The method of relieving the steam valves from undue pressure to their seats is as follows: At *m* and *n*, Fig. 233, are two inclined planes at an equal angle to the valve face, and at the back of the valve is a pressure plate P, fitting up to these inclined planes. The valve works between its seat face and the face of plate P, and beneath the latter is a screw *r*, by means of which it may be adjusted, it being obvious that raising the

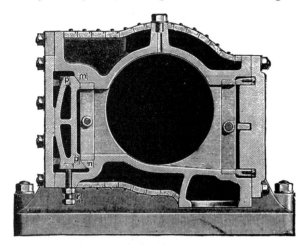

Fig. 233.

Vertical Secton through the Cylinder aud Valves.

plate P moves it away from, while lowering it causes it to approach the back face of the valve until, on meet-

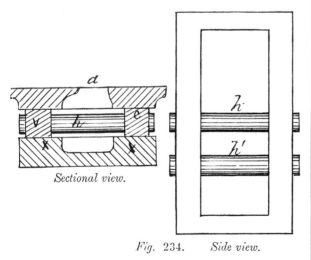

Fig. 234. *Side view.*

The Admission Valve.

ing that face, the steam is excluded and the valve relieved of the steam pressure.

The steam valves are rectangular frames, such as in Fig. 234, the valve spindle, for the valve at the crank end of the cylinder, attaching to hub *h*, while the spindle for the head end valve passes through the hub *h'* which is in line with the center of the latter valve. In the sectional view the valve is shown in its mid-position.

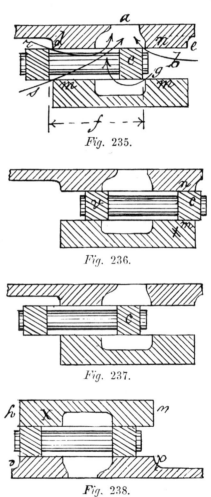

Fig. 235.

Fig. 236.

Fig. 237.

Fig. 238.

Various Positions of the Admission Valve.

The full effective width of the port is the width at *a*, it being obvious that the width of the mouth of the port is, from the point of admission until the point of cut off, covered by the thickness of the metal at *c*, Fig. 235. The valve length at *f* is proportioned that when the port opens on one end as denoted by the arrows *b* and *g*, it will also open at the other, as denoted by the arrows *r* and *s*.

In Fig. 236, the valve is shown at one end of its travel for cut-off at half-stroke, and in Fig. 237, it is shown in position at the other end of its travel, and about to reverse its motion.

The wear on the valve and its seat, and on the face *m* of the cover plate X, is equalized, because the metal at each end, *c v* of the valve, passes entirely

has not moved ⅛ inch. This is shown in Fig. 238, in which the valve is shown at one end of its shortest travel, and it is seen that it still moves up to the shoulder *e* of the valve face, and to the end *y* of the plate X.

THE ACTION OF THE VALVE MECHANISM.

Fig. 239 shows the position of the parts when the crank-pin is on its dead center D. The rod R, is

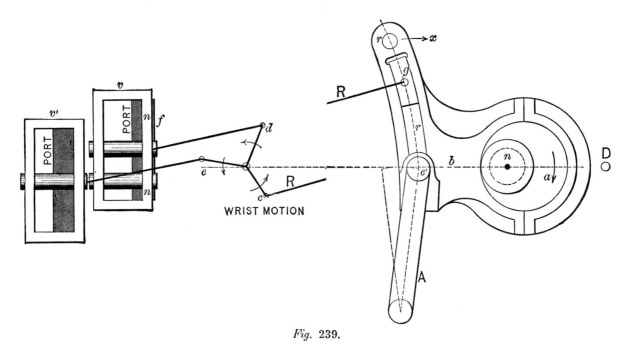

Fig. 239.

Position of the Valve-Motion at the Beginning of the Stroke.

over the surfaces *n m*, and, therefore, prevents the wear from forming a shoulder, which would occur if the ends of the valve did not travel past the shoulders *d* and *e*.

These conditions also prevail when the link-block is placed coincident with the point of suspension of the link, at which time the valve has its least possible amount of travel, the port only opening to the amount of the lead, and the cut-off occurring when the piston

shown broken and shortened, and the valve spindles are omitted for clearness of illustration. The steam ports are indicated by the shaded portions beneath the valves.

The throw lines of the crank and the eccentric, are set exactly in line one with the other, both being on the line of centers *b b* of the engine, and on the same dead center.

The valve *v* has opened the port to the amount of the

lead as shown at *f*, and it is seen that from the position of arm *d*, the valve will be moved quickly, while arm *e* will scarcely move the valve *v'*.

As the motion proceeds (the eccentric, moving in the direction of its arrow), the upper end of the link will move in the direction of arrow *x*, and it is this tipping motion that acts to open the valve for the admission. In this, and in all the figures, the arc *r* denotes the position the center-line of the link stands in when the cra..k is on its dead center, and the parts in the position shown in Fig. 239.

When the parts have reached the positions shown in center, the link-block could be moved from end to end in the link-slot without imparting any motion to the valve, the lead would be equal for all points of cut-off. But this arm moves in an arc of a circle, and the effect of this is, that if the valves were set with the crank on the dead centers, the points of cut-off could only be equalzed for some one point of cut-off, as, say, at half-stroke, because when the link-block was moved further down the link, and toward the line of engine centers, the valve would cut off earlier at the head end than at the crank end of the cylinder, the difference being considerable in the shorter points of cut-off.

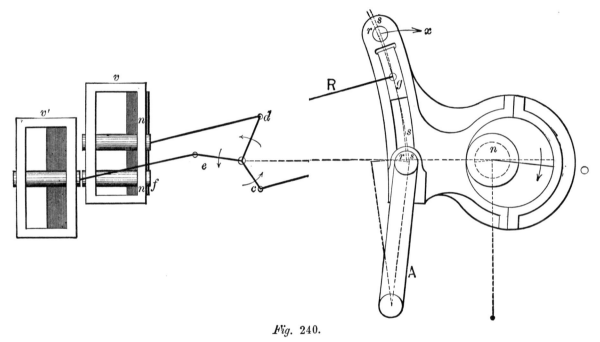

Fig. 240.

The Position of the Parts when the Lead is Set for the Head-End Port.

Fig. 240, the link-block *g* may be moved from end to end of the link-slot, without moving arm *c* of the wrist motion and, therefore, without moving the valves, and this is the position in which the parts must stand when the valve is to be set for this crank position.

If the path of the center (*e'*, Fig. 239) of the arm A, was in a straight line on the line of engine centers, and the link was so set that, with the crank on either dead

This, however, is remedied by the construction shown in Fig. 241, in which the outer dotted circle represents the path of the crank, and the inner circle *n* the path of the center of the eccentric, the line A represents the arm A in Figs. 229 and 230, and its upper extremity, *e*, represents the link-sustaining pin whose arc of motion is therefore arc *a*.

Suppose we draw, below the center C of the crank

a vertical line A′, equal in length to the arm A, and from its lower extremity, as a center, draw an arc *e″*

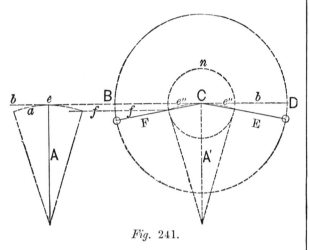

Fig. 241.

Finding the Positions of the Crank when the Valves are Set.

line *f f* drawn to the extremities of these two arcs will be parallel to the line of centers *b b* of the engine.

From the center C of the crank-shaft, we then draw a line E, cutting the intersection of arc *e″* with the dotted circle *n*, and this line E represents the position the crank must stand in when the lead is set, and, at this time, the link-block can be moved from end to end of the link without moving the valve. Similarly, a line F, drawn from C and touching the other end of arc *e″*, where it cuts the circle *n*, gives the crank position (for the other piston stroke) at the time when the link-block can be moved from end to end of the link without moving the valves, this also being the position the crank is in when the valves are set, the lead for the two ports being made equal.

It follows from the construction in Fig. 241, that these two crank positions are at an equal distance below the line of centers *b b*, and that the amount to which they are below it, depends upon the proportions of the

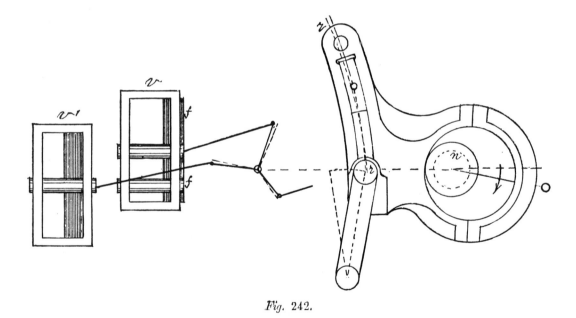

Fig. 242.

Positions when the Head-End Port is Full Open.

passing through the center C, and this arc will correspond both in curvature and length to the arc *e*, and a

various parts, which will be treated upon hereafter. Referring again to Fig. 240, the dotted arc *r* is the position the link was in when the crank was on the dead center, as in Fig. 239, and the amount of tipping of the link, in the direction of x, is shown by the distance of arc r from arc s.

When the crank has moved 10° from the dead center, and the piston has moved about $\frac{1}{8}$ inch from its dead center, the port is full open for the admission, the position of the parts being as in Fig. 242, in which the arc r represents the position of the link, when the crank was on its dead center, and shows, therefore, the amount the link has moved.

Referring now to the amount of opening there must

until the crank has moved 70° from the dead center, at which time the parts occupy the positions shown in Fig. 243, but after this the link moves forward parallel, the port remaining full open until the piston has moved 84° from its dead center, the position of the parts then being as in Fig. 244, the opening at f, equalling one-fourth of the effective area of the port (or one-fourth the width at a, Fig. 234).

The piston has now traveled 10 inches of its 24 inches of stroke, and the port, as the motion continues, begins to close for the cut-off. The positions of the parts, at the time of cut-off, is shown in Fig. 245, the

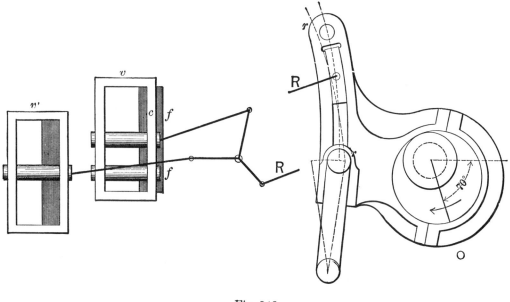

Fig. 243.

The Valve at the end of its Travel for the Head-End Port.

be at f, before the port is open full for the admission, it is seen, in Fig. 234, that the effective width of the port is its width at a, and as the steam enters in the four streams denoted by the arrows, it is obvious that when the opening denoted by the arrow b is equal to $\frac{1}{4}$ the width of a, the port is virtually fully opened.

As the motion proceeds, the link continues to tilt,

piston having moved 12 inches, and it is seen that the valve v' has scarcely moved.

When the parts have moved to the position shown in Fig. 246, the link will again be in such a position that the link-block g may be moved up and down the link, without imparting any motion to the valves. The wrist arm d has now moved to such a position that it imparts

PLATE   X——Figs. 244 & 245.

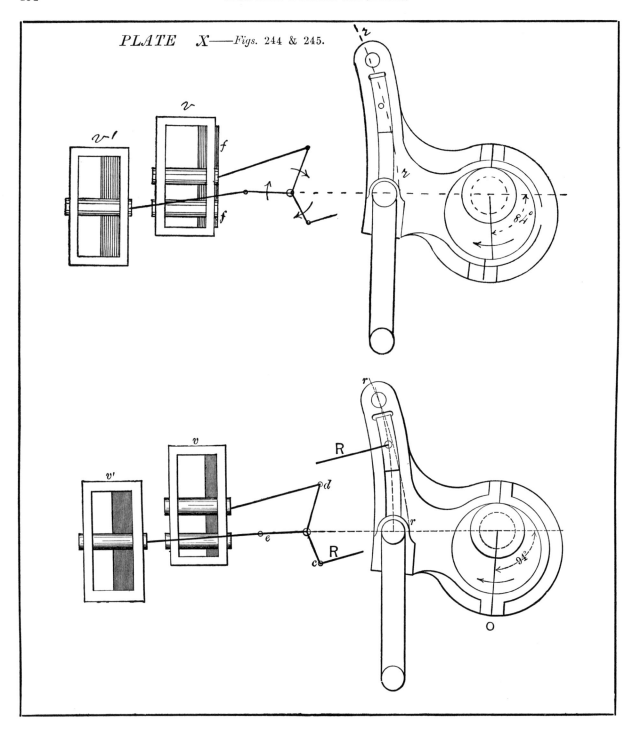

PLATE XI.

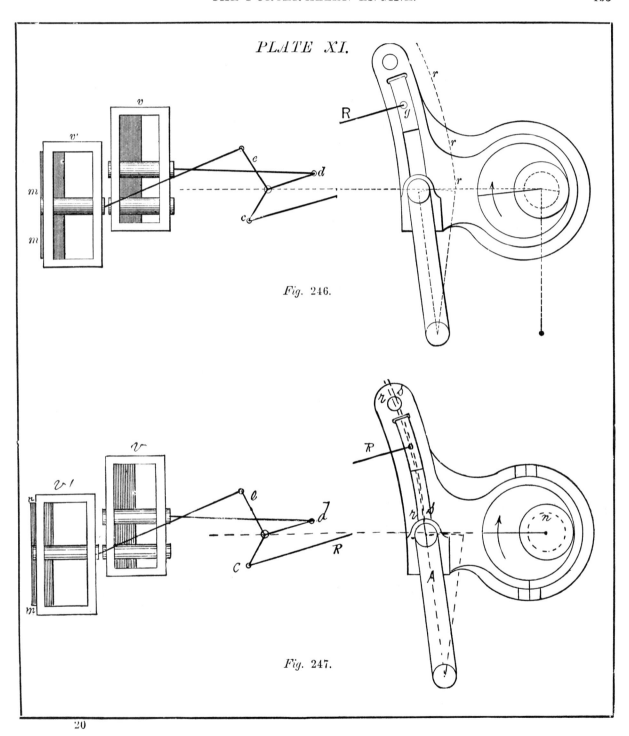

Fig. 246.

Fig. 247.

but little motion to the valve v, while wrist arm e is in position to move the valve v' quickly.

open to the amount of its lead. The valve motion has now been investigated thrughout one piston stroke.

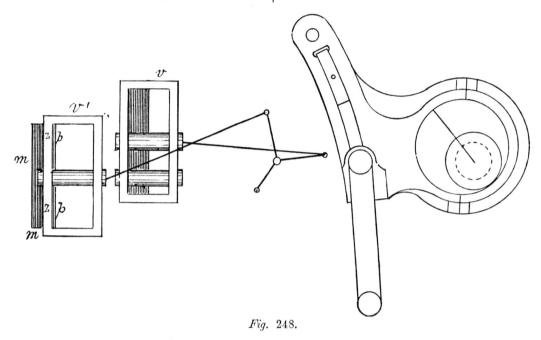

Fig. 248.

The Valve at the End of its Travel for the Crank-End Port.

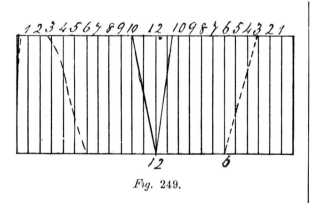

Fig. 249.

Diagram of Port Openings.

The position of the parts, when the crank arrives at its dead center, is shown in Fig. 247, the valve v' being

During the next piston stroke, the motion varies in two respects, first, the amount of valve lead is greater, for reasons which will be explained presently; secondly, the port opens quicker, and remains fully open longer, although the point of cut-off occurs at the same point, or at half-stroke, as before.

Fig. 248 shows the position of the parts, at the time the valve v' has traveled to the end of its path over the port, and it is seen that the amount of opening at m, for the admission, is greater than that at f, Fig. 243, which shows the corresponding position for the other stroke. This occurs because the end Z of the valve moves further back than does the end c in Fig. 243.

Fig. 249 represents the steam port openings during the two strokes, the link-block being set first to cut off at one-half, and then at one-quarter, of the piston stroke, the latter being marked in dotted lines, and it is seen that, for the cut-off at half-stroke, the port remains fully opened for $10\frac{5}{8}$ inches of one stroke (the head end), and for 10 inches on the other, while both cut-offs occur at 12 inches.

Similarly, for the cut-off at one-quarter stroke, the port remains full open during $2\frac{5}{8}$ inches at the crank end, and during $3\frac{1}{4}$ inches at the head end, the cut-off occurring at the sixth inch on both strokes. Furthermore, the port is full open when the crank is on the dead center at the head end, while, at the crank end, it does not open full until the piston has moved about $\frac{1}{4}$ inch for the cut-off at half-stroke, and about $\frac{3}{8}$ inch for the cut-off at quarter-stroke.

Now the piston motion is, on account of the angularity of the connecting-rod, quickest when moving from the head end, and, therefore, when the port *m* is acting to admit steam, hence the steam supply is better proportioned to the piston velocity, than it would be if both ports opened equally.

It has been observed that, in order to equalize the points of cut-off for all points of cut-off, it is necessary that the arm A vibrate in an arc that meets the line of centers, when the arm is vertical or at a right angle to the line of centers *b b* of the engine, and that with a proper adjustment of the parts, there are two positions in which the link-block can be moved from end to end in the link-slot, without imparting any motion to the valves.

These positions are those shown in Figs. 240, and 246, the crank standing, in both cases, at an equal distance below the line of centers *b b*, ready for the valve lead to be set, an equal degree of lead being given for the two valves. When, however, the crank is brought to the respective dead centers, there will be more lead at the head end than at the crank end, as may be seen on comparing Figs. 239 and 247. This is also shown in the diagram of the port openings in Fig. 249.

It is obvious, however, that the lead might be equalized, if desired, by giving to the valves an unequal degree of lead when setting them with the parts in the positions shown in Figs. 240 and 246, but it is proper to give an unequal lead when the crank is on the dead centers, and for the following reasons:

Let the piston stroke be 20 inches, and the circle in Fig. 250 being 4 inches in diameter, will represent the path of the crank-pin one-fifth full size, and its diameter on the line from 0, to 20, will represent the piston stroke. We divide the circle into 40 equal divisions, representing crank-pin positions, those from 1 to 20, being for the path of the crank-pin during one piston stroke, and those from 20 to 0. representing crank-pin positions for the other piston stroke.

To find the corresponding piston positions we set a pair of compasses to represent the length of the connecting-rod on the same scale that the circle represents the path of the crank-pin, or, in this case, one-fifth full size. Then prolong the line 0, 20 (which represents the center line of the engine), and with the compasses set as above, draw arcs from the divisions on the circle, to the line 0, 20, these arcs giving us, on that line, the piston position for each crank-pin position.

Thus arc *a* is drawn from division 10, and gives us at *b* the position of the piston when the crank is at point or division 10. In the figure this construction is carried out from division 10 to 20, for the last half of one stroke, and from division *r*, to division 0, for the last half of the other piston stroke.

We may now compare the positions of the piston on one stroke, with its corresponding position on the other as follows:

When the crank is at division 11, and the piston at *c*, the crank has 9 divisions to move through to complete the stroke (the crank being supposed to move in the direction of the arrow). The corresponding crank-pin position for the next stroke, is at *r*, because when at *r*, it will also have to move through 9 divisions before completing its stroke and arriving at 0. With the crank at *r*, the piston will be at *f*, and we may find the difference between the two piston positions, by drawing from the center C of the outer circle, a semicircle *d*, from which we see, that with the crank at the corresponding points 11 and *r*, of its path, there is a difference in the piston positions represented by the radius or distance from *e* to *f* (measured of course on the line 0, 20)· This difference is caused by the angularity of the connecting-rod, and increases in proportion, as the connecting-rod is shorter than the crank.

The length of the connecting-rod, in this example, is taken as three times the length of the piston stroke, or six times the length of the crank. By following out this method of investigation, we find that while the crank is moving through the ten divisions, from 10 to 20, the piston speed is less than it is while the crank is moving through the ten divisions from q to 0, which is clear, because, during the one period, the piston only moves

amount of linear motion at the two ends of the stroke). by means of an unvarying weight on the crank disc, it being obvious that the weight that would counterbalance the parts when the piston is at the crank-end

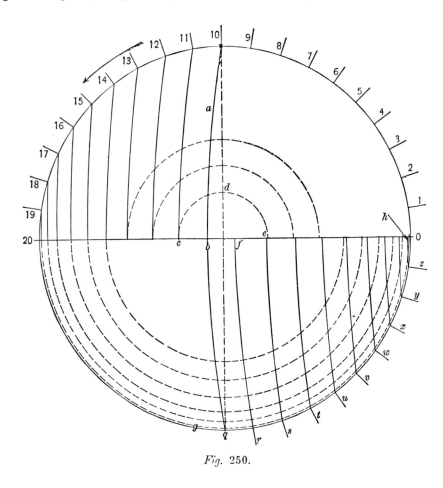

Fig. 250.

The Variation of the Piston Motion.

from b to 20, while during the other it moves from b to 0.

This assumes great importance (when the piston has a high velocity), because it renders it impossible to counterbalance the weight of the piston, of the cross-head, and of the connecting-rod (which all have a variable

of its stroke, would be insufficient to do so when it is at the head-end of its stroke.

It will be noted, that as the piston approaches the ends of its stroke, the above variation diminishes, and that, so far as the counterbalancing is concerned, we need only consider the relative speeds of the piston, when reversing its motion at the ends of the stroke. Suppose, then, we compare its velocity during the motion of the crank, while moving through its last division on each stroke, or, from 19 to 20 in one case, and from Z to 0 in the other.

From the piston position, when the crank is at 19, we draw dotted circle *g*, and we find that it passes between the end 0 and the arc *h*, and the difference between the distances of circle *g* and arc *h*, from the end 0 of the stroke (measured, of course, on the line 0 20), represents the difference in the velocity with which the piston comes to the dead centers, and to its state of rest before beginning the next stroke. This difference is about 40 per cent.

We have here assumed the velocity of the fly-wheel and crank to be entirely uniform, which, practically, it is, but as the later portion of the stroke is performed under expansive steam, and, therefore, under a reduced steam pressure, it is obvious, that the tendency is for the velocity of the fly-wheel to reduce (assuming the engine load to remain constant), but at whatever speed, or under whatever conditions, the engine may run, the discrepancy between the fly-wheel and piston velocity here described, must, if the engine has one cylinder only, exist, and the variation of velocity will be mainly in the piston, cross-head, etc., and not in the fly-wheel, indeed it is obviously desirable to maintain the velocity of the fly-wheel as uniform as possible, so as to have a uniform velocity in the machinery, driven by the engine.

In high speed engines, a uniformity of fly-wheel velocity is more easily obtained, and maintained, than in those running slower, for the reason that the period of time, during which the live steam is cut off and the fly-wheel called upon to maintain the speed (notwithstanding the diminished steam pressure, existing during the expansion period), is diminished.

It follows, from Fig. 250 and its accompanying explanation, that the fly-wheel must, in maintaining a uniform velocity, maintain the inequality of piston motion, at the two ends of the stroke, or, in other words, that this inequality is essential to a uniform fly-wheel velocity, and that the means taken to counterbalance the piston, etc., must be such as will permit it to continue until the piston is near the end of its stroke, when it

must be resisted for a length of time, merely sufficient to avoid a *knock* or *pound* in reversing the motion of the piston, etc., the period varying with the engine speed.

It has been already shown that this cannot be done for both strokes by means of a counterbalancing weight, hence, we must resort to some other means, which is found by giving the valve more lead at the head end, than at the crank end of the stroke, as has been described.

In proportioning the parts of an engine of this description, the distance from the center of the eccentric to the center of the link-slot is made equal to six times the throw of the eccentric, hence, since the length of the connecting-rod is six times the length of the crank, it bears the same proportion to the throw of the eccentric that the length of the connecting-rod does to that of the crank.

The link supporting arm is also made equal in length to six times the throw of the eccentric, and, with these proportions, the cut-off will be at half-stroke when the link-block is distant from the line of centers to an amount equal to to six times the throw of the eccentric.

The length of the arm *c* of the wrist motion may be such that the link will cause it to make one-quarter of a revolution, so that the arm *d* shall move as much past its dead point, as it lacks of moving to a position at a right angle to the line of centers, or it may be at such an angle to the arm *c* that it will move from the horizontal to the vertical position, the only appreciable effect being to diminish the amount of retardation of the valve movement while the lap is traveling over the port.

The wrist motion arms *d* and *e*, Figs. 231 and 239, may be made as much longer than *c* as the width of the port may require, it being obvious that by multiplying the motion, by making *d* and *e* longer than *c*, the throw of the eccentric may be kept at a minimum.

The positions of the eccentric, and of the link supporting arm, require to be very exact, as the least error in their alignment one with the other, or in the height of that arm, destroys the symmetry of the motion, and also, the equalization of the points of cut-off.

Another point to be considered in proportioning the eccentric, the link, and its supporting arm is, that unless the proportions are correct with relation one to the other, the port at the head-end will be apt to have no lead because the lead is set when the crank has passed the dead center, and moving it back to that center, will, unless the above parts are properly proportioned, take off the lead, while, at the other end, the lead will be unduly increased.

The exhaust valves being independent, the point of release and the compression may be regulated at will.

The Buckeye Engine.

Figures from 251 to 266 represent the Buckeye high-speed automatic cut-off engine.

Fig. 251 is a front view of the engine, showing the

is regulated by a governor attached to a wheel upon the crank-shaft. It differs, however, from other engines of this class, inasmuch as that its governor moves the cut-off eccentric *around* upon the shaft, instead of *across* it, as is commonly the case, and thus gives to both its valves an equal and constant amount of travel, which prevents the surfaces from wearing unevenly.

Fig. 253 is a horizontal section through the cylinder valves, steam chest, etc. The cylinder section is taken on one continuous central plane, while the steam chest and valves are shown upon two planes. The first, and nearest to the eye, extends from the crank end of the cylinder to the fracture at *f f f*, and shows the main valve in section through its center. From the line of fracture *f f f*, to the head-end of the cylinder, a lower plane is taken, so as to pass through the center of equilibrium rings, or balance pistons, this lower plane

Fig. 251.

cross-head, guide bars, and connecting-rod, while the back view, Fig. 252, shows the governor and valve mechanism.

This engine belongs to that class in which the speed

being on the line *b' b'*, Fig. 254. This latter figure is a sectional view on the line A A, showing the back of the main valve with the balancing pistons removed from the crank end but in place in the head end.

Fig. 255 is a cross section of the cylinder and valves taken on the line B B B, in Fig. 253, so as to pass through the center of the valve-balancing pistons, but leave the end of the cut-off valve in full view.

Fig. 256 is a sectional view on the line C C, Fig. 253, and shows the back of the cut-off valve upon its seat in the main valve.

THE CONSTRUCTION OF THE VALVES.

The main valve V is a box or chamber, fitting at each end to the cylinder port faces, and provided inside

The steam from the inlet takes the course shown in Fig. 253, by the arrows, filling the chamber D, which is in one piece with the valve chest-cover, and passes into the back of the valve through the openings *d*, of which there are four (two at each end of the valve), as seen in Fig. 254. The main valve is therefore filled with live steam, while the exhaust passes outside its ends to the outlet. The means employed to regulate the pressure of the valve to its seat are as follows:

The inner wall of chamber D has, at each end, a hub bored to receive equilibrium rings or valve-balancing ing pistons, these consisting of a spider *a* (whose arms are shown at *a'*) the follower *c*, and packing ring *e*.

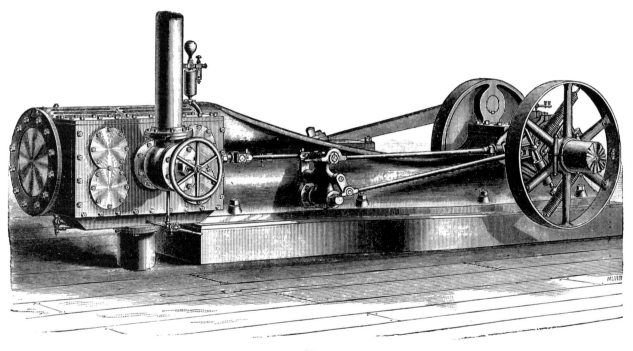

Fig. 252.

The Buckeye Engine.

with a flat surface at each end, whereon the cut-off valve *v v'* operates. The cylinder ports being shown at *p' p''*. The cut-off valve consists of two plates rigidly connected by means of the stretcher-rods *h h'*, and rides within the main valve, its spindle passing through that of the main valve.

The spider has a guiding stem attached to its hub, over which the follower slips easily. The spring S holds the follower and spider together, and confines the packing ring *e* in its proper place.

The faces of the equilibrium rings, or pistons *a*, seat upon the back of the main valve, and, therefore, tran-mit to the valve whatever steam pressure they receive. The live steam, therefore, holds the valve to its seat by acting on the area enclosed within the circumference of the four equilibrium pistons a, and this area is so proportioned as to overcome the tendency of the valve to lift from its seat. This tendency is due to the cylinder port and the port in the main valve, and is greatest at the moment of admission.

With the parts in the position shown in Fig. 253, for example, it is obvious that the steam in the port, at the head-end of the cylinder, is acting on the underneath face of the main valve, and, therefore, in a direction to lift the valve from its seat, while, at the same time, the port on that end of the main valve permits the steam to press upon the cylinder face, and the steam being within the valve causes an equal area on the in side of the valve, and opposite to the port, to be unbalanced, and, therefore, act also to lift the valve.

The area, however, of the annular pistons, is, as before stated, proportioned so as to overcome these two tendencies and hold the valve to its seat sufficiently to keep its steam tight.

It is obvious, however, that when the cylinder port is open to the exhaust, and the cut-off valve covers the port in the main valve, as is the case at the crank end of the cylinder, in Fig. 253, the pressure upon that part of the valve that covers the cylinder port is reduced to that of the exhaust, while the port in the valve is filled with steam that is enclosed between the cylinder face and the cut-off valve, hence, the annular pistons would, in the absence of any counteracting pressure, hold the valve more closely to its seat than is absolutely necessary at that end. But this is provided for by the relief ports or recesses, x x', which, during that portion of the valve stroke in which it would other wise be held to its seat with more force than necessary, receive steam, through the small hole shown to pass through the valve at r, and this steam, acting on the face of the valve, relieves it of the undue pressure referred to. The relief recess is equal in area to the cylinder port, and is so located that it will be uncovered by the heel, or inner edge, of the main valve face, just after the cylinder port, at that end, is closed against the exhaust, and before any considerable compression pressure can arise in the cylinder.

The steam hole r is located to fill the recess at the proper time, which is just after it has been covered, as above stated, by the valve heel.

It will be seen that the face m of the main valve does not quite reach the face of the hub g, and that the annular pistons, therefore, project through the bores of g, and this allows the main valve to lift from the seat to the amount of the space at m, in case the cylinder should receive a charge of water with the live steam. After the water has discharged into the exhaust, the steam pressure returns the valve to its seat, and the spring r causes the annular pistons to follow it up.

The cut-off valve is provided with an inclined plane at t, resting upon a corresponding inclined plane provided on the main valve, hence it moves of its own gravity up to its seat on the main valve and its weight acts to cause it to seat itself fairly.

The mechanism for operating the valves is as follows: Referring to the general view, Fig. 257, and the cross-sectional view of the frame and rock-shafts in Fig. 258, the main eccentric is fixed upon the crankshaft, and its rod R drives the upper end of the main rocker (having journal bearing at p', in Fig 258.) At F', Fig. 258, it affords journal bearing for the rod r, Fig. 257, which drives the main valve spindle.

The cut-off eccentric drives the rod S, which connects to the lower arm s of the cut-off rocker; s is fast upon A, which has journal bearing in the main rocker. The arm a is fast upon A, and provides at E' journal bearing for the rod that drives the cut-off valve.

The cut-off eccentric is a working fit upon the crankshaft, so that it may be moved around it by the governor, the construction being shown hereafter in connection with the governor. The construction of the valve rods is such that the spindle for the cut-off valve passes through that for the main valve.

In Fig. 259, we have the main and cut-off valves removed from the other parts of the engine, their spindles, the rock-shaft, the eccentrics, and the crank being represented by their center-lines.

For ease of illustration, the rods from the rock-shafts are shown shortened, and as if connected direct to their respective valves, which is sufficient for explaining the action of the mechanism, while it renders it easier to illustrate the movements of the parts. The rods R and

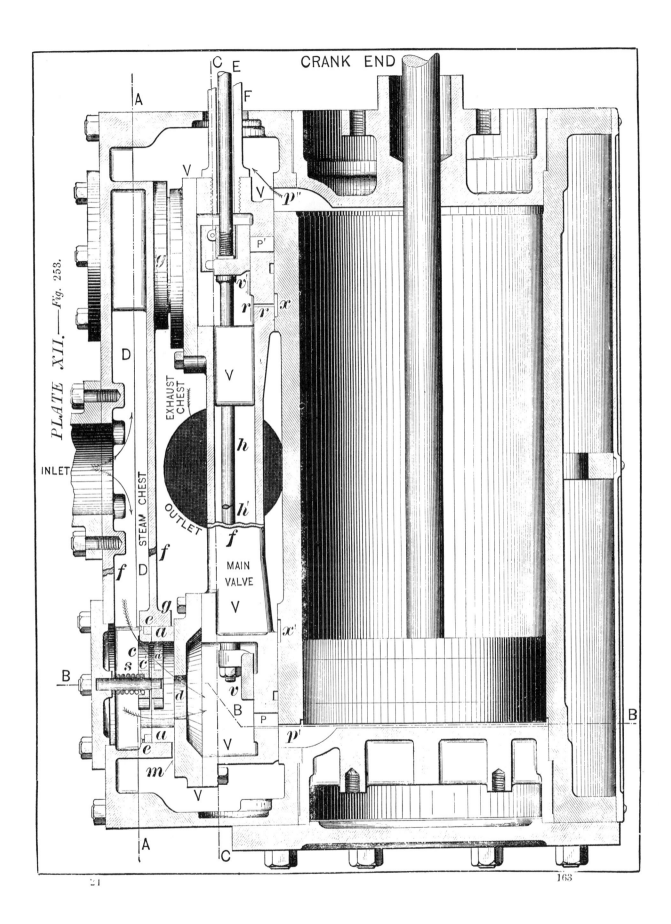

PLATE XII. — *Fig.* 253.

CRANK END

INLET

STEAM CHEST

EXHAUST CHEST

OUTLET

MAIN VALVE

PLATE XIII.

Fig. 254

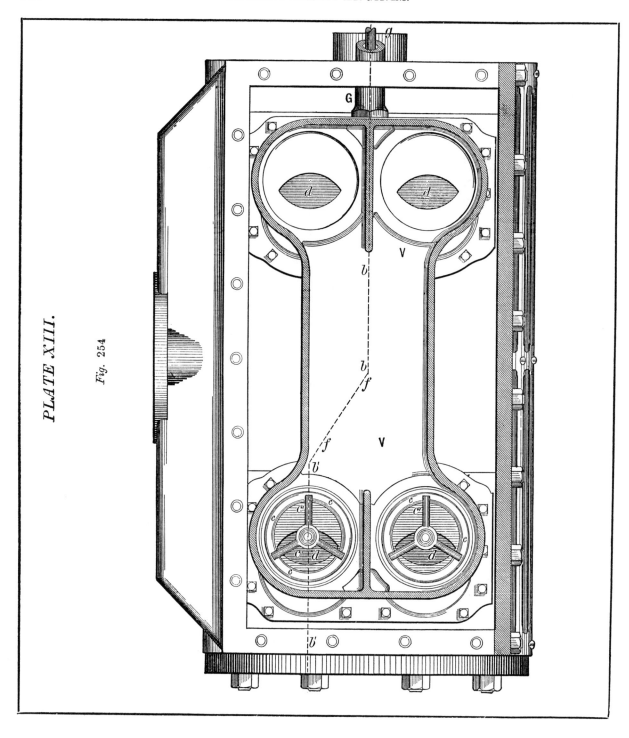

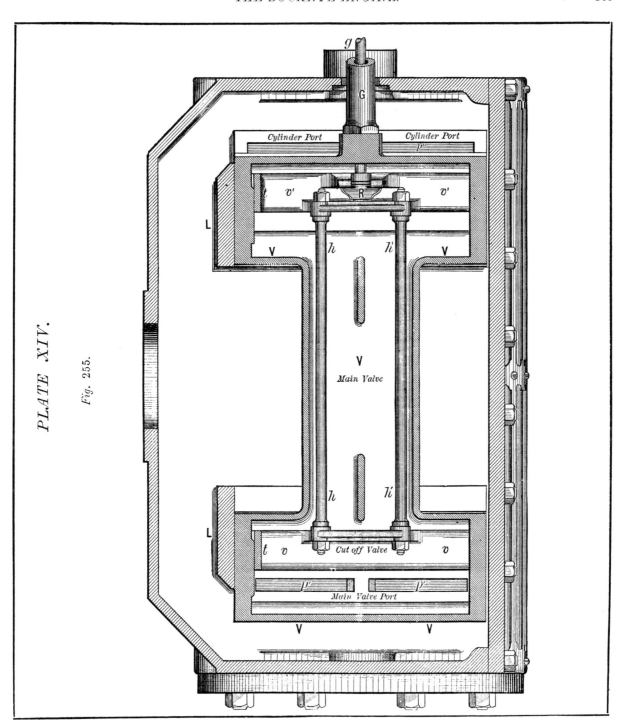

PLATE XIV.

Fig. 255.

S correspond to R and S in Fig. 257, as is also the case with the main and cut-off rock-shafts.

The crank is on the dead center B, and revolves in the direction denoted by the arrow. The main eccentric is at *f*, and the cut-off eccentric at *e*. The main valve is operated by rod F, which is driven by the upper end *p* of the main rock-shaft. The cut-off valve is operated by the spindle E, which is driven by the upper arm *a* of the cut-off rock-shaft, which is pivoted in the main rock-shaft at Λ.

Since both eccentrics follow the crank in the direction of revolution, therefore, when the crank moves from the dead center, the valves move in opposite directions, the main valve opening port *b'* for the live steam. As soon as the cut-off eccentric *e* passes the line of centers *b b* of the engine, it operates the cut-off valve in the

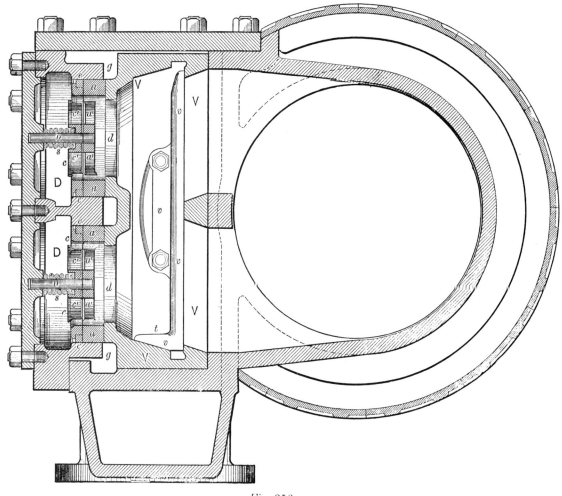

Fig. 256.

Cross-Section of Cylinder and Valves.

same direction as the main valve, and both valves move in the same direction, until such time as the main eccentric *f* has also passed the line of centers, after which the valves are operated in different directions, until the cut-off valve crosses the line of centers on the other dead center, at which time, the valves will again move together, until the main valve has crossed the line of centers at that end.

tions. But when the cut-off eccentric has crossed its line of centers, it moves rod S in a direction opposite to that in which the main rod R is moving, and its rock-shaft reverses the direction of motion of the valve, hence, both valves move together until the main eccentric, as before stated, also crosses the line of centers of the engine.

The proportions of the parts, in Fig. 259, are for an

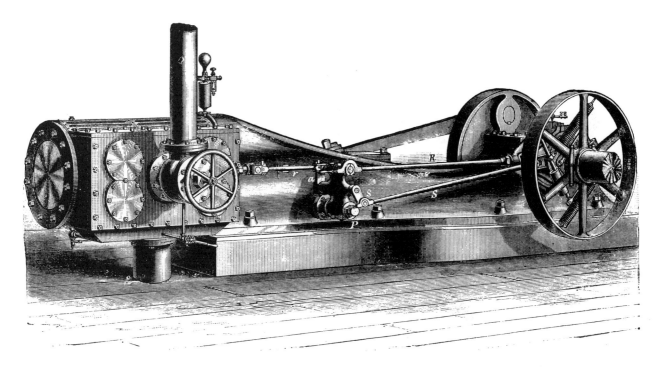

Fig. 257.

This occurs by reason of the positions of the eccentrics, and of the employment of the cut-off rock-shaft, and will be readily perceived, because, with both eccentrics on the same side of the line of centers of the engine, they are both moving in the same direction, and, as the cut-off rock-shaft *s a*, reverses the direction of motion of the cut-off valve, while the main valve moves on the same as if it were connected direct to its eccentric, therefore, the valves move in opposite direc-

engine having a cylinder of 14 inches diameter and 24 inches stroke, the throw of both eccentrics being 1½ inches, giving to each valve a travel of 3 inches. These proportions give the latest points of cut-off at five-eighths of the stroke, this being the latest the engines are designed to have. Earlier points of cut-off are effected by advancing the cut-off eccentric from its position at *e*, in the figure, to the position denoted by *e'*.

We may, however, for the present, confine our attention to the action of the mechanism, when in position for the latest point of cut-off, and, in Fig. 260, we have the position of the parts when the cut-off is effected, and it is seen that the main and cut-off valves are moving in opposite directions, and that its cut-off eccentric is near the point at which it moves the cut-off valve most rapidly, hence it follows that the cut-off is effected rapidly. This construction permits of the employment of a very short valve travel, thus reducing the duty of operating the valves.

As the main valve is driven by an eccentric that is

Suppose, now, that the governor has moved the cut-off eccentric in position to effect the cut-off at quarter-stroke, and the crank being on the dead center B, the positions of the parts will be as in Fig. 261, in which it is seen that from the positions of the eccentrics, the cut-off valve will move in the same direction, and as fast as the cut-off valve.

The positions of the parts, when the cut-off is effected at quarter-stroke, is shown in Fig. 262, and it is seen that from the position of the main eccentric *f*, its valve is at rest. When, however, the cut-off eccentric *e* has

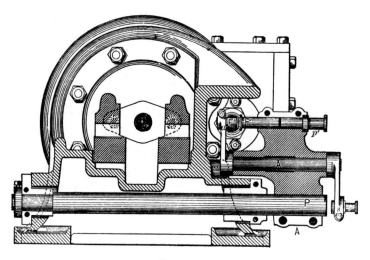

Fig. 258.

Cross-Section Through the Frame and Rocker.

fixed to the crank shaft, its amount of travel is constant, and its ends *w m*, pass over the ends *y* and *r* of its seat, and as the ends *u w*, of the valve, pass alternately over the ends *x* and *n*, of its seat, therefore, at all points of cut-off, the tendency to wear the seat unevenly is avoided (this tendency existing when the stroke of a valve is decreased).

In order that the cut-off valve may similarly pass over each end of its seat on the main valve, the shoulders on the faces *d* and *g* are provided.

nearly reached the line of engine centers, the motion of the parts is such that the two valves again move together.

Fig. 263 shows the port openings for the cut-off at $\frac{5}{8}$ stroke, and it is seen that, for the head-end B of the cylinder, the port is open full at the third inch of piston motion, remains full open up to the eighth inch, and cuts off at the fifteenth.

For the crank end, the port is full open at the third inch, remains open for $8\frac{1}{2}$ inches, and cuts off at fifteen.

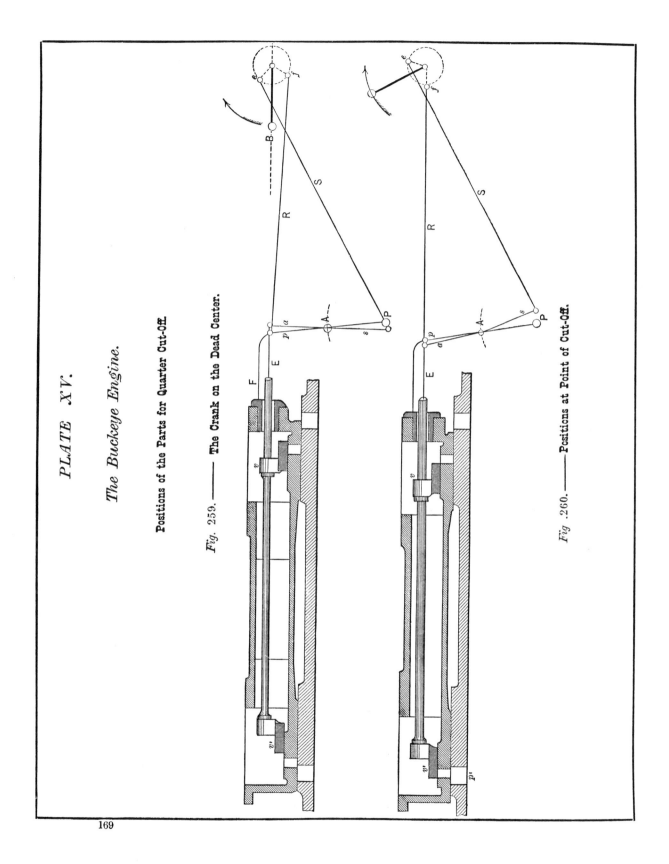

PLATE XV.

The Buckeye Engine.

Positions of the Parts for Quarter Cut-Off.

Fig. 259. —— The Crank on the Dead Center.

Fig. 260. —— Positions at Point of Cut-Off.

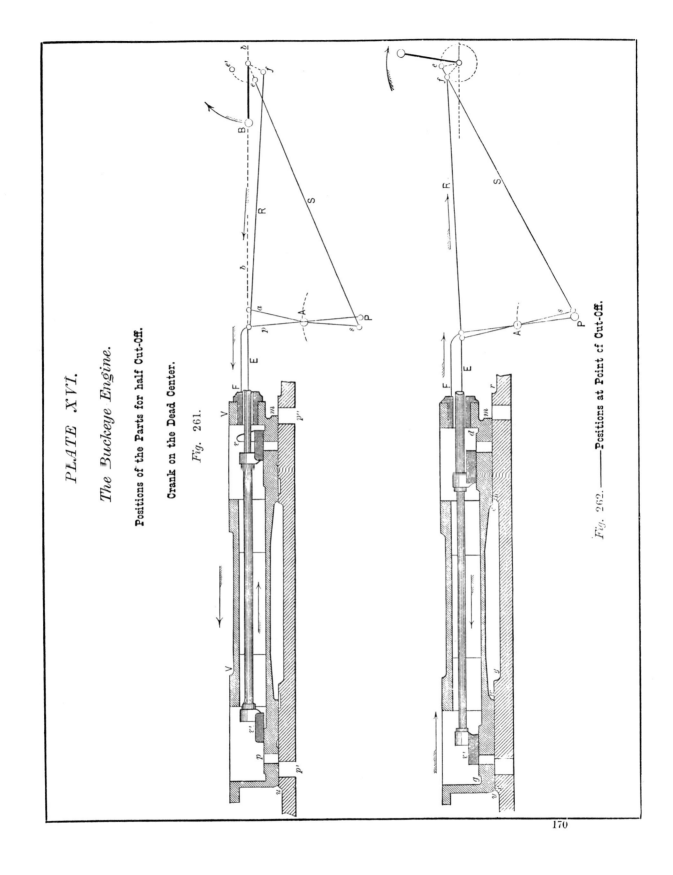

PLATE XVI.

The Buckeye Engine.

Positions of the Parts for half Cut-Off.

Crank on the Dead Center.

Fig. 261.

Fig. 262. ——Positions at Point of Cut-Off.

170

The amount of compression is 1½ inches at each end, the equilization being effected by unequal laps at the two ends (*m u*, Fig. 259) of the main valve.

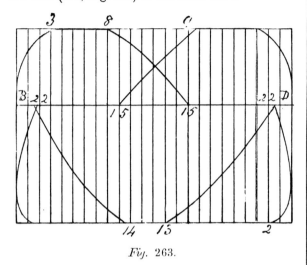

Fig. 263.

Diagram of Port Openings with Cut-off at Half-stroke.

The exhaust opening is, it is seen, greater than the steam port opening.

The lead, compression, cushion, and exhaust being governed by the main valve, having a constant amount of travel, obviously remain constant for all points of cut-off.

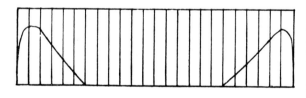

Fig. 264.

Diagram of Port Openings with Cut-off at Quarter-stroke.

The steam port openings, when the cut-off occurs at quarter-stroke, is shown in Fig. 264, and the points of cut-off are, it is seen, equalized.

22

THE GOVERNOR.

Fig. 265 illustrates the mechanism constituting the governor, or speed regulator, which advances the position of the cut-off eccentric upon the crank-shaft. Two levers, *a a*, are pivoted at their ends *b* to arms of the governor-wheel. Upon these levers, and adjustable along their lengths, are the respective weights A A. The ends of levers *a a* connect, by ball and socket joint, to the links B B, and these are attached, by ball and socket joint, to C, which is in one piece with the cut-off eccentric, the latter being a working fit on the crank-shaft. When the wheel revolves, the centrifugal force, generated by the weights A A and the lever *a a*, will cause these parts to move outwards, as denoted by the dotted lines in the upper part of the figure. This causes the links B B to advance the cut-off eccentric upon the shaft in the direction of crank-revolution, thereby hastening the point of cut-off, as has already been explained.

The outward motion of the arms *a a* is resisted by the springs F F, and it follows that when the engine is motionless, or is running too slow to enable the centrifugal force to overcome the tension of the springs, these springs will hold the cut-off eccentric at the position in which it effects the cut-off at the latest point in the stroke, the springs being under tension, and the weights A being held against the wooden buffers at *f*.

The further the weights are situated from the pivoted ends of the lever, the greater the effectiveness of the centrifugal force generated by them at a given speed of governor-revolution, and the further out the levers swing, the greater the amount of centrifugal force generated, because the weights revolve in a larger circle and, therefore, at a greater velocity.

On the other hand, however, the outward motion of the levers and weights can only occur by distending the springs F F, and the power required to do this increases in proportion as the springs are distended.

While the governor is at rest, the force of the springs is static, but as soon as the parts revolve, this force becomes centripetal, hence, we have, so far as the weights and springs are concerned, two opposing forces, one centripetal, and the other centrifugal, and the most perfect adjustment of the weights and spring tension, is that in which, at a given speed of revolution, these two forces are equal in amount, throughout the whole range of movement of the springs and levers.

nal, or in other words, it would maintain the engine speed equal under all changes of load.

In the Buckeye engine, the travel of the valve is constant in amount for all points of cut-off, and the power required to operate it is, therefore, equal. But there is another element to be considered, inasmuch as that the

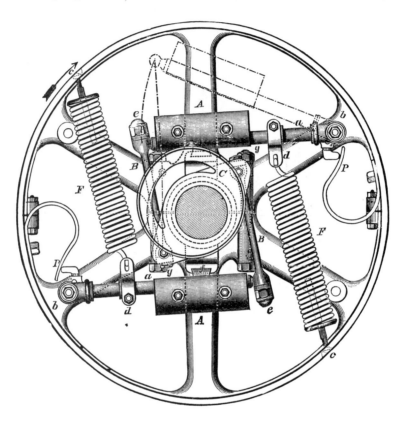

Fig. 265.

The Governor.

Now, suppose the parts are so accurately proportioned, and the adjustment so correctly made, that this equilibrium of the opposing centrifugal and centripetal forces is established, and if the power required to operate the valve is equal in amount at all points of cut-off, the action of the governor would be perfectly isochro-

weighted levers act (as they move to different positions) at a varying leverage to the eccentric with the following result:

When the weighted levers are in the positions in which they are at the greatest leverage to the points of connection on the eccentric, they will obviously exert more force in proportion to the centripetal force than they would when in their positions of least leverage, and (assuming the power required to operate the valve to be equal for all points of cut-off) the engine speed will increase, because, in order to maintain a constant and equal speed, the centrifugal force must, throughout

In the Buckeye engine, the resistance offered by the friction of the cut-off eccentric and valve gear is centripetal in its tendency. This tendency is of greatest effect when the levers *a a* are at their extremes of movement, and the auxiliary springs P P, Fig. 266, are employed to correct it.

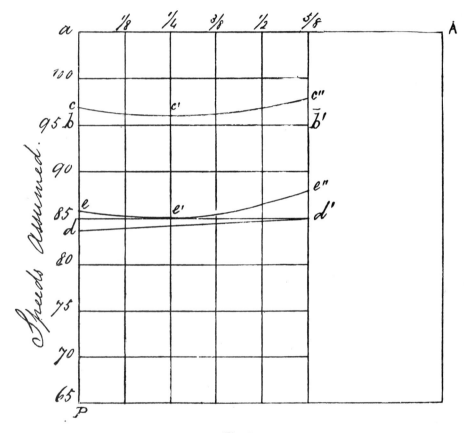

Fig. 266.

Diagram of Speed Regulation.

the whole range of lever movement, bear a constant and equal proportion to the total force resisting it, this total force including the power required to operate the valve as well as that required to resist the tension of the springs.

These auxiliary springs are intended to start the levers out at the proper speed when the tension given to the main springs is that proper to secure the best regulation during the outer half of their range of movement, where it is mainly confined when the engine is properly loaded. Under such an adjustment of tension, it was found that the centripetal tendency of the resistance offered by the cut-off valve and gear, being augmented in effect in an increasing ratio from mid-movement inwards, in proportion to the increasing acuteness of the angle formed by the links with the eccentric ears, required more than the mean working speed to overcome it, and the auxiliary springs are made of such force as to just overcome this undue frictional centripetal tendency. They leave contact at mid-movement, where they are no longer needed, the above tendency being then at its minimum. Its increase from mid-movement outwards is provided for in the tension of the main springs, which is increased on that account to obtain a diminished ratio of increase of spring force.

The range of lever motion is sufficient to regulate the speed under all changes of load under all ordinary conditions of boiler pressure, and it is obvious that the ratio of the centripetal to the centrifugal force will remain the same whatever changes may occur in either the engine load or the boiler pressure.

DIAGRAM GRAPHICALLY ILLUSTRATING SPEED REGULATION AND THE USE OF THE AUXILIARY SPRINGS.

To graphically illustrate the variation of the effective centripetal and centrifugal forces, and the use of the auxiliary springs, let the line a A, in Fig. 266, represent the length of the piston stroke divided by lines representing points of cut-off at $\frac{1}{8}$, $\frac{1}{4}$, $\frac{3}{8}$, $\frac{1}{2}$ and $\frac{5}{8}$ stroke. The vertical line a P is equally divided into lines representing speeds.

Now, suppose that while the levers move from the inner to the outer end of the range of motion, the distance of the center of centrifugal force, from the center of the shaft, is doubled, and the amount of centrifugal force will be doubled. And if the total spring tension, resisting the centrifugal force, is also doubled at the same time, the two forces will increase in the same ratio, and the speed regulation will be isochronal. Such regulation may be represented, so far as the centrifugal force is concerned, by a line b b' parallel to a A.

Now, suppose that the centripetal friction would accelerate the speed at the earliest cut-off by an amount represented by the distance from b to c; at $\frac{1}{4}$ cut-off by the distance c', and at $\frac{5}{8}$ cut-off by the distance c'' from line b b', and through these points we may draw a curved line representing, in its curvature upwards, the amount to which the centripetal friction would accelerate the speed; this acceleration being represented, at $\frac{1}{4}$ cut-off, by the distance b' c'; at the earliest cut-off by the amount b c; and at $\frac{5}{8}$ cut-off by the distance b' c''.

This shows close regulation at and near c', but near c there will be a change of speed accompanying changes of cut-off, and to remedy this, more spring tension is required.

From c' to c'', the speed variation is shown, by the upward curvature of the line, to be in the wrong direction, because, while the point of cut-off would be passing from c' towards c'', the speed would be accelerating, whereas the engine load is increasing, and its tendency is, therefore, to decrease the speed, hence, for this part, the tension is too great, because, with less tension, the diminution of spring force would be more rapid, which would compensate for the increasing effectiveness of the centripetal friction.

By the application of the auxiliary springs P P, Fig. 265, the speed may be cut down from c'' to b'. This will give stability of speed throughout all points of cut-off from $\frac{1}{4}$, or c', to $\frac{5}{8}$, or b', leaving a margin of variation for points of cut-off between $\frac{1}{4}$ and zero, which margin is represented by radius d c.

This margin may, however, be diminished, to any required extent, by giving a spring tension that would give the fastest speeds at the latest cut-offs, as represented by the line d d', and on this line we may draw the assumed frictional curve, e c' e'', corresponding to line c c' c'', and it is seen that from e to e', it is sufficiently isochronal, and by changing e' e'' to e' d', by means of the auxiliary springs, the regulation becomes sufficiently isochronal throughout the whole range of cut-off.

The Armington-Sims Engine.

In Figs. 267 and 268, we have two views of the Armington-Sims high speed Automatic cut-off engine, the construction being as follows:

Fig. 269 is a vertical section of the cylinder and valve, the latter being a piston valve and double ported, as seen in Fig. 270, in which it is shown broken. The port *a* is receiving steam through the openings, denoted

The steam, after passing through this opening, passes through the valve, and finds egress at *f*, into the annular groove *r*, and thence into port *a*. The opening, denoted by arrows *s v*, extends all around the circumference of the valve, and the steam passes directly into the annular groove and port *a*.

The exhaust passes out at the end of the valve, as denoted by the arrow at B, and is, therefore, independent of the admission.

The lap of the valve is marked in Fig. 271, in which it is seen that the valve just closes the port *a*, and will keep it closed for the period of expansion, or until the valve has moved far enough to the right to open the

Fig. 267.

The Armington-Sims Engine.

by the four arrows, the valve being hollow, and it is obvious, that the opening denoted by arrows *c d*, extends around the whole circumference, with the exception of the metal at *e*, of which there are four sections in the circumference of the valve.

exhaust, as in Fig. 272, this period of motion corresponding to the passage of the steam lap of a simple slide-valve over the port.

The compression begins when the end *m* of the valve covers the port, as in Fig. 272, and increases in amount as the valve travel is reduced and the cut-off occurs earlier, as will be seen hereafter.

The engine speed is governed, and the point of cut-off varied to suit the load, by a mechanism that varies the travel of the valve, but maintains the lead equal at whatever point in the stroke the cut-off may occur, the construction being as follows: Figs. 273 and 274, show the construction of the regulator or governor, removed from the main shaft.

The weights 1 1 are pivoted at their outer ends to the

In moving the inner eccentric C forward, its angular advance, and, therefore, the amount of valve lead is obviously increased, and in moving the eccentric ring D backward, the eccentricity of the combined eccentrics (C and D) is reduced, and thus reduces the valve travel.

The necessity for increasing the lead, in proportion as the amount of valve travel is reduced, is shown in Fig. 275, in which A represents the common throw line of the two eccentrics, the rod being at *e*, and it is seen, that if we move the point of connection *e* of the rod

Fig. 268.

The Armington-Sims Engine.

arms of the wheel, and to these weights are pivoted the links 2 2, which at their other ends are pivoted to the inner eccentric, which is a working fit on the crankshaft. To one of these weights the lever 3 is pivoted, its other end being pivoted to the outer eccentric or eccentric ring, as it may be termed, hence, as the wheel revolves, the centrifugal force of the weights causes them to move outwards at their free ends, thus turning the two eccentrics upon the shaft, and thereby altering both their total amount of throw, and the position of the common throw line with relation to the crank.

along the line A and towards C, we shall move the valve to the right and decrease the opening or lead at *c*; hence, to maintain a constant degree of valve lead, we must increase the angular advance of the eccentric, it being borne in mind that the eccentric and eccentric ring are, in effect, a simple eccentric, capable of adjusting the amount of valve travel and of maintaining the lead constant, and it may be remarked that the throw line of the two, or, in other words, their line of greatest eccentricity, always passes from the center of the shaft through the center of the eccentric ring.

In Fig. 276, for example, the eccentrics are shown with their throw line A at 55° from the crank, the center of the eccentric E being at *e*, and that of the ring R at *f*, and it is seen, that the throw line A passes

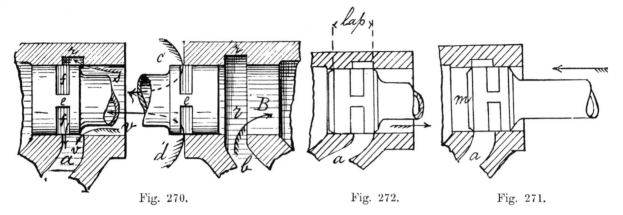

Fig. 270. Fig. 272. Fig. 271.

The Valve.

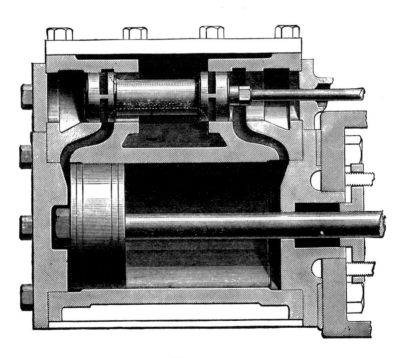

Fig. 269.

Section through the Cylinder and Valve.

from C through *f*. The dotted circle *a* is drawn from the crank center C, and shows that the line A A, is the

the crank, and it is seen that the throw-line A still passes from C through the center *f* of the eccentric

Fig. 273.　　　　　　　　　　　　Fig. 274.

The Governor.

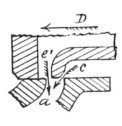
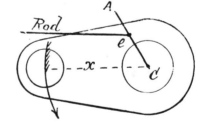

Fig. 275.

common throw-line of the two eccentrics (E and R), and also that the amount of travel the valve would have is the radius G. In Fig. 277, the eccentric E has been moved forward, and the ring R moved back, causing their common throw-line A to stand 44° behind

ring R, the amount of valve travel being reduced to the radius G'.

To find the position of the common throw-line A, for the two eccentrics for any required point of cut-off, we proceed as follows:

In Fig. 278, let the outer, or large circle, represent the path of the crank-pin center, and let it be required to find the position for the common throw-line of the

further to the left hand in order to open port *a* full for the admission of steam, and to whatever amount it leaves the edge *h* of port *a*, it must move back the

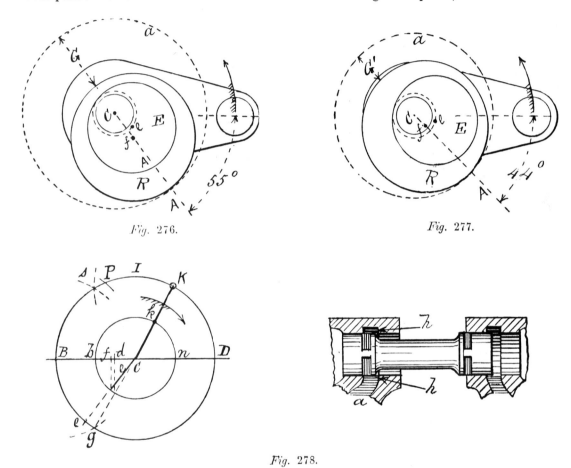

Fig. 276.

Fig. 277.

Fig. 278.

Finding the Position of the Eccentric.

two eccentrics, the cut-off to occur when the piston has moved seven-tenths of its stroke, and the crank stands at K, its direction of motion being denoted by the arrow. On the right of the figure, we have the valve in the position it would occupy when the crank was on the dead center B, and it is clear that the port *a*, being open to the amount of lead only, the valve must move
23

same distance before it can close the port and effect the cut-off. Now, when the valve is at the end of its stroke, the common throw-line A, of the two eccentrics, must be on the line B C, and it is clear that, supposing the valve to have no lead, three things must happen.

First (omitting all considerations as to the crank position), the common throw-line A will move towards *b* during the whole time the port *a* is being opened. Second, this throw-line will stand at *b* when the port *a* is full open, the valve being at the end of its travel, and third, the throw-line A will move past B, while closing the port *a*, to the same distance it did in opening it.

The amount of this distance (measured on the crank circle B I D, and supposing the valve to have no lead) will obviously be half the distance between K and B, which we obtain by arcs *s* drawn from K and B respectively. With the compasses set to the radius B *s*, we mark, from B, the arc at *g*, and it is clear that if the crank is at B and the throw-line at *g* C, then, while the crank moves from B to *s*, the common throw-line A will move from *g* to B, and the port will be opening. Then, while the crank is moving from *s* to K, the common throw-line will move from B to *s*, and the cut-off will occur. It has been supposed that the valve had no lead, but we may now take the lead into account as follows:

As the inner circle represents the path of the common center of the two eccentrics, therefore its diameter

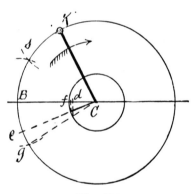

Fig. 279.

b n, represents the travel of the valve. Now, it has been shown that with no valve lead, the common throw-line of the eccentrics would stand at *g* when the crank was at B. From *g* we draw a line to C, and where this cuts the inner circle, we erect a vertical line *d*.

From *d* mark off the amount of valve lead, and draw the line *f* parallel to line *d*; from C draw a line *e e*, cutting the line *f* at its junction with the inner circle, this line *e* giving the position for the eccentric when the

crank is at B. By prolonging this line to *e'*, and marking, from *e'* to P, a distance on the outer circle equal to the distance from B to K, we get the angle the common throw-line moves through while the crank moves from B to K, taking the lead into account.

In Fig. 279, we have a similar example, the cut-off to occur when the crank arrives at K. By dividing the arc B K, we get arc B *s*, and mark B *g* equal to B *s*, we then draw *g* C, this being the position of the common throw-line when the crank is at B and the eccentric has no lead. From *g* C we mark *d*, and distant from *d* to the amount of valve lead, we mark *f*, and a line *e*, from C to the intersection of *f* with the inner circle (which represents the path of the common center of the two eccentrics), is the common throw-line corresponding to lines A in Figs. 276 and 277.

Instead of using two circles, one for the crank-pin path and one for the path of the common center of the two eccentrics, however, we may use a single circle to

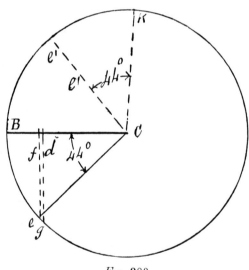

Fig. 280.

find the angle of the common throw-line necessary to cut-off at any given point in the piston stroke.

Let it be required, for example, to find the angle of the common throw-line to the crank necessary to effect the cut-off when the piston is at half-stroke, and we draw the circle, in Fig. 280, representing the path of the crank on one scale, and the path of the common center of the two eccentrics on another scale, or it may be full size if the amount of valve travel has been determined. Suppose the crank to be at B and the cut-off is to occur when the crank is at K, the piston then being at half-stroke, and we take half the arc B K, and from B mark off point *g*, representing the position of the common throw-line when the valve has no lead; from *g* we erect the vertical line *d*, and to the left of *d*, and distant from it to the amount of the lead, we draw *f*, and where *f* cuts the circle, as at *e*, we draw a line *e* C, which gives the angle of the common throw-line of the two eccentrics to the crank necessary to enable the cut-off to occur when the crank reaches K.

THE VALVE PROPORTIONS.

The amount the valve leaves the port open is obviously equal to the distance from *d* to B measured on

valve travel necessary for the point of cut-off corresponding to the angle of eccentric.

The point of cut-off having, however, been determined, and the angle of the eccentric to the crank found by the foregoing construction, we may proportion the amount of travel to suit the width of the port. Suppose, for example, that the steam port in the cylinder requires to have an opening of an inch and we may proportion the valve as in Fig. 281, the width at *c* being only sufficient to give the necessary strength, because it has no effect upon the distribution of the steam, whereas, the wider it is the wider the port must be, and this increases the clearance space.

The width of the port *d* must be not less than the sum obtained by subtracting the width of the metal, at *c*, from the width of the port and dividing the remainder by 2, so that, when the valve is in the position shown in the figure, *c* being in the middle of the width of port *a*, the opening will be equal on each side of *c*, and the port being fully opened, there will be an open-

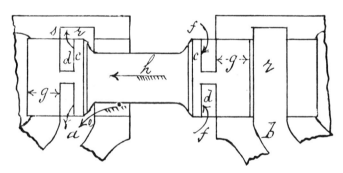

Fig. 281.

the line C B, but, as the valve takes steam at both ends, the amount of effective opening is twice *d* B. It will now be seen that there is, for each point of cut-off, a definite degree of angle of eccentric to the crank, and that this angle cannot be departed from, except in so far as concerns the amount of lead given, and it follows that the motion of the eccentric and the ring, when moved by the weighted levers 1 1, in Fig. 273, must be such as to cause the common throw-line A, Fig. 277, to stand at the proper angle to the crank, and at the same time regulate the degree of eccentricity of the center *f* in that figure so as to give the amount of

ing at the end denoted by the arrows *f*, equal to one half the effective width of the port *a*. It is obvious that if the width of ports *d* was less than this, the admission would be correspondingly diminished, but that it may be made greater without affecting or increasing the effective amount of port opening.

The dimension *g* may be proportioned so as to either equalize the point of release for the two strokes, or so as to equalize the point at which the compression will be effected.

The diameter of the valve may be made such that its circumference equals the diameter of the cylinder the length of the port equaling the piston diameter. | recess *r*, thus closing the port *d* and leaving *r* full open,

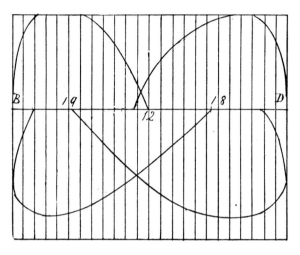

Fig. 282.

Port Openings for Cut-Off at Half-Stroke.

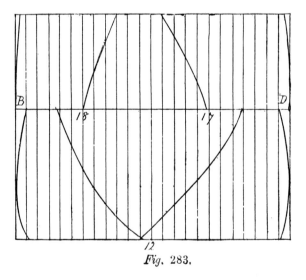

Fig. 283.

Port Openings for Cut-Off at Seven-tenths Stroke.

The valve travel, for the latest point of cut-off, is such as will, at the end of the valve travel, bring the inner edge of the port *d* coincident with the edge *s* of the

save where it is covered by the thickness of the metal at *c*.

Now, suppose the effective width of port opening is

required to be an inch, and the width of the annular recess must equal the thickness at *c* plus 1 inch.

Suppose the latest point of cut-off is to be at seventenths of the piston stroke, and we may find the necessary positions of the eccentrics, as in Fig. 276, the common throw-line being found to be 55° behind the crank.

If the thickness at *c* is ⅜, then for an effective port opening of an inch, the annular recess must be 1⅜ inches and the port *d* ½ inch wide, so that when the valve is in the position it occupies in Fig. 281, there will be ½ inch opening at *f* and ½ inch at *d*, and as the motion continues in the direction of the arrow *h*, the opening

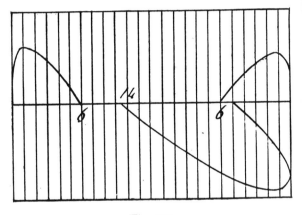

Fig. 284.

Port Openings at Quarter Cut-off.

denoted by the arrow *e*, will increase as rapidly as port *d* closes. The travel of the valve must, therefore, be equal to the effective width of the port (in this case 1 inch) plus the length of *g*, which may, as before remarked, be proportioned to equalize the points of release or of compression. Suppose the points of release are to occur at the 23rd inch of piston motion, the total piston stroke being 24 inches, and the length of *g* must be 1¾ inches at the head end, and 1 15/16 inches at the crank end, and the travel of the valve being 4 inches, the port openings will be as in Fig. 282, where it is seen that, at the crank end B, the port is open full

when the piston has moved about ⅝ inch, remains open full for 12¼ inches, while the cut-off occurs at 16¾ inches and the release at 23 inches.

For the port at the head end, the admission is full when the piston has moved half an inch, remains full open up to the 15th inch, the cut-off occurs at the 18th inch and the release at the 23rd inch. The amount of cushion is 3¾ for one stroke, and about 4¼ for the other.

In order to effect the cut-off at half-stroke, the governor must have moved the eccentrics to the position shown in Fig. 277, the common throw-line A standing at 44° behind the crank, and, in order to maintain the lead equal the valve travel must be reduced from 4 to 3 inches. The disposition of the steam is as shown in Fig. 283, the port at the crank end B being fully opened at the 2nd inch of piston motion, remaining full open for 6 inches, cutting off at 12 inches.

The exhaust begins at 21¾ inches, and the compression, on the return stroke, at 18¾ inches.

When the piston moves from the crank-end D, the point of cut-off is at 13¼ inches, the release at 22¼, and compression begins at 17½ inches. It is seen that in proportion as the point of cut-off is earlier in the stroke, the release occurs earlier, and the amount of compression is increased.

It will be observed that the exhaust does not fully open, but it is to be borne in mind that the port, when acting as an exhaust port, has a greater effective width than it has as a steam port, being, in this example, 1 inch wide as a steam port, and 1⅜ wide as an exhaust port. This occurs from the fact that the thickness of the metal on the valve at *c*, Fig. 281, must be deducted from the width of the port, when considering it as a steam port, but not when considering it as an exhaust port.

For the cut-off at ¼ stroke, the common throw-line of the eccentrics must stand at 23° behind the crank, and the travel will be reduced to 2½ inches, the steam distribution being as shown in the diagram, Fig. 284. It is seen here that the port does not open full for either the admission or the exhaust, while the compression (which is shown for one stroke only, viz., the crank-end port) occurs at 14½ inches, giving 5½ inches of compression. The exhaust is here again earlier, occurring at the 19th inch in the stroke.

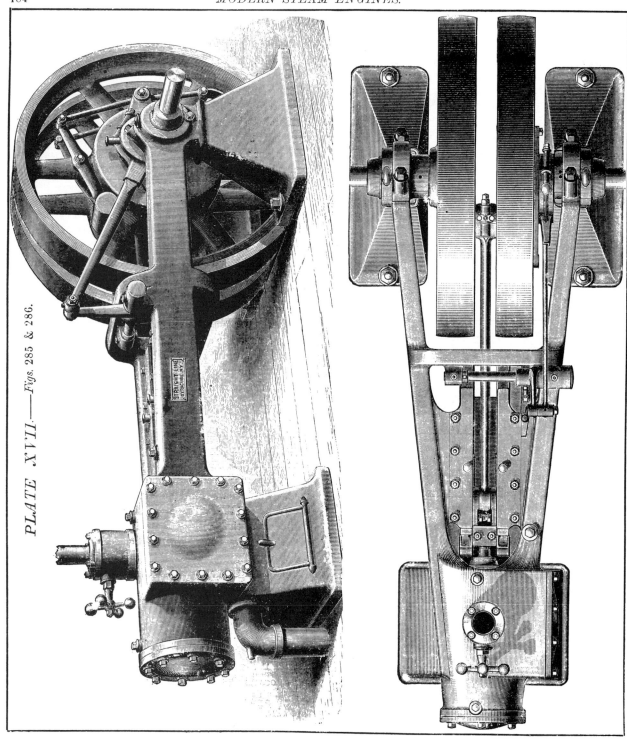

PLATE XVII.—Figs. 285 & 286.

The Straight-Line Engine.

In the figures from 285 to 302, is illustrated the Straight-Line engine, designed by Professor John E. Sweet of Syracuse New York.

The name Straight-Line is given because the outlines

features of marked individuality, designed to meet the requirements of the high piston speed that has become a marked feature of modern practice. All parts that require to be steam tight, are made so by accurately fitted parts, and are kept so by simple means of correcting, or taking up, the wear, dispensing with the packing commonly used, and, therefore, reducing to a minimum, the cost of maintaining the engine in good working order and of running it.

The motions of the working parts are all direct and

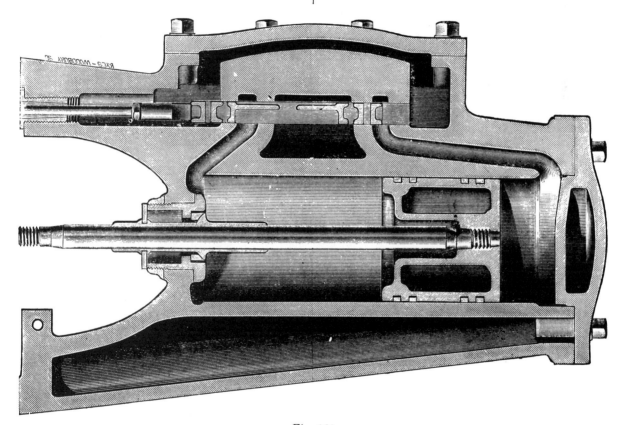

Fig. 287.

Horizontal Section through the Cylinder and Valve.

of the engine are in the direction of the strains, and are therefore, straight lines.

Throughout the details of this engine will be found

are positive, while by an original, ingenious, and simple arrangement of parts, an automatic cut-off is obtained from a single eccentric and valve, the latter giving practically equalized points of cut-off, while the lead is varied equally for the two ports.

Fig. 285 is a side elevation, and Fig. 286 a plan of the engine, and it will be seen that the frame runs direct from the cylinder to the main-shaft bearings.

The fly-wheels run between two bearings, and one of them contains the governor or speed regulator. The frame rests upon three legs or points, thus ensuring that it shall bed fairly upon its foundations, whether the same be unstable or not. The two side members of the frame are connected by a cross rib, on which are the bearings for the rock-shaft, while the guide-bars are contained within a rigid portion of the frame.

Referring now to the details of construction, Fig.

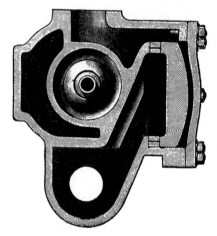

Fig. 288.

Vertical Section through Cylinder.

287 represents a horizontal section through the cylinder and Fig. 288 a vertical section through the same, while Fig. 289 is a side, and end view of the piston removed from the cylinder.

The piston is hollow, and therefore light in proportion to its length, it is secured to its rod by two taper seats and a parallel thread, which is made an easy fit, so as to prevent its influencing the fit of the taper seats.

The piston packing is constructed as follows: The rings are made in two sections, the lower of which is driven tightly in the grooves and faced off even with the piston surface. The joint openings are made very

narrow, and as the joint faces are horizontal, (as seen in the end view, Fig. 289), therefore these openings do not increase as the rings open out to compensate for the wear.

Moreover, the openings, being near the bottom of the piston which rests on the bore of the cylinder, are virtually closed by the piston. The upper parts of the rings are made $\frac{1}{8}$ inch larger in diameter than the cyl-

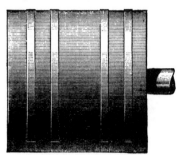

Side Elevation.

Sectional End View.

Fig. 289.

The Piston.

inder, and are closed and sprung into their places, being what are termed *snap* rings, and it follows, that being in two parts, the ring will conform itself much more readily and correctly to the cylinder bore, than is the case when the ring has a single split.

In place of the ordinary gland and its accompanying packing, a single Babbit-metal bushing is employed as shown, being a free sliding fit which is found to be sufficient to prevent leakage of steam, because of the

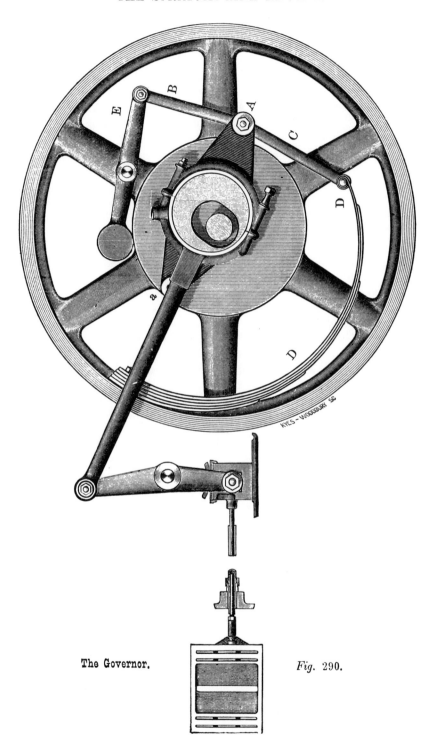

The Governor.   *Fig.* 290.

brief period the piston occupies in making a stroke being too short to permit the steam to pass through. The bushing rests in a spherical seat, being maintained therein by the steam pressure as well as by an outside collar.

The bushing is, therefore, free to move laterally with the piston rod, and this leaves the duty of guiding the piston-rod entirely upon the piston and guide bars, where it properly belongs, and thus prevents the bore of the bushing from wearing by reason of the piston-rod moving laterally as the guide-bars, etc., become worn.

The valve and its operating mechanism is constructed as follows: Fig. 290 represents the governor, rock-shaft and valve removed from the engine. The eccentric is suspended from a lever that is pivoted to the face of the wheel at *a*, and is connected, at its end A, to arms or links B and C. Link C is connected to a spring D, which, when the engine is in motion, exerts a tension acting to move the eccentric across its shaft, and, therefore, to increase the valve travel, so as to prolong the point of cut-off.

Link B is pivoted to a lever E, which, in turn, is pivoted to an arm of the wheel, and weighted at its other end. When the wheel revolves, the centrifugal force generated by the heavy end of the lever E, acts in a direction to move that end outwards, and, therefore, to move the eccentric inwards across the shaft, reducing its throw, and, therefore, the valve travel, thus hastening the point of cut-off.

These opposing forces are so regulated, in the strength of the spring and the weight of the heavy end of the lever E, that up to the time the engine has attained its proper speed, the eccentric is at its greatest throw and the cut-off is at its latest point in the piston stroke, but if the conditions (such as a sudden decrease in the engine load or duty) are such as tend to induce an increase of speed, the increased centrifugal force, generated by the heavy end of E, moves the eccentric inward across the shaft, and thereby hastens the point of cut-off. In the contingency of the spring breaking, therefore, the eccentric would be moved by lever E to its point of least throw or central upon the shaft, and the engine would stop.

Now, suppose the engine to be running at its normal speed under a heavy load, and, as the eccentric will be at its greatest throw, therefore its center will be at its greatest distance from the axis of its shaft, hence the centrifugal force, generated by the eccentric itself, will be in a direction to cause it to move still further outwards, and in opposition to the heavy end of lever E, and as the lever A is pivoted at one end only, while the other swings outward with the eccentric, it also will generate a centrifugal force opposing that generated by the heavy end of E. But on the other hand, however, the spring D, being fast at one end only, its free end generates a centrifugal force acting in the same direction as the lever E, and, therefore, to counteract the unbalanced centrifugal force of the eccentric and lever A.

Furthermore, as the conditions lead towards an increase of speed, the eccentric is moved more central to its shaft (carrying with it the lever A), hence it revolves in a path of less diameter and generates less unbalanced centrifugal force, and, as a result, there is less resistance to the prompt action of the governor in the necessary direction. The friction of the eccentric in its strap has no influence upon the spring or lever E, and does not, therefore, disturb the eccentric moving mechanism.

The equilibrium thus obtained, is sufficient to practically relieve the spring and lever E from the disturbing elements, sometimes found, of a varying unbalanced centrifugal force generated by the governor mechanism when moved to different positions to answer the requirement of varying the point of cut-off.

Another characteristic of the governor, is that the weight of the governor ball (as the weight on the lever E is termed) is such as to practically counterbalance the weight of the eccentric and its strap. The object of this is to prevent the weight of the eccentric and its strap from disturbing the proper circular path through which the eccentric center should rotate, and thus obviate the disturbance that would occur from the tendency of the eccentric and strap to fall towards the shaft when above it, and fall away from the shaft when below it, in the path of revolution, this tendency existing unless counteracted by this, or by some other means.

The joints of the levers B and C Fig. 290, of the governor are (to reduce friction and obviate the necessity of oiling them) constructed as in Fig. 291, the eye having a flat plate of tempered steel upon which the

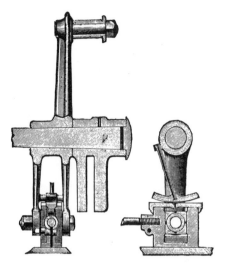

Fig. 292.

constant contact with that side only, of the eye, on which

Fig. 291.

the steel plate is placed. The pivot joint of lever E, Fig. 290, has a large oil hole drilled through the center of its pin, and from this hole smaller holes are pierced to the wearing surface of the pin. The central hole is plugged at its outer end, and thus forms an oil pocket or reservoir affording continuous lubrication.

Fig. 292 is a sectional view through the rocker-shaft, and shows the means of maintaining the valve-rod in line. The lower rocker-arm carries a pin having journal bearing in a block, sliding vertically in a box into which the end of the valve-rod is secured. This box

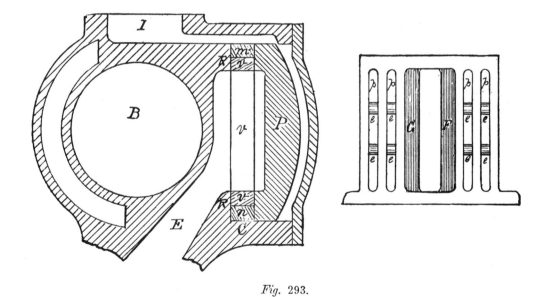

Fig. 293.

The Valve Construction.

pins roll, this construction being available because, from the tension of the spring D, the pins are held in

slides along a guide, and is kept down to its seat on the guide by means of a segment that is shown attached to the rocker hub, and seated on top of the box. The block obviously moves in the arc described by the pin in the rocker-arm, and, therefore, moves vertically in

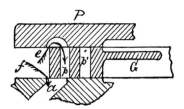

Fig. 294.

the box which is covered by a gib on which the seg-ment rolls as the rocker swings, hence, the segment not only keeps the box to its seat on the guide beneath it, but also acts to relieve the rocker bearings of the weight of the rocker-shaft.

The construction of the valve, and the means of eliminating the pressure to its seat that exists in an ordinary slide-valve, is shown in figure 293, in which on the left, is shown a sectional view of one end of the cylinder, and on the right, a side view of the valve

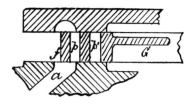

Fig. 295.

removed from the steam chest, (a top view of the valve, in section, is shown in Fig. 287). The valve is, it will be seen, a rectangular frame having four ports *p*, the ribs between the ports being kept rigid by the pieces *e*, which are solid with the valve, but do not extend to its face.

A rectangular piece *n*, rests on the lower face C of the valve chest, and on this rests the valve *v*. On the top of the valve is a rectangular piece, or bar *m*, cor-

responding to piece *n*, and at the back of the valve is a shield, or pressure relieving piece P. The pieces *m*

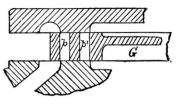

Fig. 296.

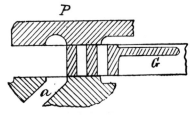

Fig. 297.

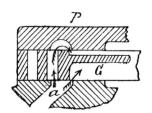

Fig. 298.

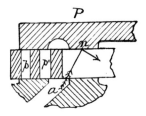

Fig. 299.

Valve Positions.

and *n*, therefore, form distance pieces, leaving between the face of P and the face, or valve seat R, an opening in which the valve may slide free from steam pressure, on its back. When, in the course of time, the valve requires to be refitted, it may be surfaced true, and the pieces *m* and *n* reduced to restore the fit of the valve in the opening in which it slides. By this means, the refitting is greatly facilitated, because all the parts may be removed from the cylinder, and surfaced either in a machine, or at the bench, instead of requiring to be surfaced while in the steam chest where it is difficult to get at them.

To fully understand the action of the valve, and the object of its peculiar construction, it is necessary to consider its action when in position for the various events during the stroke.

In Fig. 294 the crank-end port *a*, and a section of the valve and of the plate P, ts shown, and it is seen that the effective port opening is double the amount to which the valve has uncovered the port, because the live steam is admitted to port *a* in two places, as denoted by the arrows *e* and *f*.

The effective port opening remains at double that due to that amount of valve motion, until the valve reaches the position shown in Fig. 295, after which, the amount of port opening remains a constant quantity until the valve has reached the position shown in Fig. 296, after which, the opening remains the same as for a common slide valve, the maximum of valve travel being shown in Fig. 297. When the point of cut-off is at half-stroke, the valve opens the port for admission to the amount shown in Fig. 295, the opening at *f* and at *p* being equal.

It is seen, therefore, that by means of port *p*, the port opening is, for all points of cut-off up to half-stroke, doubled throughout the whole of the admission period, and that for points of cut-off later than half-stroke, it is doubled up to half piston stroke, remains stationary for a certain period depending on the point of cut-off, and for still later points, continues the port opening the same as a common slide valve.

The action is similar when the valve moves on its return stroke to effect the cut-off, the amount of port opening being doubled from the position shown in Fig. 296 until final cut-off, which prevents wire drawing the steam.

We have now to consider the exhaust, and by means of the port *p'* in the figures, this also is double that due to the motion of the valve, as may be seen in Fig. 298, in which the valve is shown in the position it occupies when the exhaust begins, the steam having two means of exit as denoted by the arrows. The double port opening continues during the same part of the piston motion, as the extra port opening does for the admission, because the port *p'* has the same effect upon the exhaust, as port *p* has for the admission.

The shield G, in the figures, is provided because it was found that in their absence, the exhaust steam flowing in the direction denoted by the arrow in Fig. 299, would, in time, cut away the face of the pressure relieving plate P, at *n*.

THE ECCENTRIC, ROCKER, AND VALVE MOVEMENTS.

We have now to consider the means by which the points of cut-off for the two strokes are maintained practically equalized for all points of cut-off, and the lead varied equally for the two strokes. In Fig. 300 let W represent the rim of the governor wheel, and suppose the crank to be at D and the eccentric (at its greatest throw) at *e*. Let the eccentric center when moved across the shaft to its position of least throw, be at *f*, and it is obvious, that the path of the eccentric in moving across the shaft, would be along the line *e f*. At a right angle to this line, and from the center C of its length, we draw a line C C', and then with the length of the eccentric-rod as a radius, and from C as a center, we mark an arc *d*, giving at *g* the position for the eccentric-rod eye corresponding to crank position D. The length of the eccentric-rod (measured from the center of the eccentric), is radius *g e*, and it is clear, that if, the eccentric-rod being pivoted at *g*, we move its other end across the shaft, it will move in a line that will, (from the great length of the rod), practically coincide with the line *e f*.

We may now turn to the lever (A Fig. 290), that moves the eccentric across the shaft, and as its fixed end is pivoted at *h* on the line C C', it is obvious, that if we move its other end across the shaft, the eccentric center will move in an arc, that from the length *h c* and for the short distance *e f*, practically coincides with line *e f*.

Now suppose the crank to move to its dead center B, and the center of the eccentric, when at its greatest throw, will have moved to *e′*, or, if at its least throw, will have moved from *f* to *f′*, we therefore draw a line from *f* to *f′*, and at a right angle to this line, and from the middle of its length, we draw a line *m m′*. Then, with the length of the eccentric-rod as a radius, we mark an arc *n*, giving us at *p* the position of the eccentric-rod eye corresponding to crank position B, and it is clear that if we swing the eccentric-rod on its pivot P, its other end will move in a line that will practically coincide with line *e f*, because of the great length of the rod and the short distance from *e′* to *f′*.

While the crank moves from B to D, the lever (A, Fig. 290) that moves the eccentric across the shaft, will move from position *h* to position *h′*, which is in line with line *m m′*, and in a position, therefore, to move the eccentric in a path that will practically coincide with line *e′ f′*.

THE POSITION OF THE ROCK-SHAFT.

We have now to find the position for the center of the rock-shaft, which may be done by taking the length of the upper arm of the rock-shaft as a radius, and from *g* as a center, marking an arc *s s*. With the same radius, and from *p* as a center, an arc *r r* is marked, the point of intersection of arcs *r* and *s* being the location for the center of the rock-shaft.

The position of the lower arm of the rock-shaft, or, in other words, its angle to the upper arm, may be found as follows:

The crank being at D, and supposing the valve to have no lead, it will be in the position shown at G, (the eccentric-rod eye being at *g*) and will be moved from its mid-position to the amount of the steam lap, the lower arm of the rock-shaft, will, therefore, have moved from its mid-position to the same amount, hence, we mark an arc *u* and arc *v*, distant from the line *t* to the

amount of the steam lap, and draw the lower rocker-arm, cutting the arc *u* where arc *v* cuts it.

As the valve is double ported, ample admission is obtained with a minimum of valve travel, and as the upper arm of the rocker is longer than the lower arm of the same, therefore the action of the eccentric is reduced, or, in other words, the path of the eccentric center, is of larger diameter than would be the case if both rocker-arms were of equal lengths. By this means, the duty of the governor in shifting the eccentric is lightened, thus enabling the employment of a minimum of weight on the end of lever A.

THE EQUALIZATION OF THE PISTON AND VALVE MOVEMENT.

Having found the positions of the various parts, we may now trace their movements as follows:

Suppose that the crank is at D, Fig. 301, and the eccentric center at *e*, and the eccentric-rod eye will be at *g* and the valve in the position shown at G. Now while the eccentric center moves from *e* to *n*, the rod end will move to position *q* and the valve to position N, the port *b* being opened to its fullest extent; and while the eccentric center moves from *n* to *w*, the valve will move back, and when at *w*, will close the port *b* and effect the cut-off, having returned to the position it occupies at G in the figure.

While the eccentric moves from *w* to *e′*, the valve will move from its position at G to its position at H, ready for port *a* to open, the piston having completed its stroke while the crank and eccentric center have made a half-revolution.

A pair of compasses are then set to the length *e g* of the eccentric-rod, and resting one point at *n*, we mark an arc, giving us at *q* the position of the eccentric-rod on the arc *x x* of motion, of the upper end of the rock-shaft.

Similarly, to find the position of the eccentric-rod eye when the eccentric center is at *w*, we rest the compasses at *w* and mark an arc, giving us at *g* the position of the eccentric-rod eye.

Turning now to the piston stroke, while the crank moves from B to D, with the crank at B and the eccen

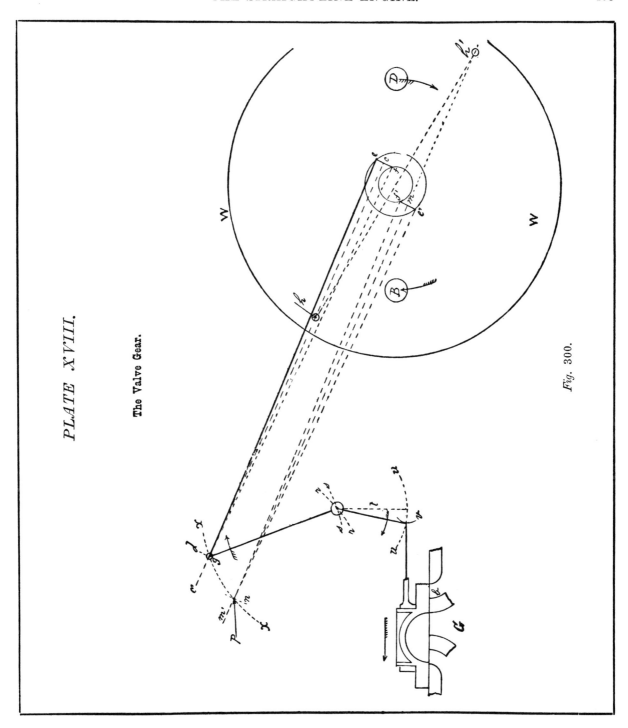

PLATE XVIII.

The Valve Gear.

Fig. 300.

tric at e', the eccentric-rod eye is at p. While the eccentric center moves from e' to n', the rod eye moves from p to s, fully opening the steam port a in the figures, and while the eccentric center moves from n' to k, the rod moves from s back to p, closing port a for the cut-off, the valve occupying the position shown at H. Finally, while the eccentric center moves from k to e, the rod moves from p to g, and the valve from the position shown at H to that shown at G, thus completing a full revolution.

Now suppose we take the path of the crank-pin, and divide its circumference into 24 equal divisions, as, a' b' c' d' etc., and let the diameter of the circle on the line D B represent the path of the piston. Then, if we take the length of the connecting-rod, measured on the same scale as the circle represents the path of the crank-pin, we may rest one compass-point on the line of centers (or line D B continued from D past B), and from the divisions, mark the corresponding piston positions. Thus, suppose the crank has moved from D to r', and the piston will have moved from D to arc 1 at the crank end; when the crank has reached division s' on the circle, the piston will have reached arc 2 on the line D B, and so on throughout the whole revolution.

When the crank has moved from D to y, it will have made a quarter-revolution, but the piston will not have moved half its stroke, as it should do to keep time with the crank motion. On the other hand, however, while the crank moves the quarter-revolution from y to B, the piston will move from arc 6 to B, which is more than half its stroke. Similarly, while the crank moves the quarter-revolution from B through divisions a' b' c' etc., to y', the piston will move through the divisions from 1 to 6 at the head end, and, therefore, more than half its stroke, and it is shown, that the piston moves faster when moving from the head end B to arc 6, than it does when moving from the crank-end D to arc 6, and it is this that causes the points of cut-off to vary in engines having a common or simple slide-valve with equal lap for each port.

Now the nature of the eccentric motion in the Straight-Line Engine, is, in connection with the line of motion of the eccentric-rod and rocker-shaft, such as to correct this evil without the employment of unequal lap on the valve, while at the same time, it main-

tains the lead equal at each end of the cylinder. Thus, on the piston-stroke when the crank is moving from D towards B, it has been shown that the port will be fully opened when the eccentric center is at n, and that the cut-off occurs when the eccentric reaches w, hence, the angle the eccentric moves through while it is closing the valve, is angle n w, or, in this example, 63° as marked in the figure. On the other stroke, when the motion of both piston and crank (on their respective paths) is from B to D, the eccentric, in operating the valve to close the port and effect the cut-off, moves (as has been shown) from n to k, which is in this example 58°, as marked, it is clear then, that since the speed of the crank is uniform, the cut-off will be effected quicker when the crank is moving from B to D, than it will be when the crank is moving from D to B. We have, therefore, that when the piston is performing the stroke from B to D, during which it is moving at its quickest when compared with the crank motion, the action of the eccentric and valve motion, is, from its peculiar construction, also accelerated, so that the valve action is timed with the piston motion. Similarly, when the piston is moving through its stroke from D to B, during which its speed is the slowest when compared with the crank motion, the mechanism delays the valve speed and again times the valve action with the piston speed, and it is apparent that a length of connecting-rod may be chosen, that will give a piston motion that will have exactly equalized points of cut-off for the two piston strokes, let the eccentric be moved to any position across the shaft that it may. This will be seen because, to whatever position across the shaft the eccentric may be moved, the periods of valve closure for the cut-off will be within the angle n w for one stroke, and n' k for the other; the latter will always be less than the former, and always less to the same amount, while the piston speed will also vary to the same amount for the two strokes.

Another feature of this valve motion is that it gives a much larger port opening at the head end of the cylinder than at the crank end, and thus gives a greater admission for the stroke, when the piston is moving the fastest. Thus, when the eccentric is at e the upper end of the rocker is at g, when the eccentric is at n the rocker is at q, giving a full port opening at that end of

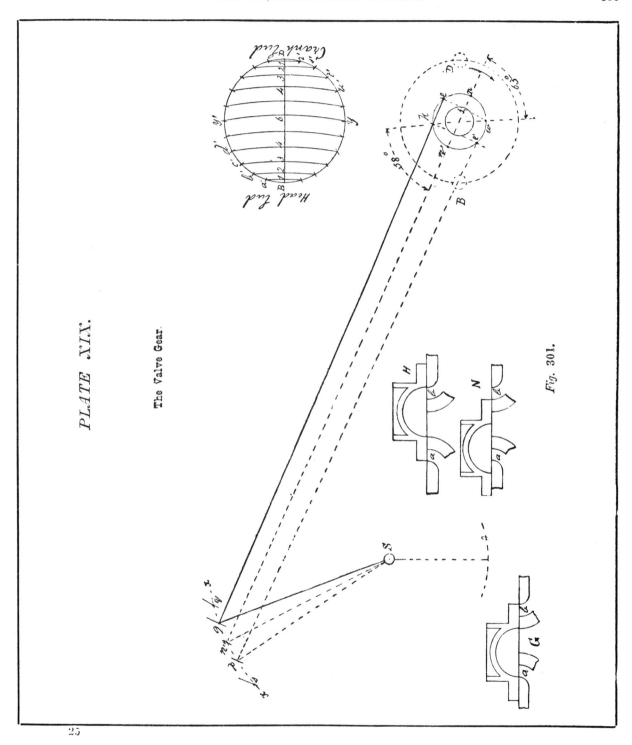

PLATE XIX.

The Valve Gear.

Fig. 301.

the cylinder. Similarly, when the eccentric is at e′, the rocker is at p, and when the eccentric is at n′, the

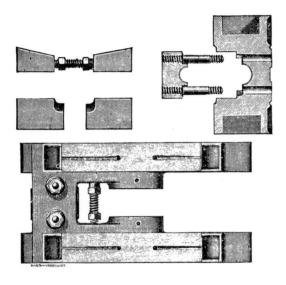

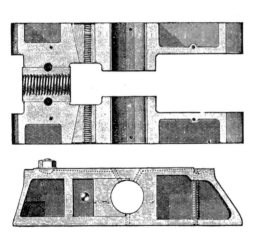

Fig. 302

rocker is at s. Now, as the distance from p to s is greater than from p to q, therefore the port opening for the end of the cylinder furthest from the crank, is larger than the port opening at the crank end of the cylinder, which is necessary, not only on account of the quicker piston motion, but also because, at the crank-end, the piston-rod reduces the steam space by an

amount ranging between from 2 to 7 per cent, hence, the head-end port requires a corresponding increase of opening to keep the steam pressure up to the same point as that obtained in the crank-end of the cylinder.

Fig. 302 shows the details of construction of the cross-head, which is made hollow and as light as is consistent with the requisite strength. The cross-head pin is held secure in the connecting-rod end, and has journal-bearing in the cross-head, by this means, a long journal-bearing is secured and the connecting-rod is prevented from moving sideways at its crank-end. The surface of the cross-head pin is cut away at the top and bottom, and the corresponding surface in the cross-head bore is recessed, which prevents the pin from wearing oval. The recess is so arranged that the bearing surface on the pin passes over the edges of the recess, thus preventing the formation of shoulders.

The Ide Engine·

Figures from 303 to 316 illustrate a High Speed Automatic Cut-off Engine, designed and constructed by A. L. Ide and Son.

Figure 303 is a side elevation of the engine, and figures 304 and 305 elevations on a larger scale, and partly in section to show the construction and the valve mechanism.

The piston is hollow, and is provided with two snap piston rings. The piston valve operates in steel bushings, that may readily be removed when worn, the valve and bushings being shown removed from the engine in Fig. 306. The bushings are each provided with openings through which the steam passes into the cylinder when these openings are left uncovered by the inner edge of the valve head. The exhaust passes through the valve to a pipe at the head-end of the cylinder.

Fig. 307 is a cross sectional view of the frame showing the construction of the rock-shaft and cross-head, and it is seen that the rock-shaft is provided with an oil pocket for lubricating the joint of the lower rocker-arm.

In Fig. 308 is a side elevation and plan, and in Fig. 309 a longitudinal section through the cross-head, while

PLATE XX.

Fig. 303.

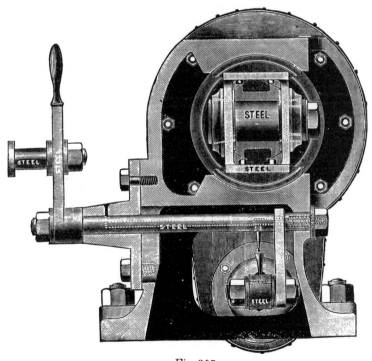

Fig. 307.

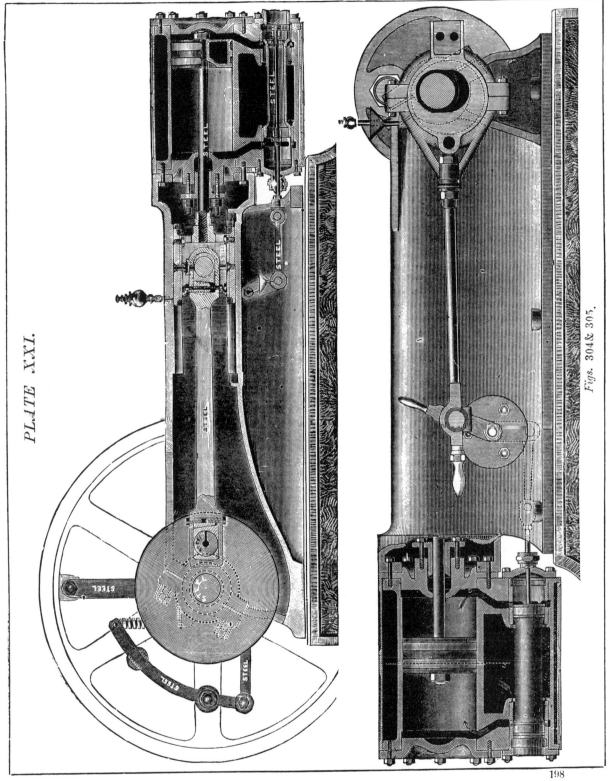

PLATE XXI.

Figs. 304 & 305.

198

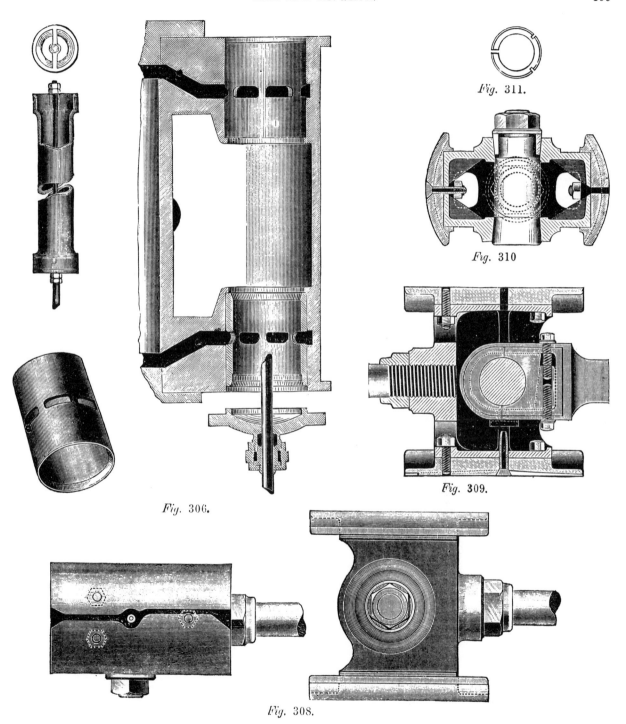

Fig. 311.

Fig. 310

Fig. 309.

Fig. 306.

Fig. 308.

Fig. 310 is an end section, and Fig. 311 an end elevation of the bush.

The pin is central in the cross-head, and is tapered at one end.

The bush is cut through in one place and nearly so in two other equidistant places, and is tapered on its external diameter, so that by means of the nut it may be drawn within the cross-head and closed upon the pin, thus firmly securing the latter, while the arrangement permits of the ready removal of the pin, and of the use of a solid end connecting-rod with wedge of full width of bearing. A constant supply of oil is supplied to the wrist-pin from the oiler on top of the engine frame. The oil being wiped from the top guide and passing through a tube in the top slide, enters a funnel in the connecting-rod, and after passing through the bearing, drips to the bottom of the cross-head and passes through a hole to the lower guide, the one oiler supplying both slides and wrist-pin bearing.

Figs. 312 and 313 represent the governor or speed regulator.

The eccentric E is attached to a hanger A, which is pivoted to the wheel at C, at L L′ are levers pivoted to the arms of the wheel at *a a′*.

Lever L is attached at one end to a spiral spring S, while L′ is attached near one end to a similar spring and at its extreme end to a dash-pot D. Each lever is provided with a weight marked respectively W and W′. To these levers are pivoted at *b′* and *b*, arms B and B′ which are pivoted at *b″* to a pin at the upper end of the hanger A. The operation is as follows:

The centrifugal force, generated as the wheel revolves by the unbalanced part of levers L L′, and by the weights W and W′, acting against the force of the springs S S, moves the eccentric along the arc *n n* (whose center is at C), and this lessens the throw of the eccentric, and therefore, the travel of the valve, causing the point of cut-off to occur earlier in the piston-stroke.

It is obvious that as the wheel speed increases, the levers open out, their extreme positions, and that of the pendulum lever, being marked in dotted lines, and the corresponding position of the center of the eccentric being at *f. e* represents the eccentric center when the parts occupy the positions shown in the full lines.

The tension of the springs S is regulated as follows:

The springs at their points of attachment to the wheel are secured in the sliding blocks *r*, Fig. 312, which may be operated along the radial slideways *g* by means of the screws whose square heads are shown at *h h′*. When the block *r* is in the position shown in the figure, the arc in which the upper end could move without further tension to the springs is denoted by the arc *m*, but if we move block *r* inwards to the other end of slideway *g*, this arc would become arc *c*, and it is plain, that as *c* is further from the outer position *x* of the end of the lever, therefore, the spring will require to be extended more, and will be placed under more tension when the block *r* is at the inner than when it is at the outer end of the slideway *g*, or, in other words, with the block *r* in the position it occupies in the figure, the spring S would, while the lever L moved out to the dotted lines, be further extended to an amount equal to the radius from *x* to the arc *m*, whereas, with the block *r* moved by the screw *h* to the other end of the slideway *g*, the spring S would, while the lever moved out to *x*, require to be distended to an amount equal to the radius from arc *c* to *x*, the radius being in each case measured in a line from *x* to *r*, as denoted by the dotted lines. The adjustment to find the required position for blocks *r* is effected by experiment with the engine under light and heavy loads, with a speed recorder and an indicator attached, so as to note the revolutions at various points of cut-off.

The dash-pot D contains glycerine, and consists of a case or barrel containing a piston and rod, the latter being pivoted at *d* while the barrel is pivoted to the lever L′, so that as the latter moves outwards towards the position denoted by the dotted lines it draws the case with it. Through the piston are the small holes *p p*, and it is obvious, that as the barrel or case is pulled outwards, the glycerine with which the case is filled must pass through these holes from one side of the piston to the other.

Now if the motion of the case over the piston is so fast that there is not sufficient time for the glycerine to pass through the holes *p*, then the glycerine on one side of the piston will be under compression, while on the other side there will be a partial vacuum, hence, the action of the dash-pot is to offer a resistance to sudden vibrations, or, in other words, to equalize and steady

the action of the speed regulator.   Referring now
to the mechanism for shifting the eccentric across
the shaft to alter the valve travel, and thereby
vary the point of cut-off, Fig. 314 represents
the parts in the positions they would occupy when the
crank is on the dead center B. The eccentric is shown
shifted inwards to its position of least throw, its center
being at f or distant from the center C of the shaft
to the amount of the steam lap of the valve which is
supposed to have no lead.

The pivoted end C' of the eccentric hanger is on the
line of centers of the engine, and if from C' as a center
we mark an arc $f f'$, then this arc will represent the
path of motion of the center of the eccentric when
moved across the shaft by the eccentric hanger. When
the eccentric is moved across the shaft to its position of
greatest throw, its center stands at e and the path of
revolution of its center is on the circle n.

To find the position the upper end of the rocker-arm
and the pivoted end of the eccentric-rod must be in, in
order that shifting the eccentric from f to e may not
move the valve, we set a pair of compasses to represent
the length of the eccentric-rod, and from e as a center
mark an arc w, and then with the same radius, mark
from f an arc y, and where w and y intersect, or at x,
is the position for that end of the eccentric-rod, the
crank being on its dead center B. Now suppose we
rest the compasses at x, and mark an arc $f c$, and it is
seen that between the circles n and n' the arcs, $f' f$,
and $c f$ practically coincide, and it becomes clear, that
moving the eccentric across the shaft will not move
the rocker, and therefore, will not move the valve.

With the eccentric center at f, its path of revolution
will be on the dotted circle n', and as the rocker-arms
are of equal lengths, the travel of the valve will equal
the diameter of circle n', or equal to twice the lap,
hence there would be no admission of live steam,
because the center of the eccentric is at the end of its
stroke, and as soon as the crank moves, the valve will
begin to move back. But as soon as the eccentric is
shifted across the shaft from f towards e, there will be
an admission of live steam, beginning when the crank
passes its dead center, and the period of this admission
depends upon the amount the eccentric has been shifted
across the shaft by the governor or speed regulator.
Suppose then, that the eccentric center is shifted from

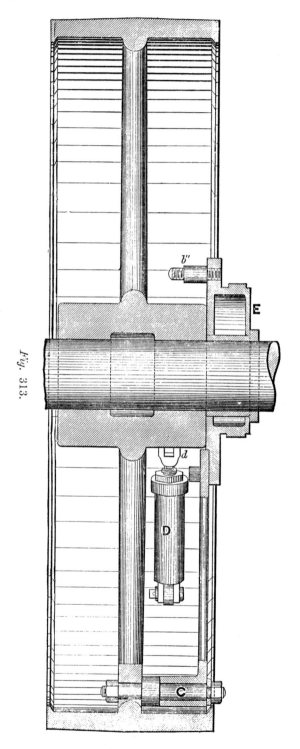

Fig. 313.

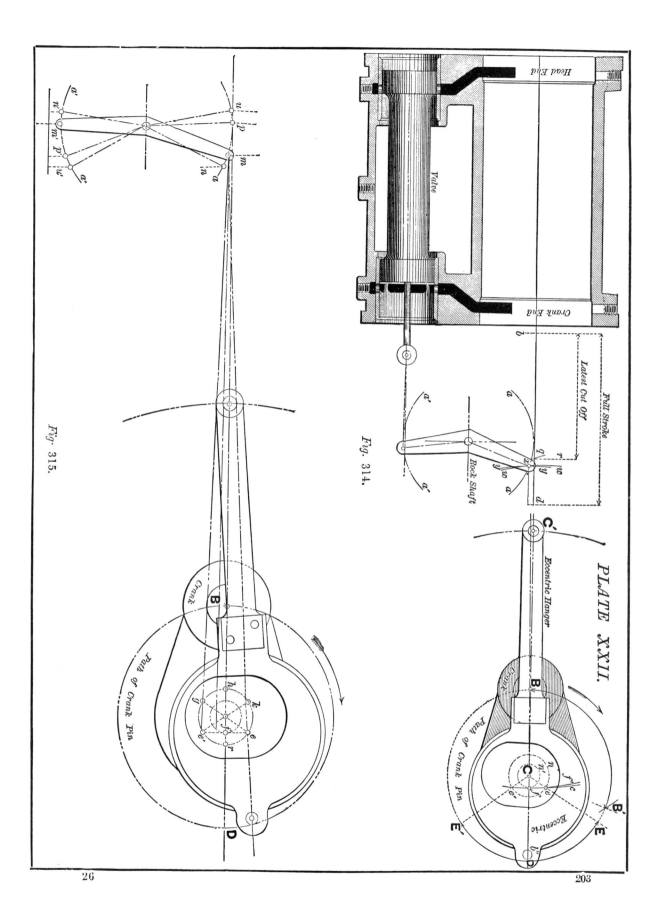

Fig. 315.

Fig. 314.

PLATE XXII.

f to *e'* so as to give the longest period of admission or latest point of cut-off, and we may find at what point in the piston-stroke this occurs, as follows: As the crank motion is in the direction denoted by the arrow, the eccentric will move from *e* to *e'*, the cut-off occurring when it arrives at *e'*. The degrees of angle it moves through, from the point of admission to the point of cut-off, is shown on the crank-pin circle at E E', and it is obvious that the crank will move through the same angle, hence we take the length of arc E E', and mark it from B, thus getting at B' the crank position at the time the cut-off occurs.

We then set a pair of compasses to represent the length of the connecting-rod, and from B mark an arc at *b*, and from D mark an arc *d*, thus marking from *b* to *d* on the line of engine centers, the length of the piston stroke; with the same set of compasses mark from B' an arc *q*, and from where *q* cuts the line of centers, erect a perpendicular line *r*, which in its distance from perpendicular line *b*, gives us the longest period of live steam, or in other words, the latest cut-off, as marked in the figure, and it is thus found that the governor, in shifting the eccentric across the shaft from *f* to *e*, varies the point of cut-off from about three-quarter stroke to no admission.

In practice, however, the rock-shaft center is lowered sufficient to bring the apex of its upper arm 7° below the line of engine centers, as shown in Fig. 316. This causes the admission to begin when the crank passes the line of the eccentric-rod, and gives $\frac{1}{16}$ inch lead when the crank is on the dead center.

Obviously therefore, under these conditions the point of admission varies for the different points of cut-off, and furthermore, in proportion as the eccentric is shifted out towards *e* for longer points of cut-off, the lead increases, because the eccentric, having a greater throw, moves the valve more during a given number of degrees of eccentric motion. The increase of lead, however, at the later points of cut-off, serves to compensate for the lesser amount of compression at the later cut-offs and thus serves to equalize the amount of cushioning on the piston.

It is obvious, however, that in this case the shortest point of cut-off will not occur until the piston has moved past the dead center enough to move the valve a distance equal to the amount of the lead. It will be observed that there is an offset in the rocker-arms, and the object of this offset is to cause the valve to move the fastest when the piston is moving the fastest, and thus proportion the admission to the piston speed. To investigate this, we proceed as in Fig. 315, in which the parts are shown in the positions they would occupy with the crank on the dead center B, and the eccentric shifted across the shaft to its position of greatest throw at *e*, for the latest point of cut-off.

Now, while the eccentric moves from *e* to *r*, the valve will move to open the port for the admission of steam, and while it is moving from *r* to *e'*, the valve will be moving to close the port, the cut-off occurring when the eccentric reaches *e'*. During this period of eccentric motion, the upper rocker-arm will move from *m* to *n*, opening the port, and then back to *m*, closing it for the cut-off, while the piston will move from the head-end towards the crank-end of the cylinder.

Now suppose the crank to be on the other dead center, and the eccentric will be at *g*, and the upper rocker-arm at *p*, then while the eccentric moves from *g* to *h*, the upper rocker-arm will move from *p* to *u*, opening the steam port, and while the eccentric moves from *h* to *k* the upper arm will move from *u* to *p*, and the cut-off will occur.

When the rocker-arm is at *n*, the lower arm is at *n'*; when the upper is at *p*, the lower is at *p'*, and so on; and if we draw a line beneath the lower rocker-arm, we may, by means of the vertical lines *n' m'*, etc., trace the movement of the valve, thus when the crank is moving from D, the valve, in opening for the admission, moves a distance equal to radius *m' n'*, whereas, during the admission for the other stroke, it moves the lesser radius from *p'* to *u'*, and we have, therefore, that during the period of admission while the crank moves from D, the valve moves faster than it does for the admission period when the crank moves from B, and it follows that since the piston moves faster during the live steam period from D, than it does for the corresponding period from B, therefore the valve motion is timed with the piston motion, giving a quicker admission for the port at the crank-end than for that at the head-end of the cylinder.

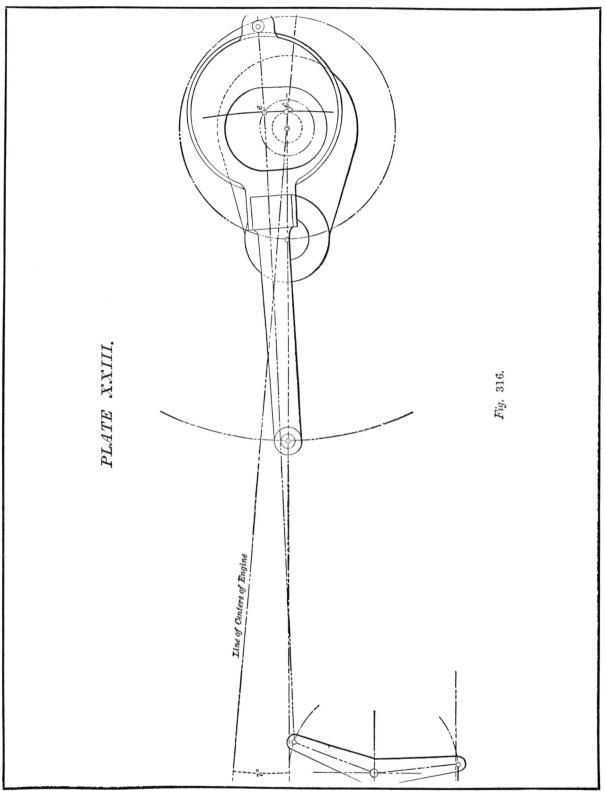

PLATE XXIII.

Line of Centers of Engine

Fig. 316.

The Westinghouse Engine.

Figs. from 317 to 324, represent the Westinghouse Automatic Cut-off Single-acting Engine.

Fig. 317 is a front, and Fig. 318 a rear view of the engine, while Fig. 319 is a sectional view on a vertical plane passing through the center of the crank-shaft

are trunk pistons, the wrist-pin *b*, for the upper end of the connecting-rod, passing centrally through the piston. Each piston has four packing-rings to maintain it steam tight, and is made long enough to form its own guide in the cylinder bore, thus dispensing with the usual guide-bars and cross-head. The upper end of each piston is chambered (as seen in the left hand piston, which is shown in section) to prevent condensation. The cranks are set exactly opposite to each other so that one piston is always in action, and the live steam

Fig. 317.

bearings, and Fig. 320 a sectional view, through the cylinder in a plane at a right angle to the crank-shaft.

There are two steam cylinders A A having covers *a a* at their upper ends only, the lower ends being open for the connecting-rod to work through. The pistons D D

period in one cylinder, corresponds to a certain portion of the exhaust period in the other, this period depending upon the point at which the cut-off occurs.

The upper end of the connecting-rod is bushed with a thimble, while at the crank-end it is provided in the upper half of the bore for the crank-pin with an anti-friction metal lining, this being the half that is (on account of the steam acting on one side only of the piston) always under pressure or compression, while the other half performs no duty.

The cranks are balanced by the over-hanging piece, or bob *x*. The cylinder covers are provided with what is termed a *pop-out* head, the construction being as follows: The center of the inside cylinder-head, Fig. 321, is a separate piece, screwed or driven into place

mentally determined) to crack out when a pressure of 200 pounds per square inch is reached.

If, therefore, the engine cylinder should receive a charge of water, the center piece will break out, thus relieving the pressure and preventing the breakage of parts that would be more costly to repair or replace.

The construction of the main bearings is seen in the sectional view, Fig. 319. The crank-journal H is taper, and works in a shell *d* lined with Babbitt metal. Between the flange of the shell *d* and the cover *d'* is a

Fig. 318.

against a shoulder, *a a*. It is prevented from any possibility of getting into the cylinder by the indicator plug *b*, which draws it up to the loose outer head *c c*. This center piece is partly cut away on the upper side by a circular grove *d d*, leaving metal enough (experi-

chamber, containing a ring wiper W, which takes up the oil as it works past the bearings, and returns it through the tube *e* to the crank case C, which is partly filled with water, upon which floats a layer of oil for lubricating the cranks and eccentrics, as will be explained presently. Collar washers *t t*, Fig. 319, of bronze form the end bearings of the cranks, and lead washers *v* prevent the taper sleeves from being forced up, so as to cause binding on the crank-journals H. A

filled with live steam. In the position the parts occupy in the figure, steam is about to be admitted to the cylinder through the annular port P, which is left open for the admission by the upward motion of the valve, the

Fig. 319.

center-bearing K bridges the crank case, and receives the downward thrust of the crank at H.

The construction of the valve mechanism is shown in Fig. 322, which is a vertical section through the central plane of the valve chest, and it is seen that a piston valve V, having four packing-rings, is employed. Steam is admitted to the central portion S, which is constantly

eccentric I moving from right to left. The exhaust for this cylinder is taking place through the annular exhaust port *p' p'*, which is also opened by the upward motion of the valve. The valve is situated between the two steam cylinders, and the one valve, therefore, serves for both cylinders, the exhaust for the other cylinder entering the chamber above the valve.

and passing through the valve to the exhaust pipe *n*.

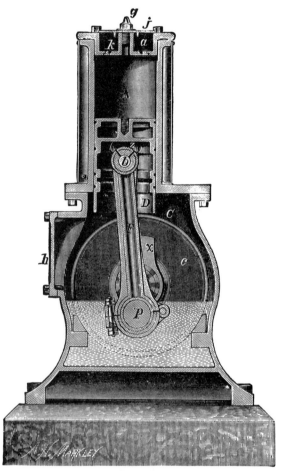

Fig. 320

The valve is guided by a piston guide J, which being

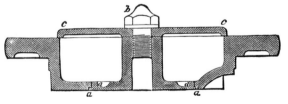

Fig. 321.

covered, prevents the exhaust from passing into the lower part of the casing in which the crank works.

The valve works in a casing *m*, which may be replaced

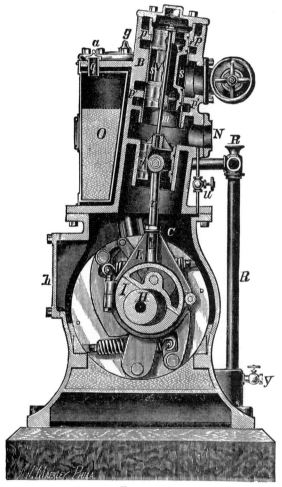

Fig. 322.

by a new one when necessary to restore the fit of the valve.

The construction of the governor for varying the point of cut-off by moving the eccentric across the shaft to reduce the throw of the eccentric, and therefore the travel of the valve, is as follows: The disc A A, Fig. 323, is cast solid, and keyed to one of the cranks.

The loose eccentric C, is suspended by the arm *c*, from the pin *d*, around which it has a motion of adjustment; B B are the Governor Weights, pivoted on the pins *b b*; one of the weights is connected to the eccentric by the link *f*, and both weigths are connected to operate in unison by the link *e*. Coil Springs D D furnish the centripetal or returning force. The eccentric encircles the shaft S, the opening being elongated to admit of the proper motion. The stops *s s* limit the motion of the weights.

In Fig. 223 the Governor Weights are shown in the position of rest, whereby the eccentric is thrown over to its position of greatest eccentricity, giving a maximum travel to the valve, corresponding to a cut-off of about

from $\frac{1}{5}$ to $\frac{1}{4}$ stroke, (at which the engine developes its rated power) the position of the parts is mid-way between the two positions shown. The position to which the governor must move the eccentric in order to cut off the steam at a given point in the stroke, may be found by the construction explained with reference to figures 278, 279 and 280, and the port openings, by the construction explained wtih reference to Fig. 225, while the valve lead and points of admission may be considered, as was done with reference to Fig. 227.

The means provided for lubricating the various work-

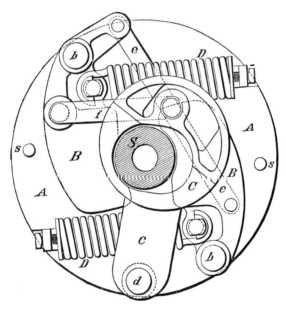

Fig. 323.

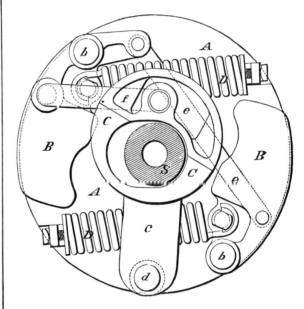

Fig. 324.

$\frac{5}{8}$ stroke. The parts of the governor remain in this position until the engine is within a few revolutions of its full speed. The centrifugal force of the weights then over-balances the tension of the springs, and the weights move outward, reducing the travel of the eccentric and valve, and consequently shortening the point of cut-off.

The extreme outward position of the weights is shown in Fig. 324. The cut-off for this position occurs so early as to barely hold the engine up to speed when running without load.

When the engine is properly loaded, so as to cut off

ing parts without the use of oil cups, are as follows:

A reservoir O, Fig. 322, contains a supply of oil, which is admitted from time to time through the cap at *q*. From the oil reservoir are pipes having the globe valves *l l*, Fig. 319, which may be operated to feed oil into the receptacles *f f*, Fig. 319, the oil passing thence to the crank-shaft journals H, and after finding its way to the ends of these journals, it is carried into the case or chamber C, Figs. 319 and 320, in which the crank works. This case is enclosed by the cover shown at *h* in Fig. 320, and contains water whose level nearly reaches the crank-shaft; floating upon this water is a layer of oil, into which the cranks and eccentrics dip during the lower part of their path of revolution, thus giving constant lubrication.

To maintain the proper level of water in the case C, a drip-pipe U, Fig. 322, is provided, which admits the water of condensation from the exhaust steam.

Similarly, either additional water or oil for the case C, may be admitted through the pipe R R, but it is ob-

319, prevents the accumulation in the case of water above the level of the top of the pipe *e*, and being connected to the bottom of the case, it carries off the surplus water only, maintaining the level of the oil at a constant height in the case.

This level is indicated by the height of the water in the funnel-head *n*, and is required to be such that the water is always in sight.

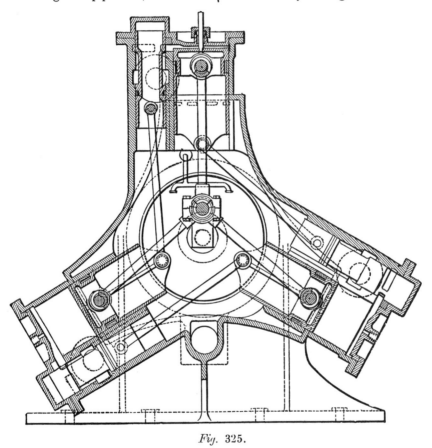

Fig. 325.

The James and Wardrope Engine.

vious, that by a proper adjustment of the valves at *l l*, Fig. 319, the oil supply may be made sufficiently constant from the reservoir O, Fig. 322, to render it unnecessary to resort to the pipes R R, for any additional supply.

A siphon over-flow, having a funnel head at *n*, Fig.

27

The Multi-Cylinder Engine.

In this class of steam engine there are two or more

steam cylinders, whose axial lines usually radiate from the center of the crank-shaft. These cylinders are single acting, or in others, receive steam at the head-end only, the other end being open, hence the connecting-rods are under compression, only being pushed by the piston and not pulled during the return piston-stroke, and the journal pressure always being in the same direction at both ends of the connecting-rod, the

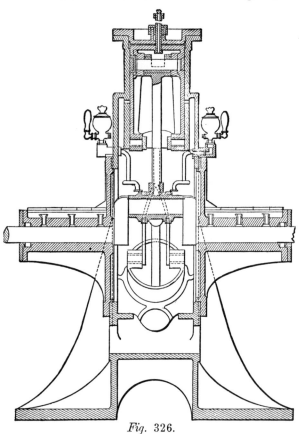

Fig. 326.

cross-head journal, crank-pin journal, and main-shaft journals are kept seated on one side only of their bearings, hence the wear does not cause play or lost motion, and delicate adjustments of fit are not necessary in order to prevent pounding or thumping.

Figs. 325 and 326 represent the James and Ward rope, three cylinder single-acting engines, these engravings being extracted from *Engineering*.

The center-lines of the three cylinders radiate from

the center of the crank-shaft, and their connecting-rods all attach to the same crank-throw, which, therefore, has no dead center. Each cylinder has a separate piston-valve which works in a line parallel to the cylinder bore, and is operated from a rod driven by the neighboring piston, as is plainly seen in the engraving.

The ends of each piston-valve are enlarged, and are provided with packing-rings. The live steam enters at the section of reduced diameter between the enlarged ends, the position of the steam pipe being shown in the dotted circles in Fig. 325. The steam enters the cylinder at its upper end, when the port is uncovered by the motion of the valve towards the head-end or outer end of the cylinder, and exhausts when the piston-valve has moved sufficiently towards the crank to leave the port uncovered, at which time the steam passes the end of the valve, and finds exit to the exhaust pipe, at the outer end of the bore in which the valve works. This part of the exhaust is therefore controlled by the valve, but there are supplementary exhaust ports which are not so controlled, these latter being shown in the dotted openings in Fig. 325, and also in section in Fig. 326; these ports are merely uncovered by the piston as it passes them and are situated so as to come into action when the piston has made about $\frac{5}{8}$ of its stroke towards the crank and permits of the escape of a large proportion of the steam.

The point at which the piston-valve effects its exhaust, depends upon the point at which the end of the valve effects the compression by closing the steam port, the action, in this respect, being precisely the same as that explained with reference to Figs. 271 and 272 concerning the piston valve of the Armington-Sim's engine, it being noted, however, that in this case the valve is single ported only. Both the exhausts enter the casing in which the crank revolves, this casing forming an enclosed chamber having the main exhaust pipe at its bottom. The exhaust steam therefore excludes the air from contact with the pistons, and thereby prevents the loss of heat which would otherwise occur.

The pistons are made long, and are trunks, the connecting-rods pivoting to their outer ends, thus giving a long connecting-rod and dispensing with the use of guides, the long pistons and cylinder bores serving for guides, as is also the case with the valves.

Fig. 327.

213

N. Y. Safety Steam Power Co's. Engine.

Fig. 327 represents an Automatic Cut-Off Engine, by the New York Safety Steam Power Co. The governor is of the usual wheel-regulator construction and operates a piston valve, which takes steam at the ends and exhausts it in the middle of its length. In order to perfectly balance the valve, the diameter at the valve-rod end is made sufficiently larger than the head-end, to make the area exposed to the steam, equal.

The Ball Automatic Cut-Off Engine.

In the Ball Automatic Cut-Off Engine, the point of

view of the engine, Fig. 329 represents the governor with the eccentric removed, and Fig. 324, the eccentric.

An arm T T is keyed to the crank-shaft, and upon the hub of this arm the pulley is a working fit, so that it can revolve a certain amount upon the arm. This amount is limited by means of two lugs which project into cavities provided in the end of the arm hub, as seen in Fig. 329. The governor balls are pivoted, by arms to the pulley arms, as shown, each of these arms being connected to two spiral springs, of which one is fast to the inside rim of the pulley, and the other is fast to the lever T. Now suppose the pulley to revolve against a steady resistance, offered by the belt to the pulley, and the balls will swing out and revolve in a circle of such a diameter as will create an equilibrium

Fig. 328.

cut-off is varied by a wheel governor shifting the eccentric across the shaft. A flat valve is used which gives a double port opening through a single port, the construction being as follows: Fig. 328 is a general

between the centrifugal force generated by the balls, and the centripetal force of the four spiral springs. Suppose, however, that the belt resistance suddenly increases, and any retardation of the pulley wheel resulting from the increase of load, will at once be com-   be moved across the shaft into position to give later

Fig. 329.

municated to the weight arms, and the weights, increas-

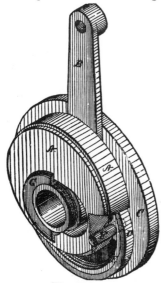

Fig. 330.

ing the centripetal force, will cause the eccentric to

points of cut-off, and increase the power of the engine to a degree corresponding to the amount of increased resistance offered by the belt to the pulley. The governor action is here, then, independent of the crank-shaft, which may go on at its regular speed. By this construction therefore, it is necessary to change the speed of the crank-shaft and fly-wheel before the governor can act, since the latter takes a short cut, as it were, and acts directly upon the valve without reference to the fly-wheel and crank-shaft.

THE ECCENTRIC CONSTRUCTION.

The construction of the eccentric mechanism is shown in Fig. 330, in which A is the main eccentric having an elongated opening, which permits it to swing across the shaft. To this eccentric is secured the arm B, which is pivoted upon the pin Y of arm or lever T in Fig. 329, thus allowing the eccentric a pendulum motion across the crank-shaft.

The amount of this pendulum motion is controlled by the rotation of a disc C. This disc has a flange D, which is eccentric to the crank-shaft, and on the inside of this eccentric flange is a ring E, which has a stud F

Fig. 331.

332 and 333. It is made in two parts T and S having rectangular flat faces, and an annular ring flange at the back.

The flange of T fits within that of S and is provided with two spring snap rings to maintain a steam tight fit. The live steam enters the inside of the valve, and pressing against the projecting lips and of the end faces of the rings, moves the two halves of the valve apart and up against the seat faces. The valve is thus relieved of pressure upon that part of its area which forms the openings through which the live steam enters.

The steam ports are so shaped, as to receive steam

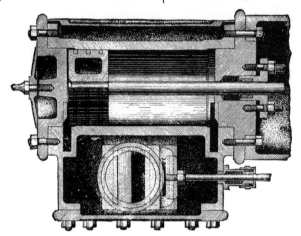

Fig. 332.

engaging with the main eccentric A, as shown.

The rotation of the disc upon the crank-shaft therefore causes the main eccentric to swing across the shaft, from the pin Y, as a center of motion. The disk D also has sleeves encircling the shaft, and projecting through the elongated bore of the main eccentric, and on the end of this sleeve is a ring-nut G which holds all the parts in place.

The disc is rotated around the crank-shaft by means of two pins in its back face, which fit into the holes shown in the ends of the links shown in Fig. 329 to be pivoted at their ends to the governor balls.

THE SLIDE-VALVE.

The construction of the valve is shown in Figs. 331,

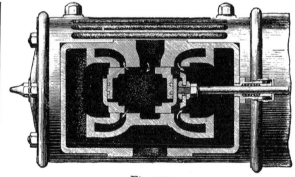

Fig. 333.

from the port openings given by both halves of the valve, the course of the steam being denoted by the arrows in Fig. 333, where it is seen that the exhaust passes out at the ends of the valve.

The Dexter Automatic Cut-Off Engine.

Fig. 334.

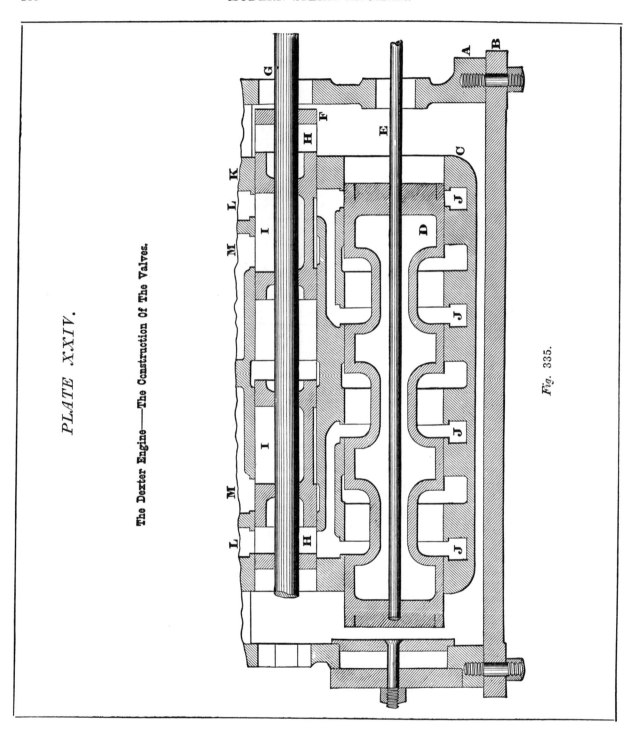

PLATE XXIV.

The Dexter Engine——The Construction Of The Valves.

Fig. 335.

The Dexter Automatic Cut-Off Engine.

In the Dexter engine (Fig. 334), a flat valve driven by a fixed eccentric, and that is balanced through the greater part of its stroke, controls the admission and exhaust, hence the lead and the points of release are equal for

rod for the main, and E that for the cut-off valve. The auxiliary steam chest C is suspended from the walls of the main steam chest, and receives steam at the center.

In Fig. 336 is given an end view of the steam chest and valves, showing the method of suspending the auxiliary steam chest within the main steam chest, and enabling it to set up to take up the wear between it and the back of the main valve.

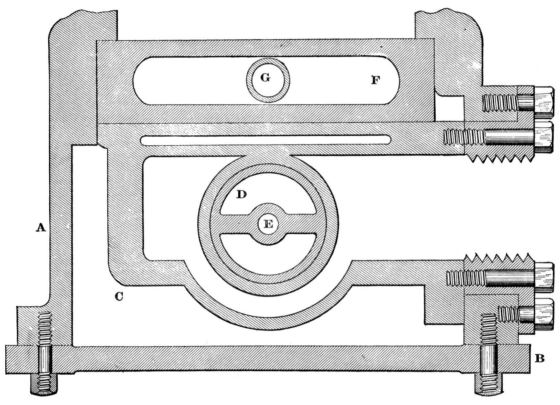

Fig. 336.

The Dexter Automatic Cut-Off Engine ——The Valve Construction.

all points of cut-off. Fig. 335 is a longitudinal section of the valves, L L are the cylinder steam ports, and M M the exhaust ports. On the back of the main valve F is a piece C, which acts as a pressure plate for the main valve, and receives a double-ported piston valve D, whose ports are shown at J, J, J, J. G is the

28

Openings lead from the central supply to the different points on the cut-off valve D, where admission occurs through the ports J, J, J, J, in the auxiliary steam chest.

It will be seen that the steam from the central openings in the cut-off valve is distributed to the main valve through two ports provided on each side of the center of the cut-off valve. The auxiliary steam chest may be moved up to the main valve to take up the wear. The eccentric C Fig. 337, is pivoted at D. The weights E E, are pivoted at G G, and are drawn inwardly by which position it will hold the eccentric rigidly against the resistance of the valve and its connections. When the weights are in their extreme inward positions, the opposite link will be in line with the points G, H, I, and the eccentric held rigidly against resistance, as in the

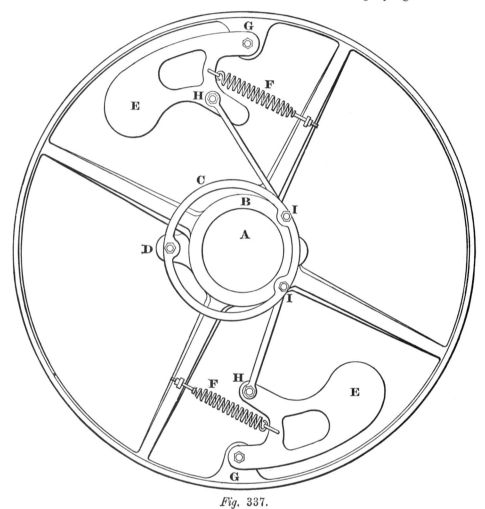

Fig. 337.

The Dexter Engine——The Construction of the Governor.

the springs F F, while they are connected to the eccentric by links with bearings at H I. The weights are shown in their extreme outward positions, and it will be seen that one of the links is in a straight line, on the points G, H, I, which is tangent to the eccentric, in extreme outward position. Between these two points the resistance is effected by the action of both links. The rigidity at the extreme inward position gives a capacity to start the valve, even though from disuse it may have become rusted or gummed to its seat.

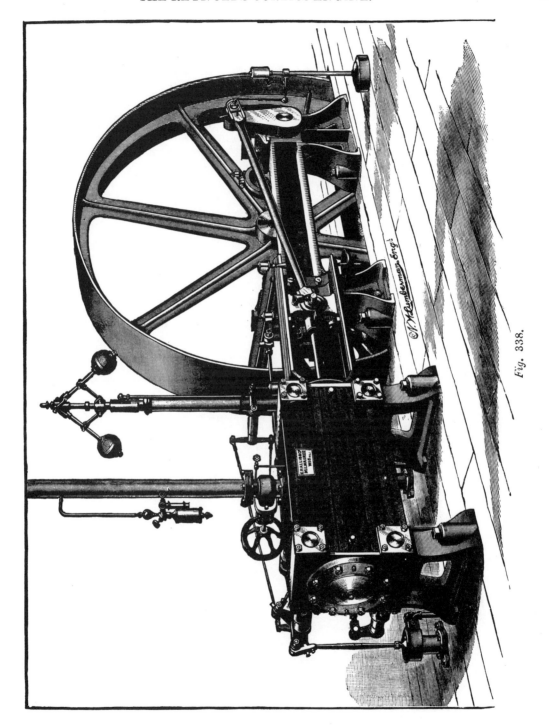

Fig. 338.

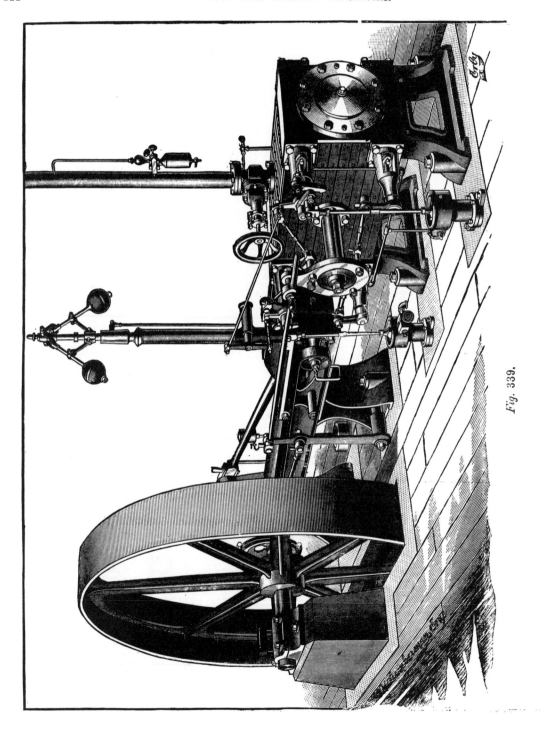

Fig. 339.

The Corliss Automatic Cut-Off Engine.

The Corliss engine is the most prominent and important of all that class of engines, in which the connection between the eccentric-rod and the valve stem is broken, in order that the valve may be closed quickly to effect the cut-off, which occurs at a point in the stroke that is determined by the governor.

The distinguishing features of a Corliss engine are the trip mechanism for releasing the valve; and its connection with the governor; the dash-pot or its equivalent for closing the valve without jar or shock; and the wrist motion which reduces the motion of the valve after it has opened the steam port.

There are two admisssion and two exhaust valves driven by a single and fixed eccentric, hence the lead and the points of release or exhaust are maintained equal for all points of cut-off.

The Reynolds Corliss Engine.

A representative of advanced practice in the Corliss type of engine, is given in Figs. 338 and 339, which represent the Reynolds Corliss Engine. Figs. 340 and 341 represent the valve gear with the parts in the position they occupy when the cut-off occurs at half-stroke, the piston having moved from the head-end of the cylinder. In Figs. 342 and 343 the parts are shown in position with the crank on the dead center, and the piston at the crank-end of the cylinder, valve v having opened its port to the amount of the lead.

Referring to Fig. 340 motion from the eccentric is imparted by the rod M to the wrist plate Y, to which are connected the rods C C' for operating the admission valves, whose spindles are seen at S S', and the rods F F' for operating the exhaust valves, whose spindles are seen at T T'.

THE VALVE GEAR.

The mechanism for the steam or admission valves may be divided into three elements; first, that for operating the valve to open the port for admission; second, that for closing the valve to effect the cut-off; and third, that which determines the point in the stroke, at which the cut-off shall occur.

The first consists of the rod M, wrist plate Y, and the rods C and C', which operate the bell-cranks r r r' r' which are fast on the valve spindles S S'. Upon the ends of bell-cranks r r, r' r', are pivoted latch links u u' which have in them a recess for the latch blocks, of which one is seen at e (the rod R' and its connection with the valve stem being shown broken away to expose e to view). During the admission the latch block abuts against the end y of the recess w and is tripped therefrom by the cam n'. The ends of arms g of the latch links abut against the hub of the arms d d', upon which are cams n, n' and at a a' are springs for keeping the ends g of latch links u u' against the hubs and cams of d d'.

Referring now to the valve mechanism at the head-end only, suppose the piston to be at the head-end of the cylinder, and latch block e will be seated in the recess provided in u to receive it, and as the bell-crank moves, the latch block will be raised by the latch link, which is carried by a crank arm corresponding to that seen at x at the crank-end of the cylinder, and as this crank arm is fast upon the valve spindle, the lifting of e will open the valve for admission. As soon, however, as the end g of the latch link meets the cam n' the latch link will be moved so that the end y of its recess will leave contact with the latch block e, and the dash-pot will cause rod R' to descend instantaneously and close the valve, thus effecting the cut-off.

THE ADMISSION.

The period of admission, therefore, is determined by the amount of motion the latch link u' is permitted to have before its end g meets the cam n' which trips the latch link, and therefore frees e from the latch link recess.

The point at which the cut-off will occur therefore is determined by the position of the cam n', because if n' is out of the way the end g of the latch link will not meet it, the latch link will not disengage from

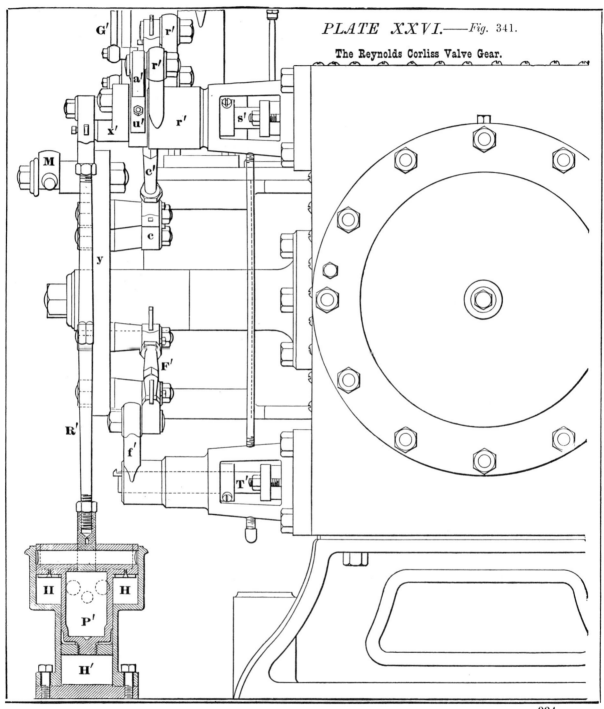

PLATE XXVI.——Fig. 341.

The Reynolds Corliss Valve Gear.

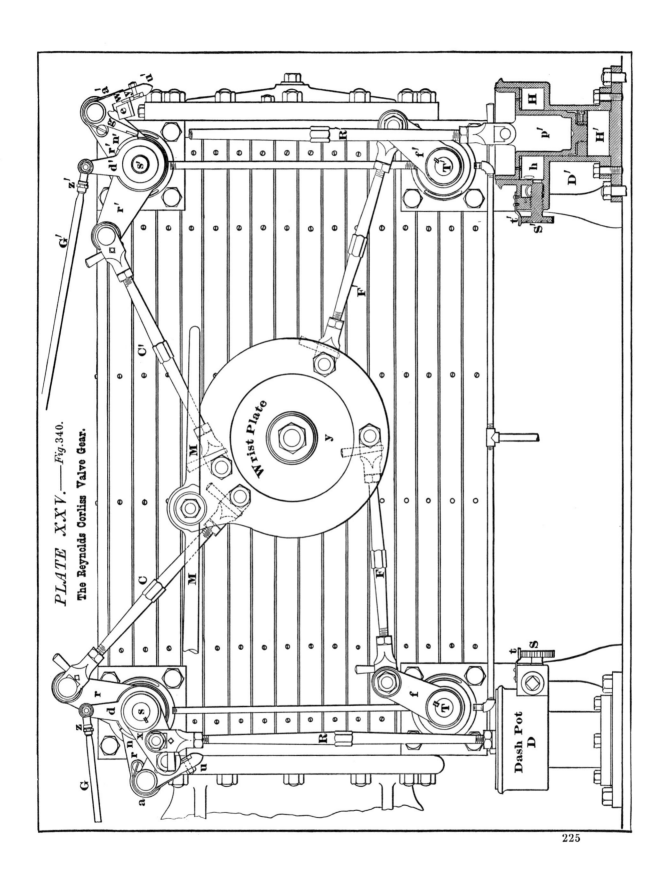

PLATE XXV.—Fig. 340.

The Reynolds Corliss Valve Gear.

Wrist Plate

Dash Pot
D

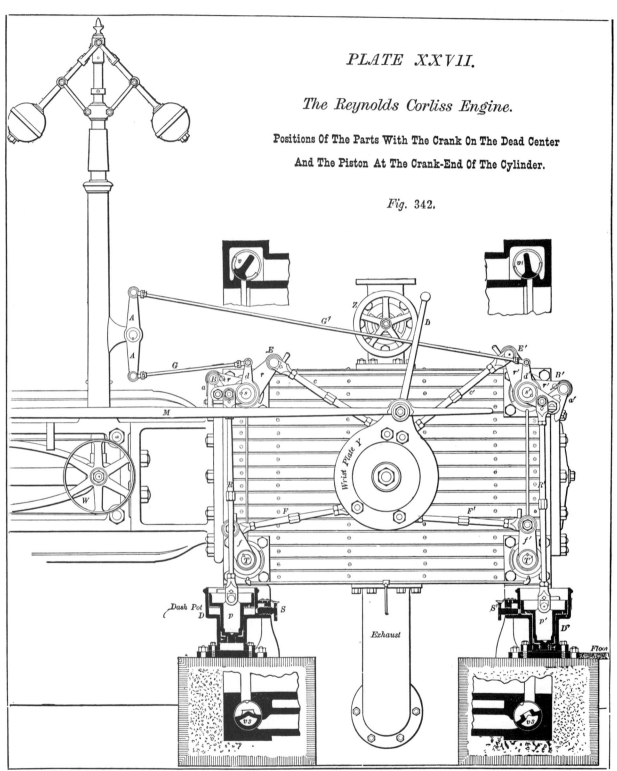

PLATE XXVII.

The Reynolds Corliss Engine.

Positions Of The Parts With The Crank On The Dead Center
And The Piston At The Crank-End Of The Cylinder.

Fig. 342.

the latch block e, and the cut-off would be effected by the lap of the valve, and independently of the dash-pot. As, in Fig. 340, the parts are shown in the positions they occupy at the instant the cut-off is to occur, therefore the cam n' has just tripped the latch link, and the end of e has just left contact with the end y of the recess w in the latch link u'.

The point in the stroke at which the tripping of u' from e' will occur and effect the cut-off, is determined by the governor, because d' is connected to the governor through the rod G'. In proportion as the governor balls rise, d' is moved from left to right, and the end of cam n' meets g earlier, or vice versa in proportion as the governor balls fall, the arm d' is moved to the left, g will meet the end of cam n' later, and the point of cut-off will be prolonged.

We now come to the means employed to close the valve quickly and without shock when the latch block is released from the latch link. Referring then to the crank-end of the cylinder, the latch block for that valve is carried upon arm x, to which is attached the rod R from the dash-pot piston (the arm corresponding to x, but at the head-end being shown removed to expose the latch block to view). We may now turn again to the head-end of the cylinder, rod R' corresponding to rod R at the other end, and it is seen that R' connects to a dash-pot piston p' having a stepped diameter, the lower half fitting into bore H', and the upper half into a bore H. The piston p' fits the bore H' and fills it when the rod R is at the bottom of the stroke, hence as p' is raised, there is a vacuum in H' that acts to cause p', and therefore R' and x, to fall quickly and close the valve the instant the latch block is released from the latch link. To prevent the descent of rod R' and piston p' from ending in a blow, a cushion of air is given in H by the following construction:

At S and S' are valves, threaded to screw and unscrew, the ends forming a valve for a seat entering H.

As the rod R' and its piston p' descends, the air in H finds exit through a hole at h, until that hole is closed by the piston p' covering it, after which the remaining air in H can only find exit through the opening left by the end of the valve S', and this amount of opening is so regulated by the adjustment of S', that a certain amount of air cushion is given, which prevents

29

p' from coming to rest with a blow. The head of valve S' is milled or knurled, and a spring t' fits, at its end, into the milled indentation, thus holding it in its adjusted position. The under surface of the upper part of p' is covered by a leather disc, while the part that fits in H' is kept air tight by a leather cupped packing.

THE GOVERNOR CONNECTION.

The connection of the cam arms d and d' with the governor, is shown in Figs. 342 and 343, in which the parts are shown in the position they would occupy when the crank is on the dead center, and the piston at the crank-end of the cylinder. The rod G' connects the cam arm d' with the upper end of lever A, which is connected to the governor, and vibrates on its center as the governor acts upon it.

Now suppose the speed to begin to diminish, and the governor balls to fall, and the direction in which A will move will be for its lower end to move to the right, thus moving d to the right, and carrying its cam away from the end of the latch link, which will therefore continue to open the port for a longer period of admission. Or, referring to Fig. 340, it is plain that if the governor balls were to lower from a reduced governor speed, G' would move to the left and cam n' would be moved away from contact with the end g of the latch link, which not being tripped, the admission would continue. On the other hand, suppose the governor balls to rise from an increase of governor speed, and d' (Fig. 340) would be moved to the right, and the cam n' meeting g earlier would trip e earlier, correspondingly hastening the cut-off.

The governor is driven by a belt from a pulley on the crank shaft to the pulley W Fig. 342, whose shaft conveys motion to the governor spindle through the medium of a pair of bevil pinions.

THE CONSTRUCTION OF THE VALVES.

The construction of the valves is shown in Fig. 342, in which v represents the steam or admission valve for the crank end port, and v' that for the head-end port, while v^2 is the exhaust valve for the crank-end, and v^3, that for the head-end of the cylinder. All four valves

PLATE XXVIII.

The Reynolds Corliss Engine.

Positions Of The Parts When The Crank Is On The Dead Center And The Piston At The Crank End.

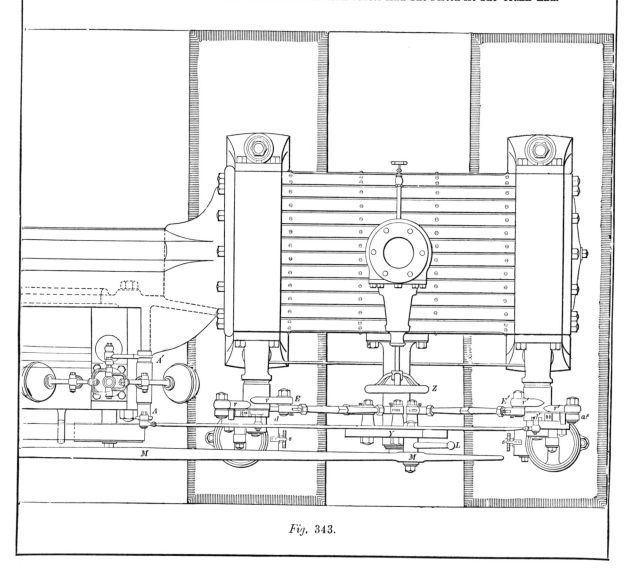

Fig. 343.

are shown in the positions they would occupy when the crank was on the dead center, and the piston at the crank end of the cylinder, hence the valve positions shown, correspond to the positions the parts of the valve motion occupy in the figure.

The faces of the valves are obviously arcs of circles, of which the axis of the shafts s s' are the respective centers. Valve v has opened its port to the amount of the lead, which in this class of engine varies usually from $\frac{1}{32}$ to about $\frac{1}{16}$ inch. As separate exhaust valves are employed, the point of release, and (as the same valve edge that effects the release also effects the compression) therefore that of the compression, may be regulated at will, by adjusting the lengths of the rods F F′, which have at one end a right, and at the other a left hand screw, so that by turning back the checknuts and then revolving the rods, their lengths will be altered.

Similarly the amount of admission lead may be adjusted by an adjustment of the lengths of rods C C′, which also have right and left hand screws.

Referring now to the admission valve v, it is seen that its operating rod C is at a right angle to bell-crank r r, hence the amount of valve motion will not be diminished to any appreciable extent by reason of the wrist plate end of rod C moving in an arc of a circle, and the point of attachment of rod C to the wrist plate is such that, during the admission the valve practically gives as quick an opening as though rod C continued at a right angle to r. But if we turn to valve v', which has closed its port and covers it to the amount of the lap, we find that bell-crank C′ and its operating rod C′ are in such positions with relation to the wrist plate, that the motion of the latter will have but little effect in moving the bell-crank r'. This is an especial feature of the Corliss valve motion, and is of importance for the following reasons:

The lap of the valve (which corresponds to the lap of a plain D slide valve) is usually, in this class of engine, such as to cut off the steam at about $\frac{1}{4}$ stroke, but the adjustment of the cam position is usually so made, that from the action of the governor, the latest point of cut-off will occur when the piston has made $\frac{5}{8}$ of its stroke, the range of cut-off being from this to an admission equal to the amount of the lead.

As the eccentric is fixed upon the shaft, the speed at which the valve opens the port for the admission is the same for all corresponding piston positions. Thus suppose the piston has moved an inch from the end of the stroke, and the valve speed will be the same, whether the cut-off in that stroke is to occur at quarter-stroke, or half-stroke, and as the valve continues to open the port until it is tripped, therefore at the moment it is tripped, the direction of valve motion must be suddenly reversed.

As the duty of its reversal falls upon the dash-pot, it is desirable to make this duty as light as possible, which is accomplished by the wrist motion, which acts to reduce the valve motion after the port is opened a certain amount for the admission.

We have, therefore, that during the earlier part of the admission, the port opening is quick, because of the eccentric throw being a maximum, while during the later part of the port opening, this rapid motion is offset or modified by the wrist motion, thus lessening the duty of the dash-pot and enabling it to promptly close the valve.

VARYING THE ENGINE SPEED.

The range of governor action, so far as the governor itself is concerned, is obviously a constant amount, because a certain amount of rise and fall of the governor balls will move the cams a given amount. But the range of cut-off may be varied as follows: At Z Z′ are adjustment nuts, by means of which the lengths of rods G G′ may be varied.

Lengthening rod G, obviously moves arm d and its cam n further from the end of latch link u, and therefore prolongs the admission period.

Shortening the rod G′ causes cam n' to move around and away from the leg g of the latch link, and prolongs the admission.

The adjustment of the lengths of G and G′, may therefore be employed for two purposes; first, to prolong the point of cut-off, and maintain the speed when the engine is overloaded, or to hasten the point of cut-off for a given engine speed, and thus adjust the engine for a lighter load.

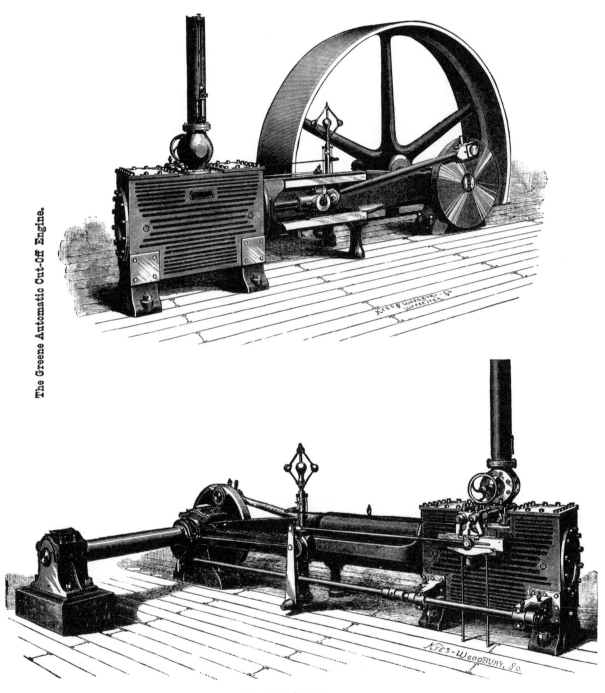

The Greene Automatic Cut-Off Engine.

Figs. 344 & 345.

The Greene Automatic Cut-Off Engine.

In the Greene Engine, of which general views are given in Figs. 344 and 345, there are two admission and two exhaust valves, each pair of valves having its own eccentric, the construction being as follows:

In Fig. 349, J represents the journal-bearing of a rock-shaft, having an arm F connected to the slide-spindle or valve-rod G, and an arm A at whose lower extremity is the toe for the trip motion. At J' is the journal-bearing for a rock-shaft, whose arm F' operates valve-rod G', and whose arm A' has a toe for the valve

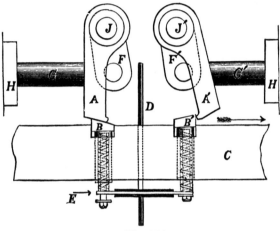

Fig. 346.

tripping mechanism. The eccentric for the admission valves operates the sliding-bar C, in which are the tappets B and B'. These tappets rest upon spiral springs that are seated upon the gauge plate E, which is raised or lowered by the action of the governor upon the rod D.

The operation is as follows: With the parts in the position shown, the sliding-bar C is moving from left to right, as denoted by the arrow above it. The toe B' is operating arm A to open the valve, whose rod or stem is shown at G', and will continue to open it until A', by moving in the arc of a circle, trips or escapes from B', whereupon the valve is closed instantly from two causes, first, by a weight attached to an arm on the rock-shaft J', and secondly, by reason of the steam-chest acting on an unbalanced area equal to the area of the valve stem C', the diameter of this stem being en-

larged to the end of obtaining a steam pressure enough to overcome the friction of the valve stem packing, and also assist the weight to move the valve back quickly, and thus effect a sharp cut-off. The point at which A' will be released ftom B, evidently depends upon the height of B' above the bar C, and this is determined by the plate E and rod D, which are actuated vertically by the governor.

By bevelling the toe ends of arms A and A', and the upper faces of the tappets B' and B, the arms are enabled to pass over the tappets on the return stroke, as shown at A B, it being obvious, that toe A will depress tappet B.

In this construction each tappet and toe will always come into contact, and open the valve for the admission at the same point in the piston-stroke, hence the amount of valve lead is maintained constant.

The exhaust valves are operated by a separate eccentric and shaft, which turns back and forth on its axis and operates the valves by a positive motion, thus maintaining the points of release and of compression constant.

The Harris Corliss Engine

Fig. 347 is a back view of the Harris Corliss Engine, in which glands for the valve stems are dispensed with by the construction shown in Fig. 348, in which A represents the valve, and a, a thrust collar whose diameter is made larger than that of the valve stem, so as to present an area large enough to receive an unbalanced steam pressure at the other end of the valve that will keep the valve seated endways against the thrust-collar and maintain a steam-tight joint without the use of shifting boxes.

The Fishkill Engine.

In Fig. 349 is shown the *Fishkill* Engine, which is of the Corliss type. In the smaller sizes of these engines an ordinary dash-pot is employed, but in the larger sizes a vacuum under the plunger is employed to assist in closing the valve after it is released. The valves are held to their seats by spiral springs, so that they may follow up the wear, and thus prevent leakage.

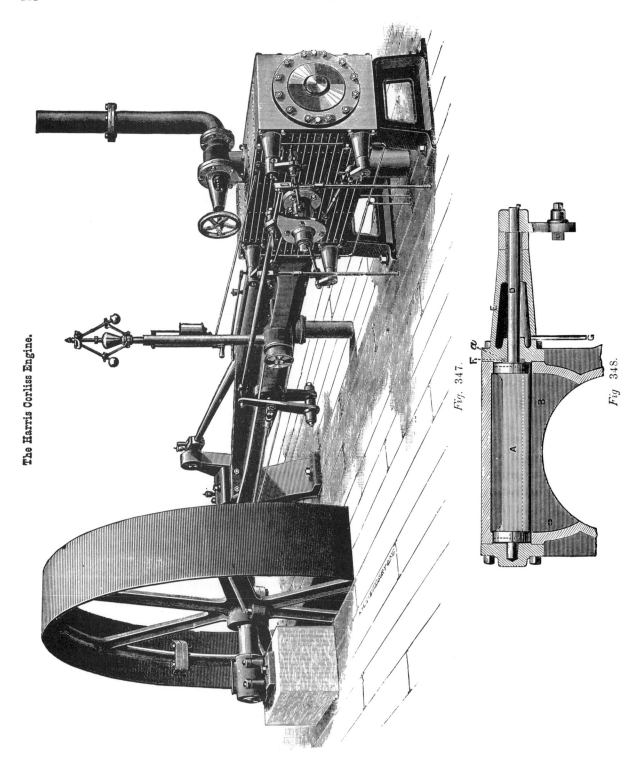

The Harris Corliss Engine.

Fig. 347.

Fig. 348.

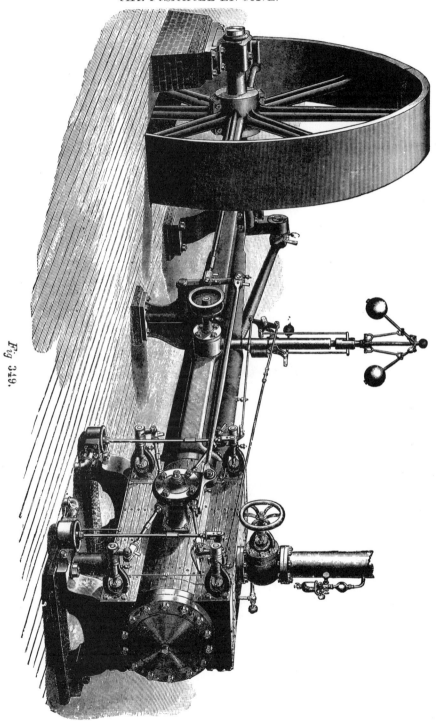

Fig 349.

The Wheelock Automatic Cut-Off Engine.

The general construction of the Wheelock Engine is seen in Fig. 350, which is a back view showing the valve gear, which is more fully shown in Fig. 351. The parts are shown in the positions they occupy at the point of cut-off at half-stroke, the piston moving from

pass through and form guide pins for the square latch blocks which are shown in dotted lines.

The latch blocks are attached respectively to the admission links D and D′, by cylindrical stems seated in the admission links, which are fast upon the stems V V′ of the admission valves. The lower arm of each admission link is pivoted to the top of the weight W of

Fig. 350.

The Wheelock Automatic Cut-Off Engine.

the head-end to the crank-end of the cylinder.

At E and E′ are the stems of the exhaust valves to which are fixed the links F and F′, which are connected together by the rod G.

Hence the rod Z from the eccentric operates both exhaust valves by a direct and positive motion. At h and $h′$ are nuts for adjusting the length of rod G.

The latch links L and L′ are pivoted to the exhaust links L L′, at m and $m′$, as are also the tongues which

the dash-pot, the pivots being shown by dotted circles p $p′$. The rod from the governor attaches at f, to the arm of the trip piece or trip cam e, while the rod R R connects arm $f′$ with f. Upon $f′$ is a pinion engaging with teeth upon the trip cam or trip $e′$, which is free to revolve upon V′. The latch link L′ has, by contact with the trip cam at point r, been lifted, thus throwing the end of the steel strip a (with which the latch link is faced) out of contact with the latch block, hence

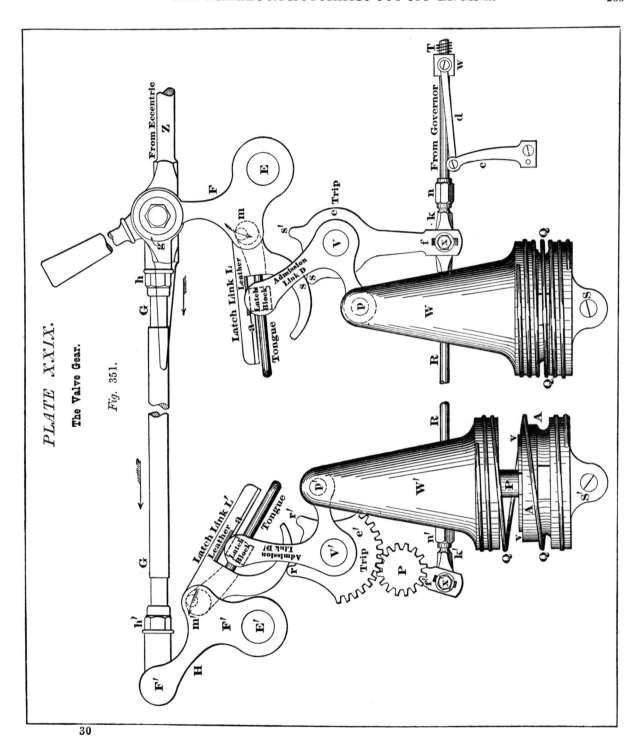

PLATE XXIX.

The Valve Gear.

Fig. 351.

the weight W' of the dash-pot is at liberty to fall and operate the valve stem V', and thus effect the cut-off.

The weight is assisted by a spiral spring Q, hence it closes the valve quickly.

THE CONSTRUCTION OF THE DASH-POT.

The weights W are air-cushioned by the following means: A sectional view taken through the center of the dash-pot on a plane at a right angle to the cylinder, is shown in Fig. 352. The base A is pivoted on the

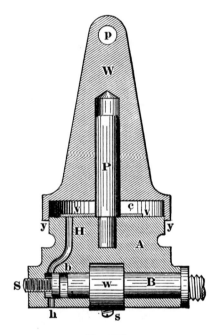

Fig. 352.

supporting stem B, a washer W, secured by a set screw *s*, preventing end motion, while leaving the dash-pot free to vibrate upon B as a pivot. Fixed in the base A is the pin P, which acts as a guide to the weight W.

At Y Y, the base fits into the recess *c*, shown in the bottom of the weight W. Suppose then, that the latch-block of the valve gear has been released, and that the weight W has fallen to the position it occupies in Fig. 351, and has therefore closed the valve and effected the

cut-off, and the recess *c* will be filled with enclosed air, which acts as a cushion because it cannot escape freely past the section Y Y, of the base, and is therefore momentarily compressed, allowing the weight to seat quietly down to the leather valve V V. The function of this valve is as follows: Suppose the weight to be seated down as it is at W, in Fig. 351, and lifting it quickly for the next admission would cause a partial vacuum in the recess, *c* and thus increase the duty of lifting the weight W, and opening the admission valve.

This however, is obviated by the leather valve *v v*, which covers air holes H *h*, this valve lifting when W is raised, and admitting air into recess *c*.

To regulate the quantity of air thus admitted, a screw S is provided, its head fitting the bore *b*, and it is obvious that it may be screwed to the left and caused to close communication between the air holes H *h*, or to the right, leaving a more free communication between them.

By pivoting the latch links upon the eccentric pins *m m'* in Fig. 351, a certain amount of adjustment is obtained. Considering, for example, latch link L in Fig. 351, and revolving its eccentric pivot *m* to the right will act to shorten the latch link with relation to the valve stems E V, and this would cause the latch block to engage earlier in the stroke, and therefore hasten the admission and increase the valve lead. Or turning *m* to the left, would have the opposite effect.

Furthermore, shortening the latch links by means of these eccentric pivots, causes the weights W and W' to lift higher, and this by increasing the tension of the spring Q Q, has the same effect as increasing the weights W W'.

THE COMPRESSION.

The amount of compression is regulated by altering the lengths of the rods Z and G, the former having an adjustment nut (not shown in fig. 351) and the latter having adjustment nuts *h* and *h'*.

EQUALIZING THE POINTS OF CUT-OFF.

The points of cut-off may be equalized as follows: Suppose the adjustment nut *n'* is operated to lengthen rod R, and arm *f'* will be moved to the left. This will move the trip cam, so as to bring its tripping point *r* more nearly vertical, and hasten the point of cut-off, or vice versa, employing nut *n′* to shorten rod R, will move point *r* to the left, and delay the point of cut-off.

Similarly for the crank-end valve, whose stem is shown at V, operating adjustment nut *n* to move K, and therefore *f*, to the right, will cause the latch link to meet the trip cam at *s* quicker, and hasten the cut-off, while employing *n* to move K (and therefore *f*) to the right, will cause contact at *s* to be delayed, and prolong the point of cut-off for valve V.

The two links F and F′ have, from their peculiar shape, a motion corresponding to the wrist motion of the Corliss Engine, opening the valves quickly during admission, and slowly while the lap of valve is passing over the port, thus the position of F is such that its motion transmits a very slight degree of motion to the latch link L, while F′ is in position to move its latch link L′ quickly.

In the later forms of this engine a flat griddle or multiported valve and valve seat is employed, the stems E and E′ moving the valve by a short link that also has the effect of a wrist motion in retarding the valve movement during that part of the stroke when the cut-off is to be effected. The latch blocks seat against a piece of leather, and being narrower than the latch notch have, after passing it, a slight motion along it. Thus while the latch link L is moving downwards, as denoted by its arrow, the latch will remain stationary while the tongue slides through it. On account of the position of the point of suspension *m* of the latch link, F can have considerable motion without altering the position of the latch block upon the leather.

The Twiss Engine.

Fig. 353 represents an Automatic Cut-Off Engine, constructed by N. W. Twiss. The cut-off eccentric-rod is here pivoted to the upper end of a link, that is pivoted at its lower end, and therefore vibrates in an arc of constant length. The link-block is attached to the rod for operating the cut-off valves, and the governor is attached to this same rod. It is obvious, that according as the governor moves the rod, and therefore the link-block down, the travel of the cut-off valve is reduced, and the point of cut-off hastened, until upon the center of the link-block, becoming in line with the center of oscillation of the link, the cut-off valve would remain motionless, and their would be no admission.

The steam chest is pierced at each end directly underneath the bore of the cylinder, for the reception of four circular slide valves, one main valve, and one cut-off valve at each end of the cylinder, thus securing the least possible amount of clearance. Steam is admitted and exhausted by the main valves. The cut-off valves are located inside of the main valves and concentric herewith, they are made double opening by means of a cavity in their centers, thus reducing the amount of travel nearly one-half. These valves have the prsssure of steam upon them and are made to compensate for wear upon their seats. The main valves receive their motion by means of drivers, having hollow stems passing through long bonnets secured to the steam chest; to these stems cranks are keyed. These cranks are connected together by a pitman which receives its motion from an eccentric on the main shaft, the eccentric-rod is made to disengage on the larger sizes by which means the engine may be worked backward or forward at the will of the engineer. The cut-off valves receive their motion by means of stems which pass through the hollow stems of the main valve drivers, at the extremities of which cranks are keyed. These cranks are also connected together by a rod, and receive their motion from another eccentric on the main shaft. The range of cut-off being from zero to five-eights stroke, and the speed of the engine may be changed while the engine is running, by adding or removing weights from the governor.

PISTON   STEAM   EXHAUST

Fig. 353.

CHAPTER VIII.

THE COMPOUND ENGINE.

In a Compound Engine there are two or more cylinders, each partly utilizing the steam. The first, which recieves steam from the boiler, is the high pressure cylinder, in which the steam performs a certain amount of duty before being exhausted into a receiver, or into the steam pipe of the low pressure cylinder, as the case may be. The receiver is a chamber from which the second cylinder receives its supply of steam.

The live steam period, in a compound engine, is confined to the period of admission of the H. P. (or high pressure) cylinder, since after the point of cut-off in the H. P. cylinder, the steam performs work from its expansion only.

The object of compounding is to enable the use of a higher pressure of steam without increasing the pressure of the exhaust, or, in other words, to enable the steam to be used more expansively, and, in some cases, to enable the piston power to be transmitted to the crank with greater uniformity.

Furthermore, by dividing up the expansion between two cylinders, there is less variation in the temperature of the cylinder at the beginning and end of the stroke,

and, therefore, less condensation during the admission period. As the exhaust steam from the H. P. cylinder is that which drives the L. P. piston, it follows that there is, on the exhaust side of the H. P. piston, a back pressure that is theoretically equal to the pressure of steam admitted to the L. P. cylinder. In practice, however, there is found to be a loss of two or three lbs. per square inch, while the steam passes from the H. P. cylinder, and through the reciever to the L. P. cylinder.

There are two principal methods of compounding. In the first the exhaust from the high pressure cylinder passes through a pipe leading direct to the steam chest of the L. P. cylinder. In the second the H. P. exhaust passes into a reciever or chamber, intermediate between the two cylinders.

When a second L. P. cylinder is employed the engine is termed a *triple expansion* engine.

In stationary engines the prevailing method of compounding is to place the low pressure cylinder in line with the high pressure, both pistons being on one rod. This method is also employed upon some of the smaller sizes of marine engines, as those used for yachts. Fig.

239

354 is an example of this arrangement, H. P. is the high pressure cylinder, the exhaust steam passing from the exhaust side E of the piston through the pipe P into the steam chest C of the low pressure cylinder. Direct passage of the exhaust steam from one cylinder to the other without passing into a reciever is only permissible when the two pistons are on one rod, as in the figure, or else when each piston connects to its own

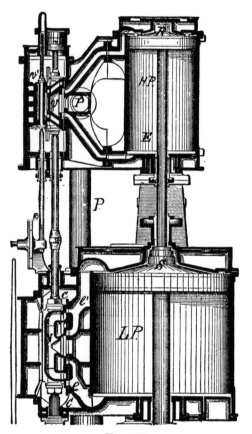

Fig. 354.

crank and the two cranks are opposite to each other so that the pistons will reach opposite ends of the stroke simultaneously, and the low pressure piston will be in position to recieve steam at the same time that the high pressure cylinder begins to exhaust.

Thus in the figure the L. P. piston is in position to recieve the exhaust from the H. P. cylinder, both pistons being at the ends of their strokes. Obviously,

however, if the two cylinders were independently connected to the crank (each having its own piston-rod and connecting-rod) the L. P. piston might be at the other end of the cylinder, and still be in position to recieve the exhaust steam from E, because both pistons would still begin and end their strokes together, the point of release of the H. P. corresponding to the point of admission of the L. P. cylinder.

In the engines of ocean steamships, the H. P. and L. P. cylinders are separately connected to the crank, whose throws are at a right angle one to the other, so that when, as in Fig. 355, the H. P. piston is at the end of its stroke and its exhaust opens, the L. P. piston is near the middle of the cylinder. These relative positions render a reciever necessary, so that a supply of steam may be at hand for the L. P. piston at the commencement of its stroke. When the high and low pressure pistons are upon the same rod as in Fig. 354, the back pressure

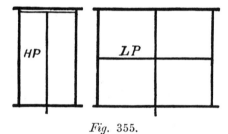

Fig. 355.

on the H. P. piston fluctuates more than it does when they are independent, and the cranks are at a right angle. Thus, suppose the high pressure piston to be at the end of its stroke, as in Fig. 354, and the back pressure will be at its greatest, because the exhaust has just begun. Its reduction will, however, proceed regularly with relation to the piston motion, because both pistons travel in unison throughout every point in the stroke. But suppose the cranks are at a right angle, and when the H. P. piston is at the end of its stroke, as in Fig. 355, and the exhaust opens, we have the following conditions:

First, the exhaust steam from the H. P. cylinder will perform a certain amount of expansion in the receiver, which will reduce its pressure, and secondly, the L. P. piston is in that part of its stroke during which it moves the fastest, and is therefore drawing most rapidly upon the steam in the receiver, while the H, P, piston is in that part of its stroke during which it moves the slowest, hence the relative speeds of the two pistons (as well as the receiver), acts to diminish the pressure during the early part of each low pressure piston stroke, and this obviously acts to equalize the receiver and L. P. cylinder pressure throughout the stroke. The amount of power developed by the engine, however, is not influenced by the fluctuation of back pressure in the high pressure cylinder, or of the pressure during the admission period of the low pressure cylinder, but is determined by the diameter of the high pressure cylinder, the pressure of the live steam, the point of cut-off for the high pressure piston, and the number of times the steam is expanded in the low pressure cylinder, or, in other words, the power developed is determined by the volume and total pressure (pressure above a perfect vacuum) of the live steam and the unbalanced pressure on the low pressure cylinder at the point of low pressure exhaust. In average practice, the diameter of the high pressure cylinder is twice that of the low pressure, and the points of cut-off are at about $\frac{5}{8}$ stroke for the high pressure, and at about $\frac{1}{4}$ stroke for the low pressure cylinder, these proportions being chosen so as to have the power about equally divided between the two pistons.

To find the total range of expansion, (which is not affected by the method of compounding), we divide the capacity of the low pressure cylinder, including the contents of one steam-port and steam passage, by the space moved through by the high pressure piston, up to the point of cut-off including the space in one steam-port and passage, and the quotient is the ratio of expansion, this is clear, because we have merely found how many times the total space occupied by the live steam has been increased. The amount of power developed by the engine may be varied by altering the point of cut-off for either the high or the low pressure cylinder, or for both.

When the cranks are at a right angle, and a receiver is used, it is usual to vary the power, by altering the point of cut-off for the high pressure cylinder only. But when both pistons are on one rod the two valves may be on one rod, and the valves being given equal lap and travel, the points of cut-off correspond in the two cylinders, and will be altered alike and simultaneously

In Fig. 354, as the two pistons are on one rod, and therefore at corresponding points in the stroke, the following arrangement is permitted. The low pressure cylinder has a single valve V, and on the same stem is a main valve *v* for the high pressure cylinder, which is provided with a cut-off valve *v'*, each valve being shown placed in mid-position. As the two valves V *v* are on the same rod, and therefore have equal amounts of travel, their amounts of steam lap must be equal, if both valves are to have equal lead, and the points of cut-off will, if effected by these valves, be at corresponding but fixed points in the stroke. By means of the cut-off valve, however, the point of cut-off of the high pressure cylinder may be varied, thus varying the admission period and the amount of expansion, and therefore the power of the engine. In order to proportion the amount of steam-port opening to the diameters of the cylinders, the high pressure cylinder and valve is single ported, while the low pressure cylinder has a double ported valve and steam-port

Farcot's Compound Engine.

Fig. 356 represents Farcot's Single-Acting Compound Engine, in which one valve serves for both cylinders. The high pressure piston receives steam on its upward stroke only, and the low pressure piston on its downward stroke only, and as there is no steam pressure between the two pistons, therefore there is no back pressure on the high pressure piston.

The live steam pipe enters at the end *n* of the valve, and therefore acts to counter-balance the weight of the valve, etc. The pistons are shown at the end of the downward stroke, the live steam-port being open to the amount of the lead. The low pressure cylinder is open to the exhaust which passes through the port in the valve, and finds exit at C. The valve action may be understood from figs. 357, 358, 359, 360, and 361.

In fig. 357 the valve *v* is shown in the position it would occupy with the crank on the lower dead center, and therefore corresponding to the position of the parts in fig. 356 the valve being supposed to have no lead. A, represents the port for the high pressure, and

b' that for the low pressure cylinder, d is an exhaust port for the low pressure cylinder, while C corresponds to the exit C in the vertical section, fig. 356.

In fig. 358 the valve is shown at the end of its stroke, the high pressure cylinder port a, being full open, and

close the low pressure exhaust at C, this being the edge that effects the compression for the L. P. cylinder.

Continuing the motion, when the edge f of the valve reaches the edge a of the high pressure port, as in fig. 360 the high pressure exhaust, and, therefore, the low

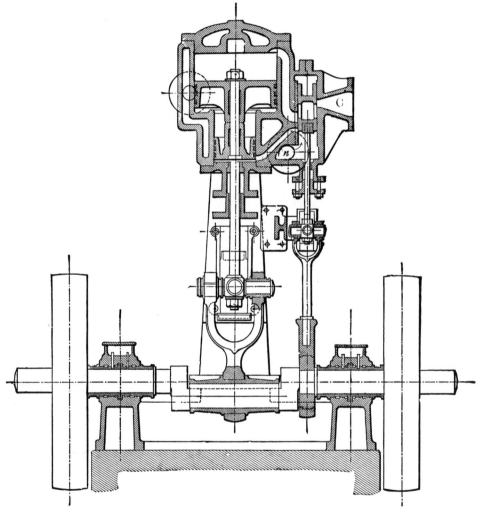

Fig. 356.

the exhaust from the low pressure cylinder passing from b through d to C. Fig. 357 also shows the valve at the point of cut-off for the high pressure cylinder, and it is seen that the admission, the cut-off and the compression is effected by the end e of the valve, and that after the cut-off is effected, the edge g of the valve begins to

pressure, admission begins the exhaust at C, being closed by the edge g of the valve. Fig. 360 shows the valve at the end of its stroke, port a being full open, and the opening at b equal to that at a. From the position in fig. 360 the valve moves to that shown in fig. 357. and we have now followed the motion throughout a full revolution of the engine. It will be seen that the ports *a* and *b* correspond to the ordinary steam and exhaust ports of a common slide-valve engine, and the

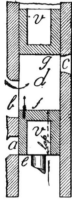
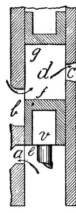
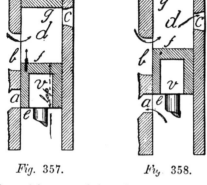

Fig. 357.   *Fig.* 358.

dimension *e f*, fig. 357, of the valve, corresponds to the lip of an ordinary slide-valve, while the port *d* corresponds to the exhaust cavity of a slide-valve, and all these parts may be proportioned by the means already described

31

with reference to diagrams for designing valve motions, all that remains, therefore, is to determine the position of the port C, which must be so located that when the

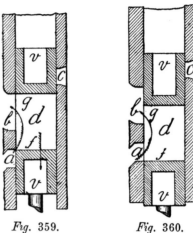

Fig. 359.   Fig. 360.

high pressure exhaust occurs, and the valve is in the position shown in fig. 359 the edge *g* of the valve, must lap over port C sufficiently to maintain a steam tight joint.

CHAPTER IX.

The Condensing Engine.

A steam engine is termed a condensing engine when it is provided with a condensing apparatus, in which the exhaust steam is condensed by water, instead of passing directly into the atmosphere.

The object of employing a condenser, is to form a vacuum on the exhaust side of the piston engine, and thus remove from it the resistance of the atmosphere, amounting to an average of about $14\frac{7}{10}$ pounds per square inch. The effective power of the live steam, and therefore that of the engine, is thus increased to an amount answering to the degree of perfection of the vacuum.

Condensers may be classified under two general heads, viz.; jet and surface condensers, this distinction arising from their construction, and the manner in which the condensing water is used. In the ordinary jet condenser, the water is (by passing through many small openings) divided into sprays, which condense the steam by direct contact, the condensed steam and condensing water being then discharged by an air pump. This air pump also discharges the air which is liberated from the water in the boiler during the process of evaporation (and separated from the steam by the process of condensation), and that which may enter

244

from leaks in the joints of the engine, or in the pipes, or from a leak in the stuffing-box of the engine piston, together with any other uncondensible gases which may be present.

In the Ejector Condenser, the condensing water is introduced through a nozzle in a solid jet, and is ejected against the pressure of the atmosphere by the combined action of the exhaust steam, and a natural or artificial head or pressure of water, thus enabling an air pump to be dispensed with.

The Siphon Condenser also operates without the aid of an air pump, provided that the condensing water can be had at an elevation of not less than 10 feet, the exhaust pipe of the engine being extended to a height of 34 (or more) feet above the hot-well. The condenser is attached to the top of the exhaust pipe, while one or more vertical air-tight discharge pipes connect the condenser to the hot-well below. The air in the condenser and pipes is first expelled by the exhaust steam, and the water injected into the condenser forms a partial vacuum. The condensed steam and condensing water pass down to the hot-well, because the column in the discharge pipe is higher than can be sustained by the pressure of the atmosphere. The air and gases are, more or less, perfectly exhausted by the column of water falling through the discharge pipe, thus maintaining the vacuum. Unless, however, a head of 20 feet can be had for the condensing water, the vacuum must be first formed or started, either by a pump or its equivalent, or by the special arrangement of pipes, which will be described hereafter with reference to the Bulkley Injector condenser.

In a surface condenser, the exhaust steam is condensed by metallic surfaces, in the form of tubes, cylinders, or plates, which are kept cool by a circulation of water. Ordinarily thin brass tubes are employed, the condensing water being circulated through them by a pump, the steam is condensed on the external surface of the tubes, and falls into the condenser, from which it is discharged by the air-pump. Thus the condensing water and the water of condensation are kept entirely separate and the latter may be returned to the boiler, while the condensing water, which may be salt or otherwise impure, passes away. Ocean steamships, and other vessels plying on salt water, are provided with surface condensers, as are also some land engines, when it is not desirable to use the condensing water in the boilers, or where a proper quantity of water for boiler use is scarce. A surface condenser requires therefore, a circulating as well as an air-pump.

The Bulkley Injector Condenser.

Bulkley's Injector Condenser is arranged on the siphon principle, but has, in place of the ordinary condenser, an injector containing an exhaust nozzle, shown at C in Fig. 361. The condensing water enters around this nozzle, and passes down into the condenser through a narrow annular orifice. The exhaust steam entering this film of water, is instantly condensed, and imparts its power to the injection water in a direct line down the discharge pipe. The speed of the water through the contracted neck of the condenser (which it completely fills) enables it to draw out the air and uncondensible vapor, into the enlarged discharge pipe below.

The general arrangement of the condenser and its pipes is shown in Fig. 362. The exhaust pipe of the engine is carried up to a height of about 34 feet above the hot-well, which should be placed at the lowest point convenient for draining it. A feed water heater, if airtight, may be set in the exhaust pipe between the engine and condenser. If the condenser has a natural head of not more than 10 feet (or what is the same thing, a pressure of about 4½ lbs. per square inch), the condenser requires the service of a pump merely to start it, because after the vacuum is once started the atmospheric pressure will cause the condensing water to flow into the condenser, thus dispensing with the employment of either an air or water pump. If the condensing water has a head of 20 feet (corresponding to a pressure of nearly 9 lbs. per square inch), the condenser may be so constructed as to start itself independent of any pump, and an example of this arrangement will be given presently. This obviously saves the power required to operate an air pump. In cases when a less head of water exists and the service of a pump is required continuously, the pump has but little duty to perform, because the vacuum will permit the atmospheric pressure to raise the condensing water about 25 out of the 34 feet. Furthermore, the exhaust steam from the pump may enter the condenser giving the pump piston the benefit of the vacuum, hence the consumption of steam by the pump will be reduced to its lowest limits. The vertical discharge pipe from the condenser passes below the surface of the water in the hot-well, and is therefore sealed by the water in the same, hence the condenser and the discharge pipe form a barometric column about 34 feet high, and the water falls from the condenser to the hot-well, because even a perfect vacuum in the exhaust pipe would be insufficient to sustain a column of water of that height. The construction therefore effects a virtually positive action in extracting the air and uncondensible gases from the condenser, nor can the condensing water pass over into the cylinder of the engine, because the water is directed downwards, and the area of the contracted throat of the condenser is greater than that of the annular inlet opening, afforded by the nozzle C, in Fig. 361.

A relief valve is placed on top of the exhaust pipe, so that the engine may work without the condenser, if the vacuum is lost from any cause. It opens automatically, and is closed air-tight by atmospheric pressure as

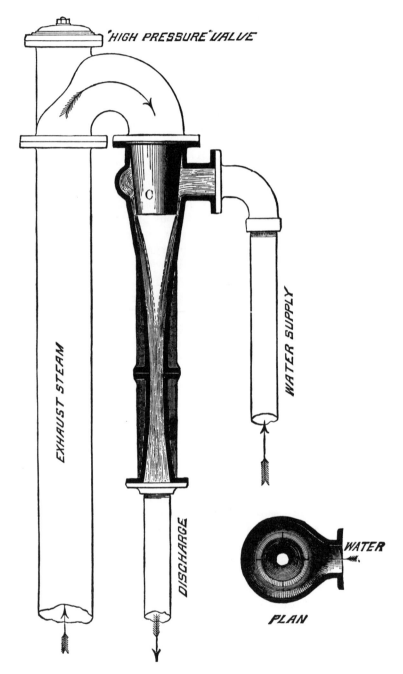

Fig. 361. Bulkley's Siphon Condenser.

PLATE XXX.

Application of Bulkley's Injector Condenser.

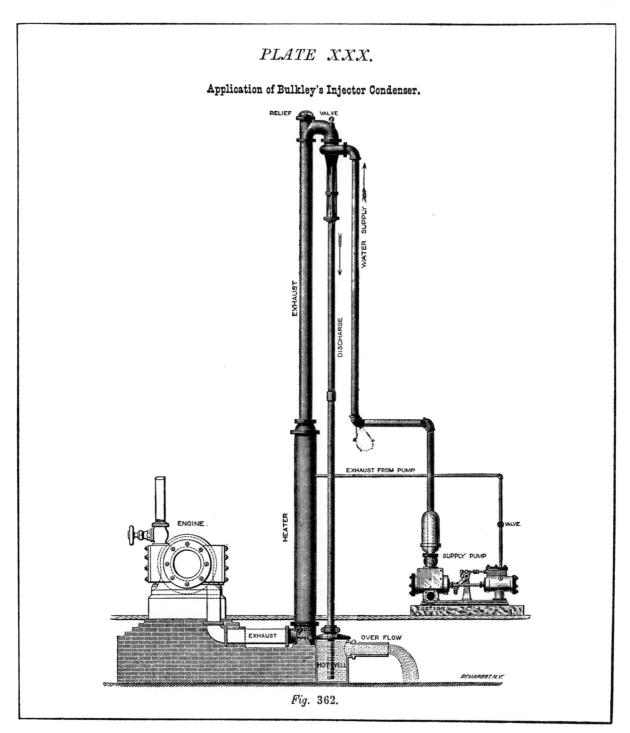

Fig. 362.

soon as the vacuum is formed. As before stated, this condenser will siphon over the water, from a head of 9 or 10 feet after starting, but the vacuum must first be formed by drawing out the air by an ejector, or elevating the water by a pump, to the condenser. With a natural head of 19 or 20 feet, the condenser may be started by the use of a valve connecting the vertical discharge and injection pipes, just below the water level. This is shown in Fig. 363. The natural level of the water, is shown in the flume, or source of supply, at about 20 feet above the level of the water at the over flow of the hot-well. Now, by expelling the air from the engine and pipes by exhaust steam, and opening the connecting valve V, the water will flow over from the flume, and down the vertical discharge pipe, condensing the steam, and thus form a partial vacuum. The water in the supply pipe will then be forced up to the condenser, by the atmospheric pressure, thus perfecting and maintaining the vacuum. The valve V is closed as soon as the vacuum is formed.

There is required for condensation, an amount of water equal to from 20, to 25 times the weight of the exhaust steam, and as the condensing water has a temperature of about 100 degrees in the hot-well, it may be used for feeding the boilers, or for any other purpose that may be required.

Several engines may exhaust into one condenser, and a single pump can be used to supply several condensers. The injector condenser occupies very little room, and may be attached to the engine, in the most convenient situation. By dispensing with the air pump used with ordinary jet condensers, a saving in power is obviously effected.

A stoppage of the water supply merely causes the exhaust steam to open the relief valve on top, and thus pass away, but cannot endanger the safety of the engine, by allowing water to enter the cylinder of the same.

KNOWLES' INDEPENDENT JET CONDENSER.

Fig. 364 represents a Knowles Independent Jet Condenser attached to a Corliss Engine. The exhaust passes through the pipe x into the heater H, and thence into the condenser C which is upon the top of the air pump, which discharges the products of condensation through the pipe D.

The air pump is driven by a steam cylinder, whose action is described with reference to the Knowles steam pump. The exhaust from the steam cylinder of the condenser passes though the pipe x x into the condenser itself, hence this cylinder also receives the benefit of the vacuum. The injection water is supplied through the pipe I near the top of the condenser, the gate valve for opening and closing the same, being shown at G.

The feed water for the boiler is drawn through the suction pipe S, and is forced through the pipe F to the heater H where it passes through a coil of pipe that is surrounded by the exhaust steam, that is on its way from the main cylinder to the condenser, and that enters at one end of the heater and passes out at the other into the condenser. The feed water thus extracts heat from the exhaust steam, and thus helps the condenser. From the heater the feed water passes through the pipe F F to the boiler; at T is the valve for admitting steam to the condenser steam cylinder, hence all the valves necessary to operate the condenser are contiguous to the engine. Pipe V is for permitting the air and uncondensed gases exit from the air pump being carried up beyond the level of the pump discharge water.

Fig. 365 represents a Knowles independent jet condenser with safety-valve and float, the construction being as follows: The exhaust steam enters at I and the injection at F. There are two perforaied discs or water dividing plates, one of which is above the steam inlet I, and the other below it, so that the exhaust steam meets at once a number of sprays of water. The lower plate has at its center, an opening bounded by a vertical flange, as shown in the figure, and steam passing through this opening meets the sprays that fall from the lower perforated disc. The water, air, and gases are drawn off by the air pump from the bottom of the condenser. At Z is a valve held to its seat by the globe shown suspended, which is hollow and light.

Now suppose that from some unusual and unexpected cause, the condenser should begin to fill with water, and the float would lift the valve Z and destroy the vacuum, thus stopping the inflow of injection water, while permitting the exhaust steam to pass out at L.

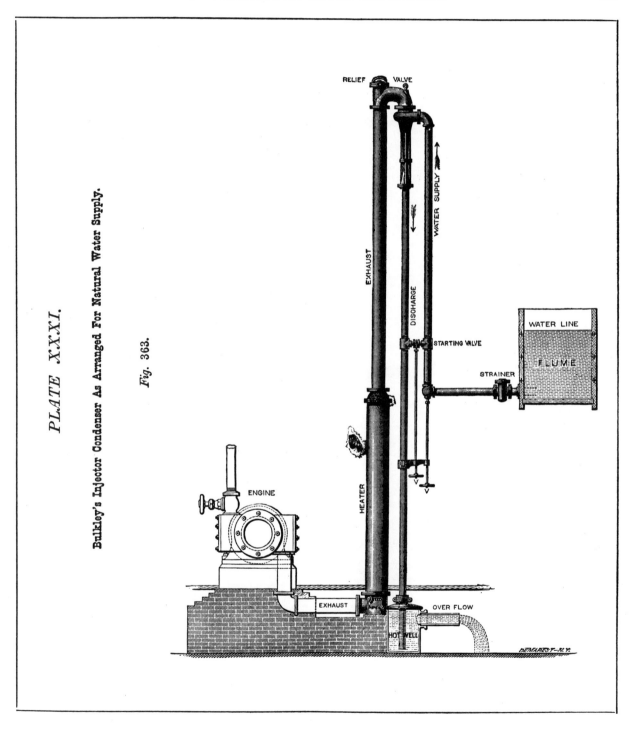

PLATE XXXI.

Bulkley's Injector Condenser As Arranged For Natural Water Supply.

Fig. 363.

Fig. 364.

Application Of The Knowles Condenser.

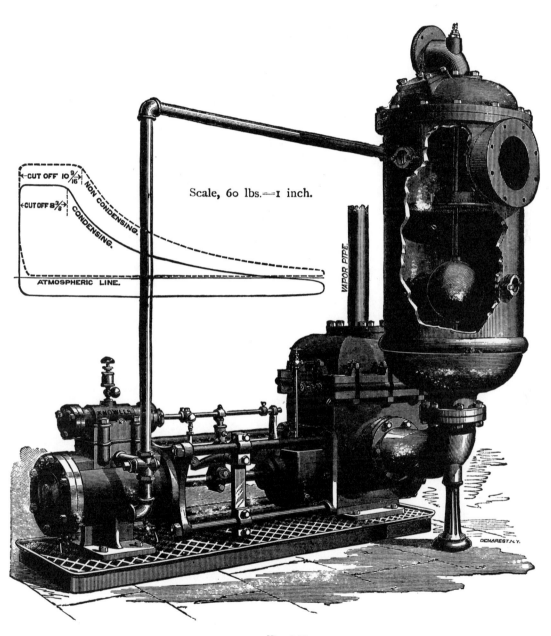

Fig. 365.

Knowles Condenser With Safety Valve.

32

A general view of Wheeler's Independent Surface Condenser is given in fig. 366, and a sectional view of the same in fig. 367.

The condenser is mounted horizontally at one end on

The circulating pump forces the injection water through the tubes, the construction being as follows:

The pistons are supposed to be moving in the stroke from left to right, hence the air pump is drawing out the condensed water and air through B, and through the left hand suction valve S. On the other side the air pump piston is forcing the discharge through the right hand delivery valve V. Similarly, the circulating

Fig. 366.

Wheeler's Independent Surface Condenser.

the circulating pump, and at the other on the air pump. Both pump pistons are on the same rod, as is also the piston for the steam cylinder that operates both pumps.

The exhaust steam enters at A and is distributed along the condenser by the perforated plate O, beneath which the tubes are arranged.

pump is recieving the injection water through its left hand suction valve S, and forcing it through the right hand valve V into a chamber F, which is separated from the chamber H by a partition E, thus giving two separate systems of tubes through which the injection water passes.

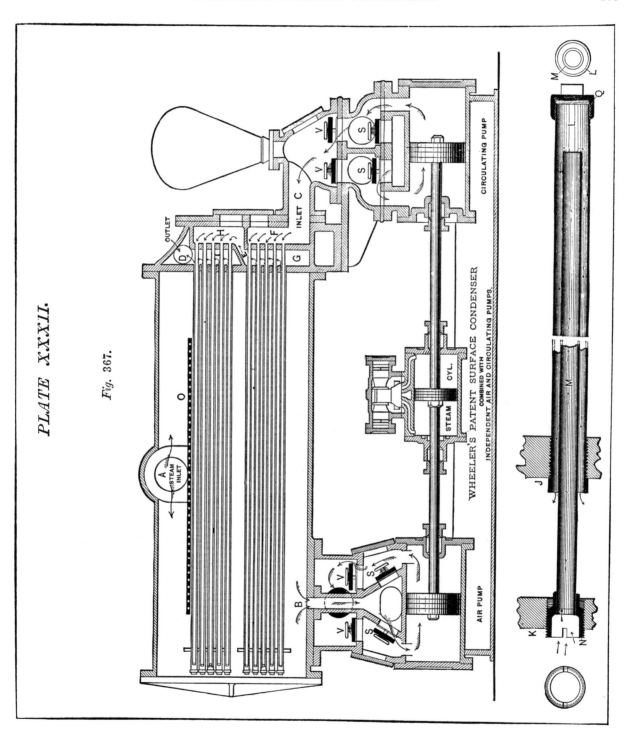

PLATE XXXII.

Fig. 367.

WHEELER'S PATENT SURFACE CONDENSER
COMBINED WITH
INDEPENDENT AIR AND CIRCULATING PUMPS.

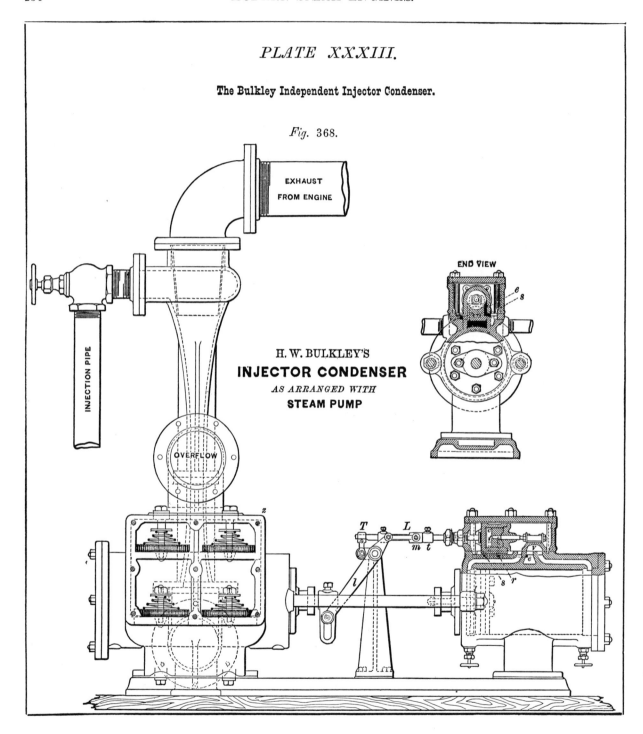

PLATE XXXIII.

The Bulkley Independent Injector Condenser.

Fig. 368.

EXHAUST FROM ENGINE

END VIEW

INJECTION PIPE

H. W. BULKLEY'S
INJECTOR CONDENSER
AS ARRANGED WITH
STEAM PUMP

OVERFLOW

The construction whereby the injection water is circulated from end to end of the tubes, is more fully shown at the bottom of Fig. 397, where a pair of tubes are shown removed from the condenser and turned end for end, and it is seen that there are two tubes, one within the other.

The plate K corresponds to the inner wall of chamber F, in the sectional view, while plate J is the inner wall of plate G, in the sectional view. The injection water enters at N and passes through the inner tube M into the outer tube L, and takes the course shown by the arrows, emerging from the end J, and passing through the opening at E in the sectional view, from the lower to the upper system of tubes, which correspond to those in the lower system, and finally passing out at D. At Q is a cap that may be removed when the tubes are to be cleaned.

The tubes are rigidly held at one end by the plates J K, Fig. 367, while at the other they are supported by plates, shown at T T in the sectional view. This permits them to expand and contract freely, and obviates the necessity of employing the paper ferrules that are necessary when the tubes are rigidly held at both ends, in order to prevent the tubes from leaking, and from creeping through the plates.

Bulkley's Independent Injector Condenser.

Fig. 368 represents Bulkley's Injector Condenser, as arranged with air and injection pumps. The constructtion of the condenser corresponds to that shown in Fig. 355, while that of the air and steam pumps is as follows: The steam pump is shown in section, and the air pump with the valve chest cover removed, to expose the pump valves. Both pistons are on one rod, which vibrate the lever *l* and through the medium of the rod connection at *m*, the rod L which is supported at T and has tappets *t*.

The valve V has a flat face, and therefore follows up its wear, and operates over three ports in the usual manner. To the valve spindle is attached a small piston C, operating in a steam chest cylinder attached to the end

of the rod L, at *r* and *s* are two small ports, which connect with the exhaust port *e*, and in the steam chest cylinder are similar ports. In the face of the seat upon which the chest cylinder slides, are cut two recesses, through which the ports in the chest cylinder take steam, as they pass over them. Thus in the end view, it is seen, that two ports denoted by *e s* are in communication. Now suppose the next piston stroke to proceed, and through lever *l* and rod L the chest cylinder will move to the left until it meets the tappet, whereupon it will move the chest cylinder to the right. The latter will carry with it the chest piston, both moving together, until such time as the port in the chest cylinder comes opposite to port *r*, whereupon steam will pass through *r*, and between the chest piston and cylinder, and the chest piston will move suddenly to the left hand end of its cylinder, opening the port at one end of the main cylinder for the exhaust, and the port at the other end for the admission, and thus reversing the motion.

The Reynolds Condenser.

In the Figures from 369 to 371 is shown the condensing apparatus, employed by E. P. Allis & Co., in connection with the Reynolds' Corliss engine.

The exhaust from the engine condenser passes first into a feed water and purifier, and thence to the jet condenser.

THE HEATER.

The construction of the feed water heater is shown in Fig. 369, and that of the condenser in Figs. 370 and 371. In the heater are two tube-plates *p p* near the top, and *p p* near the bottom.

The water from the feed pump enters the pipe A, passes up it into pipe B, and descends through the annular space between the two pipes, into the space below the lower tube-plate *p p*, whence it passes up the tubes to the top of the heater, where it enters the feed pipe that passes to the boiler. In the upper chamber of the heater is the scum pan, in which the scum of

PLATE XXXIV.——*Fig.* 369.

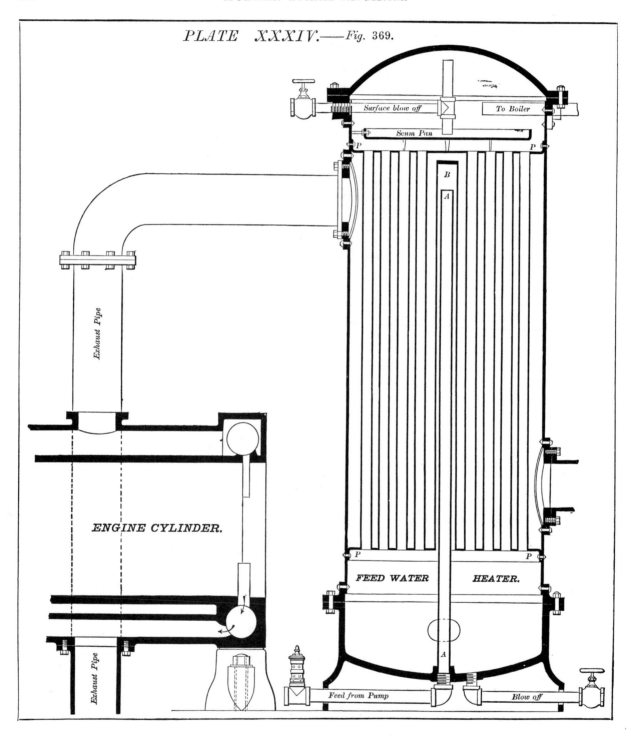

Surface blow off

To Boiler

Scum Pan

P   P

B

A

ENGINE CYLINDER.

Exhaust Pipe

Exhaust Pipe

P   P

FEED WATER   HEATER.

A

Feed from Pump   Blow off

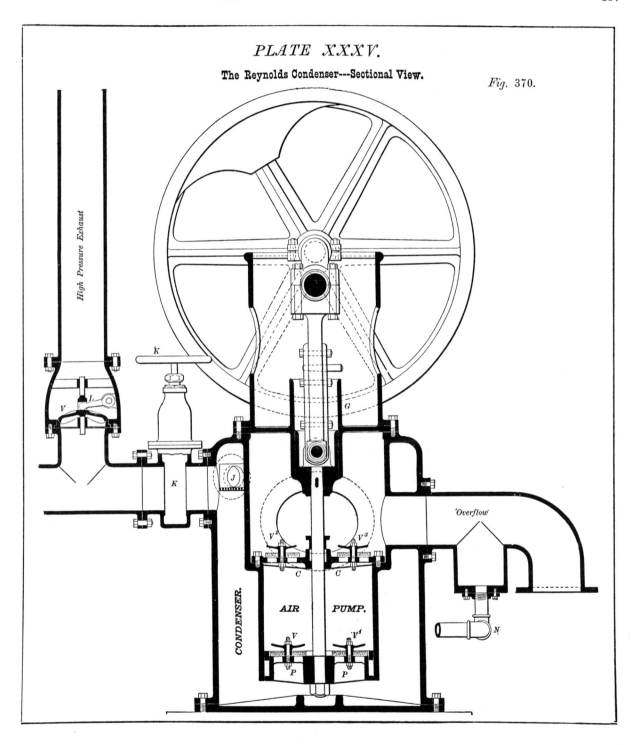

PLATE XXXV.

The Reynolds Condenser---Sectional View.

Fig. 370.

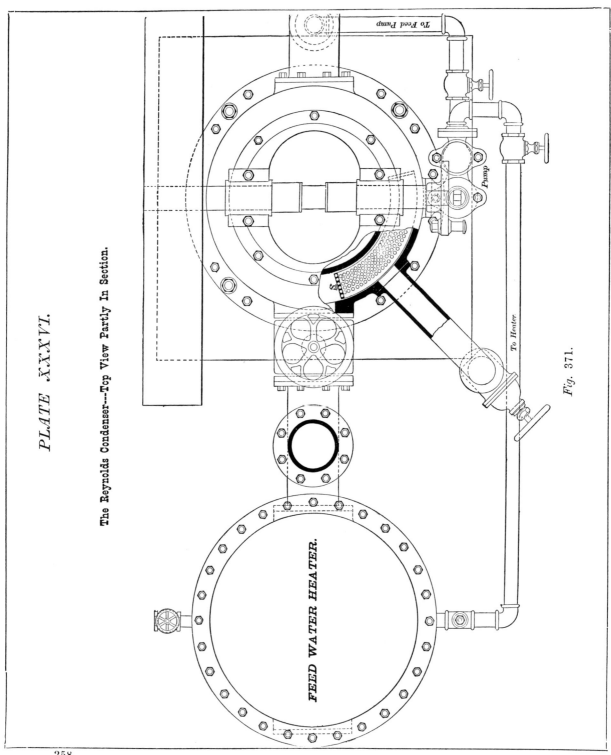

PLATE XXXVI.

The Reynolds Condenser---Top View Partly In Section.

FEED WATER HEATER.

To Feed Pump

Pump

To Heater

Fig. 371.

258

the water accumulates, and from which it is occasionally blown off through the pipe shown, the lower end of which enters the scum pan.

The settlings from the feed water may also be blown off through the blow-off pipe, shown at the bottom of the heater.

The exhaust steam from the engine cylinder enters on one side beneath the upper tube-plate p, and leaves through the pipe T above the lower tube-plate, passing thence into the condenser.

THE CONSTRUCTION OF THE CONDENSER.

A sectional view of the condenser is shown in Fig. 370, and a plan partly in section in Fig. 371.

Referring first to the sectional view, Fig. 370, the pipe T corresponds to pipe T in Fig. 369, in which is a valve K for stopping off communication with the condenser, if it should be required to use the engine for non-condensing; F is a relief valve that when the condenser is in operation is held closed by the vacuum, but that may open by means of the lever L, to permit the exhaust from the engine to pass up the high pressure exhaust pipe when K is closed, or in case, from some cause or other, the vacuum should be lost, or the condenser flooded with water.

33

The spray plate (shown in section, at S in Fig. 371) extends partly around the condenser, the injection water entering above it at J. This plate therefore divides the injection water into fine streams, which fall down the condenser and condense the steam. The water of condensation falls to the bottom of the condenser, and the air and gases remain above it until removed by the air pump. The capacity of the air pump is sufficiently great to discharge the injection water, the water of the condensed steam, and also the air and gases, as fast as they are separated from the steam, about $\frac{1}{5}$ of the pump capacity being filled with water at each upward or drawing stroke.

The suction valves V V' are in the piston, or bucket of the air pump, and the discharge valves V^2 V^3 in the cover at the top of the pump. These valves consist of flat rubber discs upon a central stem, having at its top a dish shaped shield, which limits the amount to which the valve can open. A spiral spring at the back of the valve closes it to its seat. The same form of condenser is used whether it be driven by belt connection to the pulley wheel, or whether the same be driven by a steam cylinder attached to the side of the condenser.

The weight of the crank, etc., is counter-balanced by a weight on the pulley wheel.

CHAPTER X.

THE COMPOUND CONDENSING ENGINE.

Figs. 372, 373, and 374 represent the Worthington Compound Condensing Engine, for the water-works of towns and cities. It consists of a pair of engines and pumps placed side by side, but forming virtually one engine because the valves for one engine are operated from the piston-rod of the other, while both pumps discharge through a common delivery pipe.

The distinguishing features of this engine are, first, that the water is given a continuous and as nearly as possible, a uniform flow through the suction and delivery pipes, its path of motion being kept as nearly straight as possible, and secondly, that the valves are permitted to close without the violence that is found to prove destructive in some kinds of pumps.

This is accomplished by the peculiar construction and arrangement of the steam valve gear, and by the arrangement of the pump valves. The valve gear is such as to cause the pistons to pause at the end of each stroke, thus permitting the pump valves to seat themselves quietly, and therefore without inducing reverse currents in the water, or checking its continuous flow through the main.

The construction of the steam end of the engine is shown in section in Fig. 373, H. P. is the high pressure piston whose rod attaches direct to the pump plunger,

260

L. P. is the low pressure piston, which has two rods R connecting to the cross-head C', at R.

The stretcher rods or tie rods, S R, tie the steam and water cylinders together and keep them in line. At G is the guide-bar for cross-head C', which (by means of a short rod) drives the bell crank B', which operates the pair of air pumps that are shown in section. Cross-head C drives bell crank B, which works the air pump for the other engine. From bell crank B', a rod b' actuates a rock-shaft, which operates the valve motion for the back pair of cylinders. The valve motion for the H. P. and L. P. cylinders of the engine that is shown in section may be traced as follows: The bell crank B actuates the rod b which operates the rock-shaft r, which connects at W to the rod for the two steam valves. These two valves are balanced by a pendulum p, pivoted at its upper end, and having at its lower end a piston fitting a bore in the back of the valve. Beneath each balancing piston there is an opening leading into the exhaust, hence the valves must be balanced, notwithstanding any wear that might in time occur in the valve balancing pistons.

In each cylinder the end ports as S S, are the steam ports, the inner ports e e and e' e' being for the exhaust. The exhaust from the high pressure cylinder

Fig. 372.

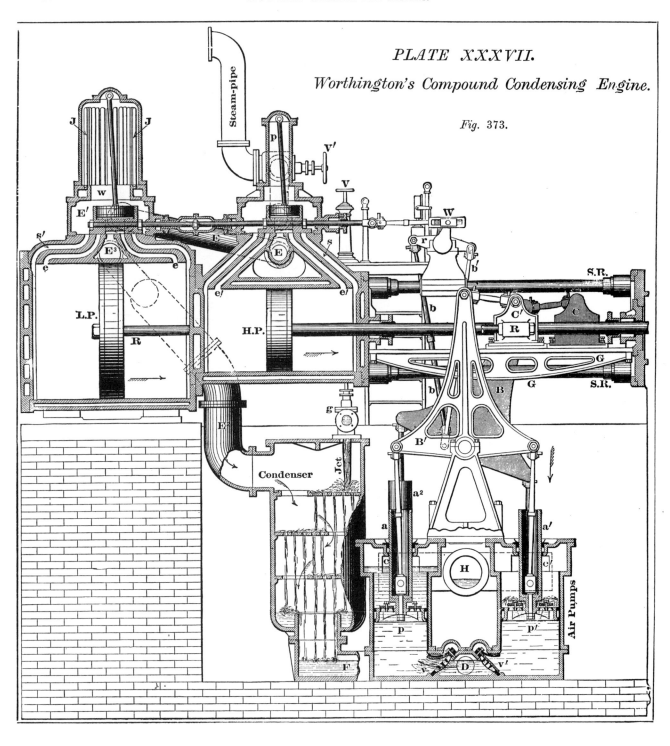

PLATE XXXVII.

Worthington's Compound Condensing Engine.

Fig. 373.

passes through pipe E to the steam chest E′ of the low pressure cylinder, where it is superheated by the pipes at J, which contain live steam at or near boiler pressure. The exhaust from the low pressure cylinder passes out at E³, and through the pipe E² to the condenser, where it meets the injection water, the jet being divided up by means of perforated plates, which separate it into small streams, as shown. From the bottom F of the condenser, the water, air and gas pass to the air pumps at D. At v and v′ are the foot valves, e being shown open to admit ingress to air pump p which is ascending. The air pump piston p′ is descending, hence its foot valve v′ is closed, while the valves in p′ are open, the air, gas and water passing from the lower to the upper side of the piston. The air pump discharge occurs through the openings at c and c′. The trunk piston-rods render guides for the air pump pistons unnecessary, and enable the use of long rods from the bars B B′, thus avoiding side strains.

Having described the general construction of the engine, we may now pass to the construction whereby the pistons are caused to pause at the ends of their strokes.

The steam valves have no lap, hence the steam follows full stroke, while the exhaust passages being closed as the pistons pass over them, a certain amount of compression is obtained. As both valves are alike, and are operated by the same rod, we may follow the motion with reference to the L. P. cylinder, only the description answering for both. The cross-head C, on reversing its motion, operates (through the medium of B and b) the rocker r, which connects at W to the valve rod. There is, however at W, a certain amount of lost motion, so that the motion of r may continue for a certain period without operating the valve, and during this period, steam port S′ remains fully opened.

When the lost motion or play at W is taken up, the valve will move to the left, and by the time the piston has nearly completed its stroke, the left hand steam port s′ will be closed, and its adjacent exhaust port e, opened.

Simultaneously, at the other end of the cylinder, the steam port will be opened, and its exhaust port e closed.

On account of the position of the attachment of the rod b on bell crank B, and of the small amount of valve travel, the port opening is so regulated, that it takes a certain amount of time, after the piston has stopped, before there is steam pressure enough to start the piston on its return stroke, and this gives the pause, or period of rest, before referred to.

We have here considered the action of one engine only, and it will be readily perceived, that as cross-head C′ is at half-stroke when C is at the end of its stroke, therefore one engine takes up the motion before the other pauses, this action following in regular rotation, so that the water is drawn up the suction pipe, and forced through the delivery pipe, with a constant and unchecked flow.

The duration of the piston pause at the end of the stroke, may obviously be regulated, by adjusting the amount of lost motion at W, because the greater the amount of lost motion, the later the port is opened, and the longer the pause. In addition to this, however, valves are employed, by means of which the steam and exhaust ports, at each end of the cylinder, may be placed into communication, so that after the piston has closed the exhaust port, the compressed steam may pass from the steam port into the exhaust, to relieve the compression if necessary, and thus permit the piston to travel the full length of stroke.

The construction of the pump is shown in Fig. 374. The length of the water cylinder is divided into two divisions, A and A′, the plunger working through a bushing, as shown at r. The suction chamber extends beneath the full length of the pump. The suction valves extend along the bottom, and the delivery valves along the top of the pump cylinder. The plunger is supposed to be moving from left to right, hence division A is receiving water, and division A′ delivering it, as denoted by the arrows. The water, it will be observed, can pass through the pump in a straight line, except in so far as it is deviated therefrom, by passing around one side of the plunger circumference. The valves are flat rubber discs guided by a central stem, and seated by means of a spiral spring at the back of each valve.

Hand holes are provided, to afford access to the pump valves. The delivery pipe from each pump connects to a central pipe, on which the air chamber is situated, so that the one air chamber serves for both pumps.

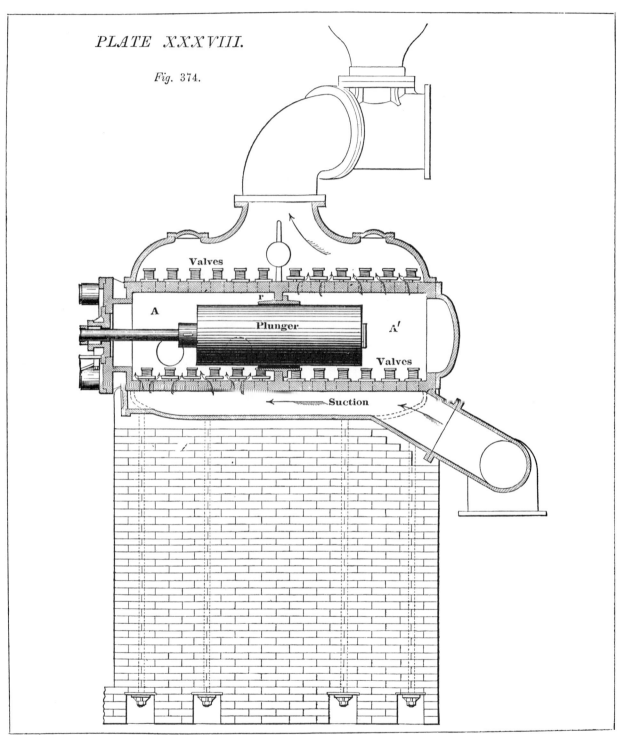

PLATE XXXVIII.

Fig. 374.

Valves

A

Plunger

A'

Valves

Suction

Vertical Compound Condensing Engine.

Fig. 375, (which is extracted from the *Engineer*) represents a compound condensing engine with surface

In this engine a double acting bucket pump serves for both the circulating and the air pump, the construction being as follows: The condenser C is in the base of one frame, the pump being worked from the engine cross-head, by means of beams A, and rods D.

The pump is divided into two compartments, each

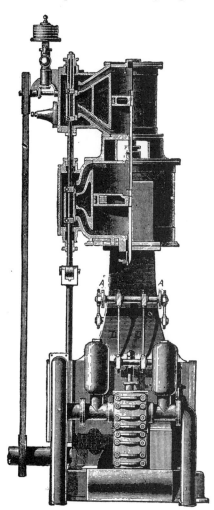
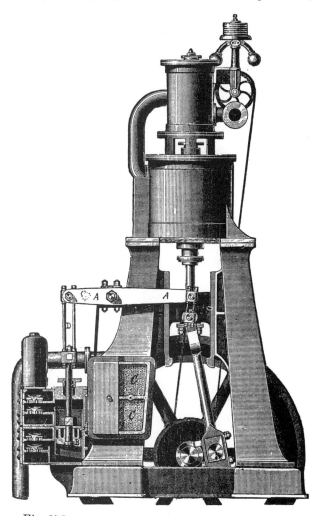

Fig. 375.

condenser, constructed by Worth, Mackensie & Co.

The two cylinders are in line with their pistons on one rod, and are separated so that the upper gland of the L. P. and the lower one of the H. P. cylinder, may be accessible for adjustment.

having its own pair of valves, of which the lower of each pair is the suction, and the upper the delivery valves, the lower pair are for circulating, and the upper for the air pump, hence each of them is single acting, while the pump bucket B is double acting, effecting on each stroke the suction of one, and the delivery of the other pump.

VERTICAL COMPOUND CONDENSING PUMPING ENGINE.

Figs. 376 and 376^a represent a vertical compound pumping engine, built by E. P. Allis and Co., for the Allegheny City water-works.

These are two engines side by side, each having a single pressure cylinder, flanked on two sides by low pressure cylinders.

The axes of the three cylinders of each engine are in line, and each operates a single acting solid plunger pump placed directly beneath it. The H. P. cylinders are jacketted, and exhaust into a receiver, from which both the low pressure cylinders recieve their steam.

The crank-shaft is over-head, the three throws being 120 degrees apart, so as to be disposed equi-distant around the shaft, and each shaft has two fly-wheels.

The jet condensers are placed between the two engines, and are worked by beams driven from the steam cylinder cross-heads. The valve gear of all the steam cylinders is of the Corliss type, that for the H. P. cylinder being controlled by the governor, while that for the L. P. cylinders is adjustable for the point of cut-off, by setting the trip pieces or cams by hand.

The plungers of the main pumps are so weighted, that together with the reciprocating parts that operate them, about one-half of the water column is counterbalanced.

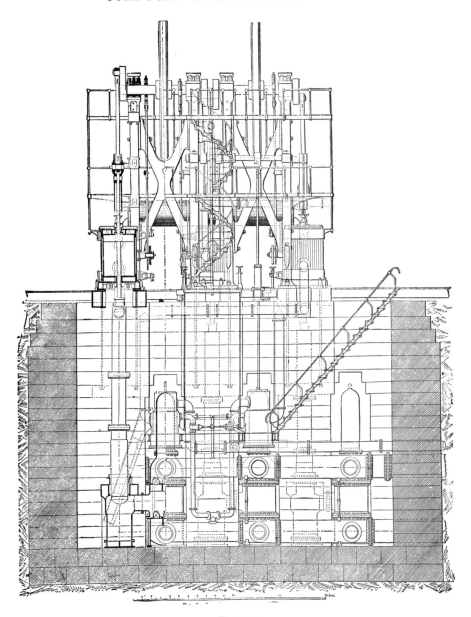

Fig. 376.

34

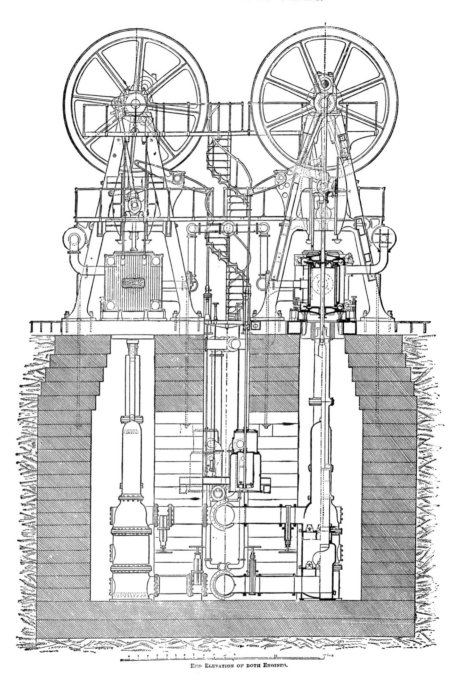

End Elevation of both Engines.

Fig. 376a.

CHAPTER XI.

THE MARINE ENGINE.

The term *Marine Engine* is applied in a general sense to the engines of vessels for service upon salt water, but by a more strict definition, it applies to vessels that make voyages by sea.

An example of a marine engine, as made in the smaller sizes, is given in Fig. 377, in which H. P. is the high, and L. P. the low pressure piston, and S the receiver, the cranks C' C' being at a right angle one to the other.

The exhaust from the H. P. P. steam passages e, passes direct into the receiver S, and is distributed to the L. P. piston by the valve V'.

The H. P. cylinder is provided with a double ported main valve V and a Meyer's cut-off valve v, and the L. P. cylinder, with a double ported single valve V'.

The valves V' for the high, and v for the low pressure cylinder, are provided with pressure relieving plates on their backs. The two main valves V and V' respectively, are driven by a link motion, which is used for reversing purposes only, the point of cut-off being varied by the cut-off valve of the H. P. cylinder, which is driven by an eccentric D set opposite to the crank C. This eccentric operates at G, an arm, which is pivoted at one end.

The cut-off valves are moved upon the rod to vary the point of cut-off by a right and left hand screw, as shown, the rod having journal bearing in a sleeve S, so that it may be revolved in S by a hand wheel operating a bevil pinion that engages with the bevil pinion at J.

The link motions for the main valves V V' are moved for reversing, by arms from the shaft H, which has an arm that is operated by the small steam engine shown at K. This small reversing engine is also provided with a link motion, which is reversed by a hand lever, whose rod is shown at r.

The weight of the valves and their connections are counter-balanced, by means of small pistons at P and P', which receive the pressure of the steam on their lower faces, the cylinders or cases in which they work, being open at the top.

At R R', and R'' are relief or snifter valves, which to open permit the escape of any water, with which the respective cylinders may become charged. These valves are usually so adjusted by either a weight and lever, or a spiral spring, that they only open under a pressure greater than the highest steam pressure the cylinders receive.

The surface condenser M, is shown at the back of the engine, its pumps being operated by beams worked from the rods N, which receive motion from the cross-

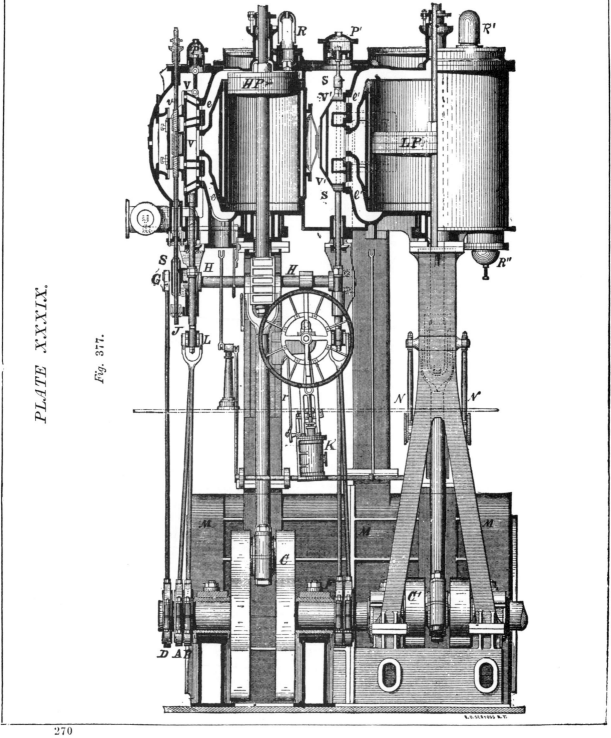

PLATE XXXIX.

Fig. 377.

Fig. 378.

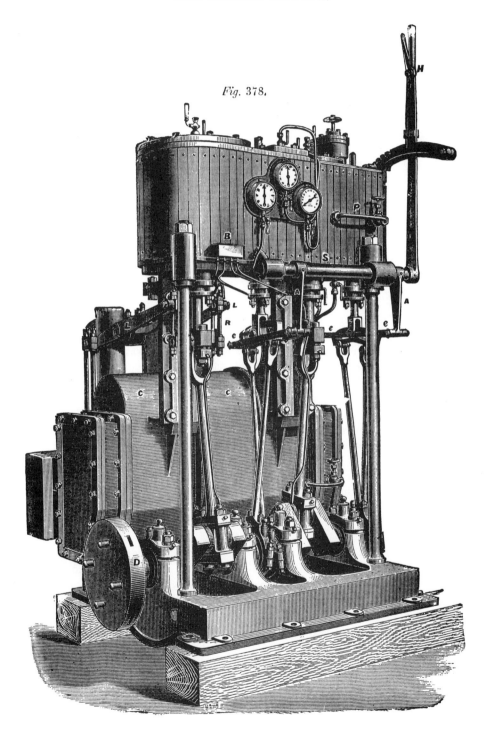

head of the low pressure piston as shown in the figure.

Fig. 378 represents a small Marine Engine, in which the power required to operate the link motion is small enough to permit its being shifted by a hand lever H, which operates a shaft S, having arms A A that shift both links simultaneously through the medium of the straight links *e*. The condenser is shown at C, the air and circulating pumps being operated by the levers H which are worked by short rods R from the low pressure cross-head.

At P is a by-pass, pass-over, or starting valve, that is opened to admit live steam into the receiver, and thus use the low pressure cylinder as a high pressure one, until the engine is started.

The starting valve enables the engine to be started when the high pressure piston is at the end of its stroke, or when the high pressure piston alone would not be powerful enough. Moreover, when the air pump is driven by the engine, and not independently, the vacuum becomes lost after the engine stands still.

These conditions render a pass-over or starting valve necessary. The starting valve is, in this example, placed outside of the cylinder, and passes the live steam from the H. P. steam chest into the receiver, but in some cases there is also a pass-over valve in the passages of the H. P. cylinder (as well as into the L. P. cylinder), so as to admit steam to the same when the engine stops in such a position, that the H. P. valves have effected the cut-off.

Obviously, however, under this latter condition alone the engine might (with a link motion easily operated by hand) be started, by operating the link motion at first to move the engine in the wrong direction, and enough to open the steam-port at the other end of the cylinder, and by again operating the link motion, start the engine in the required direction, but it is necessary to provide means by which the engine may be started quickly, and in the required direction, and these ends are accomplished by means of the pass-over or starting valve. If the vacuum is kept, and the engine is in such a position, that the live steam can be admitted, the starting valve need not, in some cases, be resorted to.

At B is an oil box from which pipes lead to the various working parts of the engine. The slots D are for the insertion of a bar to move the engine by hand, for set-

ting the valves, or for other purposes. The high and low pressure cylinders are provided with waste water cocks operated separately by hand, whereas in larger engines all these cocks are connected to one lever, whereby they may be opened or closed simultaneously.

Fig. 379 represents a Marine Engine, such as is used for coasting vessels. Each cylinder has its link motion which are shifted by a common tumbling shaft, which is operated by the hand wheel shown, The link motions are employed to vary the points of cut-off, as well as for reversing, flat slide-valves being used.

The right hand handle A is for the throttle valve V of the high pressure cylinder, while the left hand handle B operates the rod shown connected to the shaft T, an arm from which works the starting, pass-over or by-pass valve at D.

The middle handle C operates the shaft S, that works the waste water cocks for the two cylinders. The link motions are shifted simultaneously as follows: The wheel W operates, by a worm and worm-wheel connection. the lever L, which has at its upper end rods passing to the back of the engine, where they operate a shaft S, upon which are arms M connected to links or arms N, which shift the links.

Two beams, one of which is seen at A, operate the air and circulating pumps which are placed at the back of the condenser. The worm and worm-wheel, shown at W, are employed to move the engine by hand when there is no steam, as when the vessel is in port, the worm obviously being thrown out of gear with the wheel when not in use.

Fig. 380 represents, partly in section, a Compound Condensing Marine Engine, for an ocean going steamship. The main valve V for the H. P. cylinder is single ported, and is operated by a link motion F that is used for reversing only.

A Meyer's cut-off is used, the two cut-off valves being adjusted to vary the point of cut-off, by means of a right and left hand screw-thread. To permit of the operation of this thread, the upper end of the valve rod is provided with a hand wheel W, while its lower end has journal bearing in the foot piece R. The rod Z for operating the cut-off valve is driven by a pin in a crank disc D.

The low pressure valve V' is double ported, and is

Fig. 379.

driven by a link motion G that is used for reversing purposes only. Both link motions are shifted by a small steam cylinder, whose end is shown at X. The piston-rod of this small cylinder connects at its outer end with a screw of coarse pitch, so that when steam is admitted to the cylinder, and the hand wheel T is revolved, the end pressure on the screw will (from the coarse pitch of the latter) cause the wheel T to continue in motion. To the coarse screw is fitted a nut upon an arm, that connects to the tumbling shaft that shifts both link motions at once. When the link motion has been shifted, steam is shut off from the small cylinder. The speed with which the shifting is effected and the parts come to rest, is regulated by hand pressure on the wheel T.

Obviously the connection between the coarse screw and the piston-rod is such as to permit the screw to revolve without carrying the piston-rod around with it. In other forms of steam reversing gears, cataract cylinders are used, the general principles of construction being such as were explained with reference to figs. 142 and 144.

The surface condenser is shown at the back of the engine, the air, circulating, and bilge pumps being operated by the respective levers B B and B' B', which recieve motion through short rods C C, from the crosshead. Relief or snifter valves S. S. S. S. are provided both top and bottom of the cylinders. To enable the engine to be turned when there is no steam, the crankshaft is provided with a worm-wheel M, operated by a worm on the rod N, at whose upper end is a second worm-wheel and worm, the latter being operated by a hand ratchet lever.

Engine with the "Joy" Valve Gear.

Fig. 381 represents an engine with the *Joy* valve gear. This gear is used to vary the point of cut-off, and for reversing purposes. Its action is similar to that of a link motion, except that it keeps the amount of valve lead equal for all points of cut-off, whereas with a link motion the valve lead varies, as the link is moved from full gear towards mid gear, as has been fully explained with reference to link motions.

The beam L, for the air pump, is driven from the engine cross-head in the usual manner. The valve motion is constructed as follows: A lever attached to the connecting-rod one end, is pivoted at the other to the journal or pivot of the box J, which slides in a guideway in the drum N. The lever C is pivoted to the same pin or journal of box J, and is supported at its other end by the rod B, which is pivoted to the pump beam L.

The rods F for operating the valve-spindles and therefore the valve, are attached to the lever C, as shown. When the slide-way in the drum N stands horizontally level, the parts are in mid gear, and the steam port opens to the amount of the valve lead only, while when the slide-way is at an angle, the travel of the valve is increased, the direction of engine motion being governed by the direction of inclination of the slideway in N. This direction is adjusted by means of the handle H, by which the drum N may be rotated, the drum being secured in its adjusted position by the hand screw x. The principles of construction of the Joy valve gear, may be more fully explained with reference to fig. 382, in which A is a lever worked at one end by the connecting-rod, and suspended at the other by a rod B. To A is pivoted a vibrating lever C, pivoted in a block B, and working at its right hand end a rod F, that works the valve stem or valve spindle T. The segment E on which block D slides, is pivoted at a point that is, in the figure, directly behind, and in line with the pivot of the vibrating lever C.

The action of this gear may be divided into two elements, first, the vibration given to the block D, by the motion of the connecting-rod, which gives to the valve a motion sufficient to take up the lap, and give to the port the required amount of lead, and second, a motion derived from the movement of block D, along the arc or segment E, which governs the amount of opening of the steam-port.

The position of segment E is determined by the rod G, from the reversing lever H. In the position in which E is shown in the figure, it is seen that as block D moves along it to the left, it will (from the motion of D alone, and independently of line C) pull rod F, and therefore the slide spindle T, downwards, thus opening

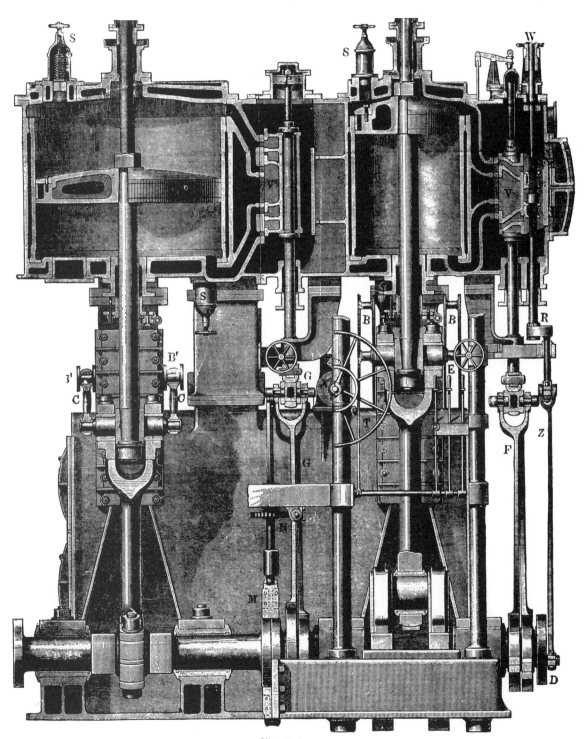

Fig. 380.

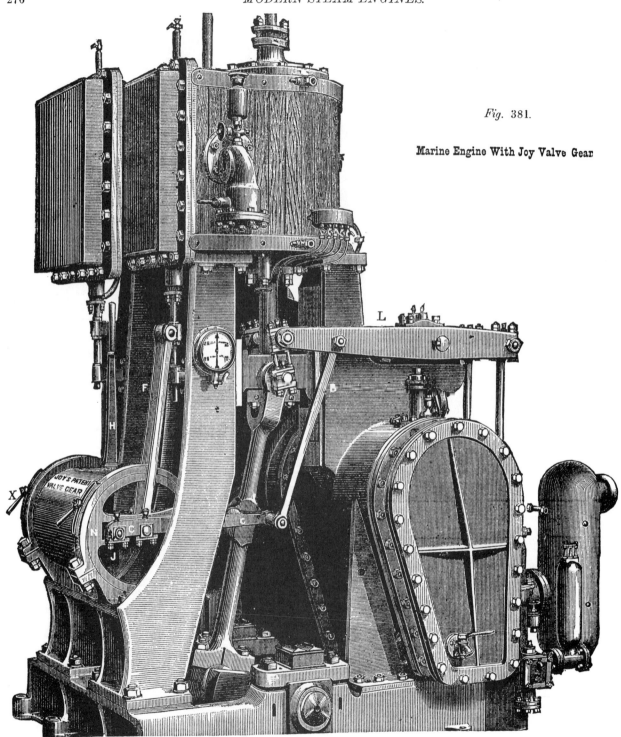

Fig. 381.

Marine Engine With Joy Valve Gear

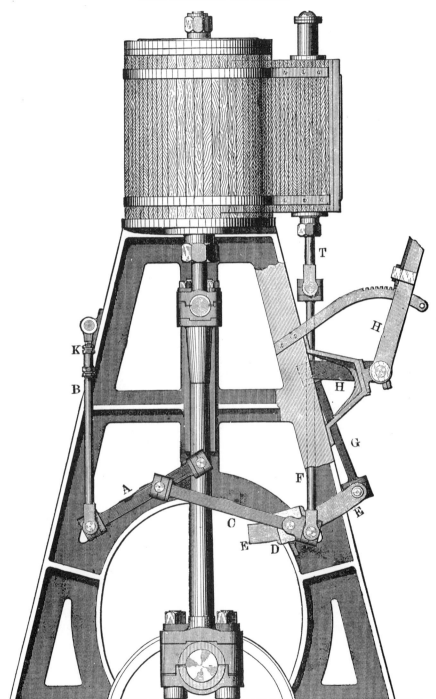

Fig. 382.

the head end port for the admission. But the motion of vibrating lever C has already moved the valve until the port is open to the amount of the lead. The curve of segment or arc E, is an arc of a circle, having a radius equal to the length of the rod F, so that if lever C were detached, block D might be moved from end to end of E without imparting any motion to the valve spindle T. It follows, therefore, that the lever A, being attached at such a point on the connecting-rod, it gives to the rod F an amount of motion only sufficient to open the port to the amount of the lap and lead of the valve, therefore when E is set (by the reversing lever) in mid-position, its arc being concentric to the upper end of rod F, the valve will open to the amount of the lead only.

It is obvious, that as the amount of valve motion necessary to take up the lap, and give the desired amount of lead, is derived from the connecting-rod, and that as the path of the connecting-rod remains the same for all points of cut-off, therefore the valve lead is equal for all points of cut-off. The amount of lateral motion of the equalizing lever A, at its point of attachment on the connecting-rod, must be as a minimum equal to twice the valve travel, which would give no lead, the increase beyond this, reduces the amount of angularity of the segment E, required for a given point of cut-off, but renders it necessary to make the segment longer, in order to obtain a given amount of port opening. The pivot of lever C (in the block D), must move, along the segment E, an equal distance on each side of a vertical line, passing through the center on which E is hung, and this is regulated by the location of the point of attachment of the vibrating lever C, on the equalizing lever A.

The adjustment nut shown at K, is for lengthening or shortening rod B, and giving more lead for one port than for the other, when it is desired to do so. The point of attachment of vibrating lever C, on the equalizing lever A, must move an equal distance above and below the pivot, on which segment E is suspended, and so long as this is the case, rod A may be connected from any point that has a motion coincident with that of the connecting-rod, thus in fig. 381 it is shown attached to a rod B, from beam L, which is also employed to drive the air pump.

Engine with Bryce Douglass' Valve Gear.

Fig. 383 represents a Marine Engine fitted with a valve motion, designed by Mr. Bryce Douglass. A short beam A is pivoted at its center to the connecting-rod, and at its lower end to a link attached to the engine frame. A rod T, from its upper end, operates an arm on the shoe S, which is pivoted at its center, so as to vibrate after the manner of a link motion.

The shoe S is mounted on the beam L, for working the air pump, and the amount of vertical motion given to the shoe by the air pump beam, is so regulated as to move the rod F, and therefore the valve, enough to take up the lap, and also give to the valve the required amount of lead. while the amount of rocking motion of S on its center, gives the port opening. The rod F has at its lower end a block sliding in a guide-way in shoe S, the curve of this guide-way being an arc of a circle, having a radius equal to the length of the rod F, so that if the shoe S is stationary, and in mid-position the lower end of F can be moved from end to end of S, without imparting any motion to F, and it follows therefore, that at mid-gear (S then being horizontal) the valve opens to the amount of the lead only, and the lead is equal for all points of cut-off.

Reversing is effected by moving the block on the lower end of F, to one on the other side of the pivoted center of S, while the point of cut-off is determined by the distance the block on the foot of F stands from the pivoted point or center of S, it being obvious, that the nearer this block is to the end of S, the greater the valve travel, and the later in the stroke, the point of cut-off will occur.

Rod F is shifted for varying the point of cut-off, or for reversing the direction of engine motion as follows:

At R, is a segmental rack, connected by rod to F, as shown. The pinion for this rack is on the same shaft as wheel W, this pinion is driven from the steam cylinder C, beneath which is a cataract cylinder D for regulating the speed at which the rack shall be moved, and enabling the engineer to stop the motion at the desired point.

Fig. 383.

For moving the engine without the aid of steam, provision is made as follows: First by a worm-wheel and worm, the latter being operated by a ratchet lever as shown; and secondly, by a train of gear wheels connecting the lower end of the worm shaft with the wheel e, in whose face are slots for the reception of a hand lever.

The main difference in the action of this valve motion in comparison with the Joy valve motion, is, that here the block at the foot of rod F, in the shoe S, has an amount of motion which varies with the point of cut-off, and which for all reduced points of cut-off is less than is the case with the Joy gear, and the wear will therefore be more uneven and greater. In the Joy gear, the length of motion of the sliding block along the segment, is the same for all points of cut-off, hence, not only is the wear equal, but the lost motion may be taken up, whereas, when the wear is unequal at different places in the same bar or guide, the fit must, in taking up the lost motion, be made to suit the least worn part or spot.

CHAPTER XII.

___

Various Applications of the Steam Engine.

___

THE TRACTION ENGINE.

A traction engine consists of an ordinary engine and boiler, mounted upon two pair of wheels, the front pair of which are used for steering purposes, and the rear pair for propelling the engine and hauling loads. The engine is provided with a reversing gear and a pump for feeding the boiler, and in some cases with a belt wheel for driving Agricultural Machinery. To obtain sufficient power without the employment of a large piston and a slow piston speed, the crank-shaft delivers the power to the traction wheels, through a train of gear wheels, which reduce the revolutions of the latter in the proportion of 20 to 30 revolutions of the crank-shaft, to one revolution of the traction wheels, the proportion of the gearing depending upon the class of work the engine is intended for, and agreeing with a speed of 2 or 3 miles per hour. The driving pinion of this gearing is thrown out of gear when the engine is to be used for driving threshing, or other farm machines. The boiler of a traction engine should be free to expand and contract, without being resisted by the engine frame, which would otherwise strain the boiler and hasten its destruction. The boiler should have a spring or cushioned seating upon both axles.

The Frick Traction Engine.

The manner in which these requirements are fulfilled in the Frick Company's Traction Engine, is shown in the following illustrations, in which fig. 384 is a perspective view of the engine, fig. 385 a plan of the engine removed from the boiler, and shown partly in section, and fig. 386 a side elevation of the engine removed from the boiler, and shown partly in section.

The frame of the engine is one continuous casting, having at its cylinder end, an eye e, through which passes a pin P secured in a bracket bolted to the top of the boiler, so that as the boiler lengthens or shortens, from expansion or contraction, the pin may pass through the eye e, leaving the boiler and the engine connection free from expansion strains; at the other end the engine frame is bolted to two side plates S and S', fig. 385^a which clear the boiler, and connect to two channel irons which pass to the front end of the fire-box, curve inwards, and thence connect to the saddle block of the front axle. This saddle block is provided with an expansion joint, permitting the boiler to expand and contract without being resisted by this part of the framing, and also with a spring, or elastic seat, upon which the boiler may ride easily.

281

Fig. 384.

The Frick Traction Engine.

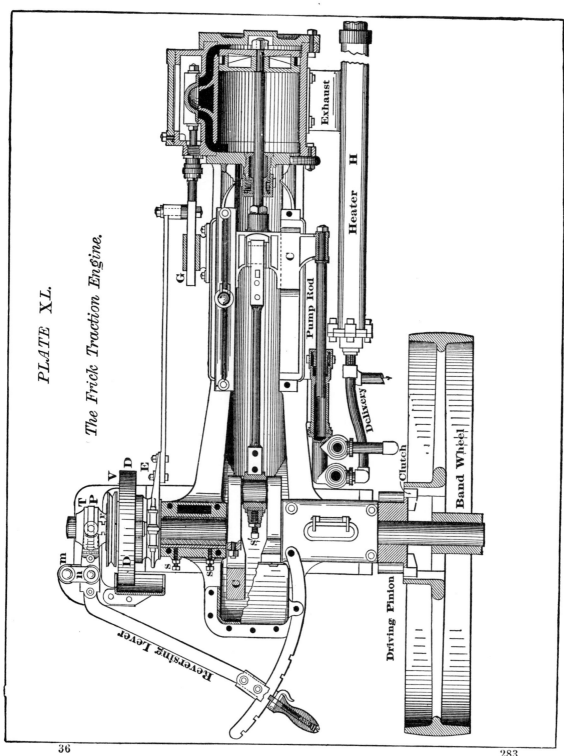

PLATE XL.

The Frick Traction Engine.

36

283

The steering is done by swinging the front axle, by means of the chains operated by the worm and worm-wheel shown in the perspective view fig. 384. To provide an elastic connection to this part of the mechanism, and thus avoid breakage, the chain is provided with a spiral spring, which is clearly seen in the perspective view.

The engine has a plain slide-valve and a reversing motion, in which the eccentric is shifted across the shaft, both for varying the point of cut-off and for reversing purposes.

The feed pump is operated from the engine cross-

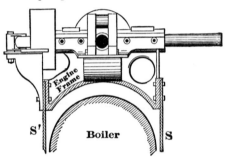

Fig. 385*ᵃ.*

head, and forces the feed water through a heater placed lengthwise of the engine frame.

Referring now to figs. 385 and 386, the cylinder and band wheel are shown in section, and the pump and heater partly in section.

The connecting-rod has a key at the cross-head end, and an adjusting screw at the crank-end, so that the length of the rod is kept as constant as possible, because while the passage of the key through the strap acts to lengthen it, screwing up the set screw acts to shorten it. The end of this set screw *s′* abuts against a plate upon which the back brass beds, the plate preventing the set screw end from indenting the brass.

The crank is balanced by the balance weight, or bob *c.* The wear of the main bearing is taken up by set screws *s s,*

The driving pinion has a clutch connection with the band wheel, so that when used for driving agricultural machines, it can be moved endwise upon the crank-shaft, and out of gear with the train of gearing used for traction purposes. When the pinion is in gear with the traction gears, it is still in gear with the band wheel which serves as a fly-wheel for the engine.

THE REVERSING GEAR.

The engine is reversed, or the point of cut-off varied by means of shifting the eccentric across the crank-shaft, the construction being as follows:

In fig. 387 D is a disk (shown also at D in fig. 385) which is fast upon the crank-shaft, and to which the eccentric E is pivoted at *b*, so that from *b* as a center, the eccentric can be swung across the crank-shaft. At *b′* is a stop pin, which is threaded into the eccentric flange, and moves in the slot *a*, which is an arc of a circle of which *b* is the center. The limit of eccentric motion across the shaft, is determined by this pin *b′* seating against the end of slot *a*, and it follows, that when the eccentric is moved into position for full gear, for either the backward or forward motion, it is driven by the two pins *b* and *b′*. The limits of eccentric motion across the shaft, is denoted by the dotted lines *x x′*, the mid-position being denoted by the dotted line *y*.

The method of shifting the eccentric across the shaft, is as follows: Fixed to the eccentric is a cross rack F meshing with a sliding rack J, which may be moved endways and parallel with the crank-shaft. The teeth of these two racks are at an angle of 45° across the width of the rack, so that by moving the sliding rack endwise, the cross rack F shifts the eccentric E.

Th cross-rack F is operated by a piece *r v*, that is fixed to the sleeve T, which may be moved endways upon the shaft, by means of a yoke P, whose two trunnions *t t′* connect to the forked or double-eye end of the reversing lever, At *n* is a link to which the reversing lever is pivoted, this link being pivoted at *m* so as to permit of the lever accomodating itself to the motion of the sleeve T, along the shaft.

The pump rod or plunger, is driven from the cross-head, the pump barrel being bolted to the engine frame. The suction valve *v* is provided with a wheel W, fig. 389, whose spindle end adjusts the height to which the suction valve can lift, and thus governs the rate of boiler feed.

From the delivery valve *v′* the feed water passes

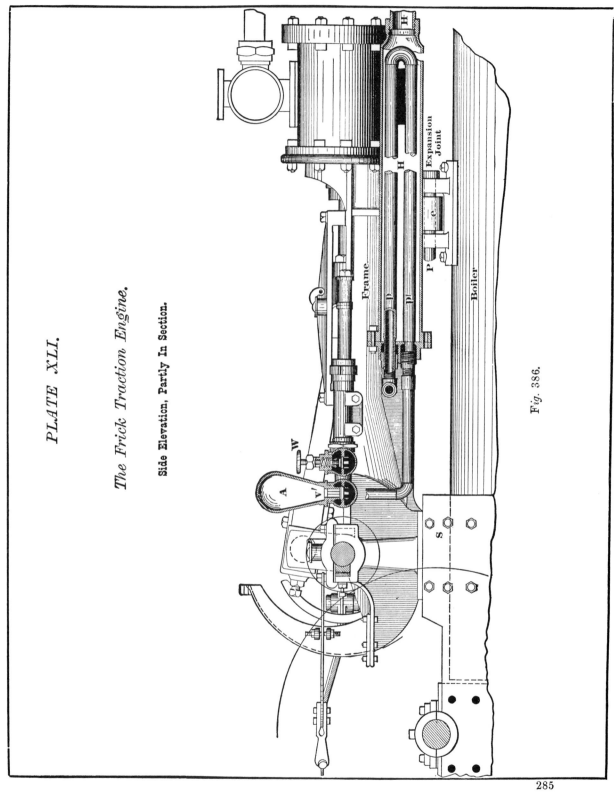

PLATE XLI.

The Frick Traction Engine.

Side Elevation, Partly In Section.

Fig. 386.

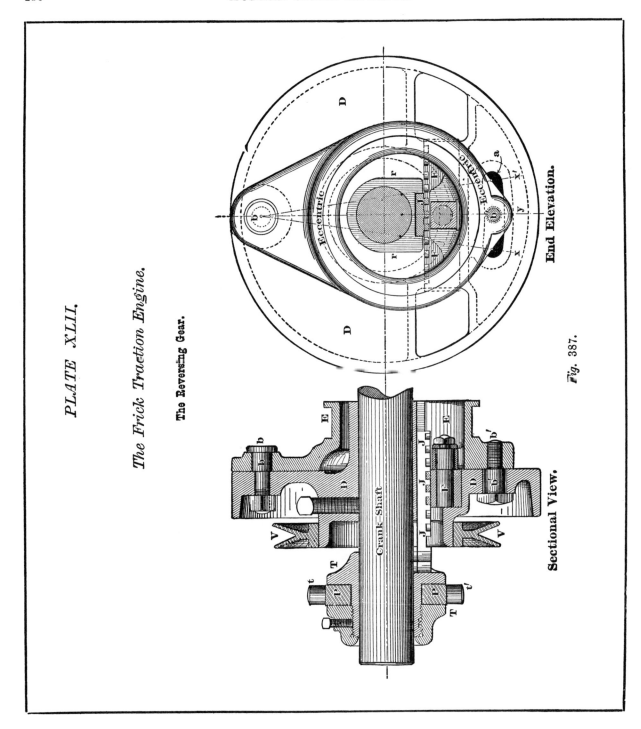

PLATE XLII.

The Frick Traction Engine.

The Reversing Gear.

End Elevation.

Fig. 387.

Sectional View.

through pipes p' and p, in the heater H, and out at u, whence it passes into the boiler. The exhaust steam passes through the heater to an exhaust nozzle in the smoke box, and thus assists the draught of the boiler.

The connection between the driving gears and the traction wheels, is as follows: To the arms of the traction wheel is secured an annular ring provided with lugs, to which are attached spiral springs, the other ends of which are attached to similar lugs upon the arms of the driving gear, hence these springs act as an

passing into the cylinder is made as follows: The crown sheet of the fire-box is inclined to the rear, as seen in fig. 388, and a dry steam chamber is provided, as shown, The dry pipe takes steam at the top of the fire-box. and discharges it into the dry steam chamber.

When the engine is descending, as in the figure, the top of the fire-box remains covered with water, and the steam passes from the fire-box through the dry pipe into the dry steam chamber, while when the engine is ascending, the steam fills the dome, and what little

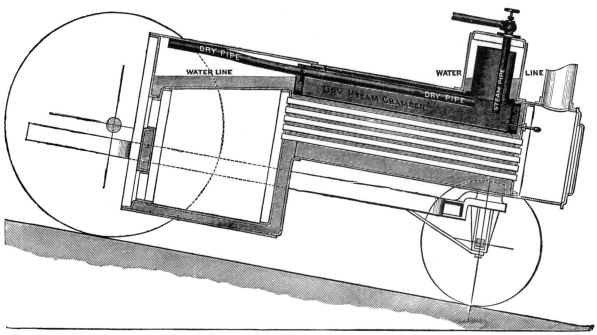

Fig. 388.

elastic connection, which permits the traction wheel to move within certain limits, either vertically or horizontally, provision being made in the construction of the wheel and axle connection, to permit of such traction wheel motion, without changing the position of the driving gear with relation to the pinion that drives it. This not only enables the boiler to ride easily when the engine is passing over rough roads, but it avoids the breakage of gear teeth, that is apt to occur where the connections are too rigid.

When the engine is to be used in very hilly districts provision for preventing the water in the boiler from

water that may pass through the dry pipe will be evaporated in the dry steam chamber.

The pipe at B is a blast pipe for forcing the draught by a steam jet.

THE PORTABLE ENGINE.

A Portable Engine is one that is mounted upon wheels, and a semi portable engine one that can be moved from place to place without requiring to be erected upon a foundation. In portable engines for agricul-

tural purposes, this is usually accomplished by mount-
ing the engine upon the boiler, and the boiler upon
wheels. For the work rooms of the small businesses
carried on in cities, portable engines are sometimes
attached to the sides of vertical boilers. Generally
however, the vertical boiler is mounted upon an iron
base plate or frame, to which the engine is bolted
independently of the boiler, this being the plan also
adopted for hoisting engines.

box to the shell of the boiler are threaded at the ends
and rivetted over.

At *b c* are angle irons, for staying the tube sheet,
and at *f*, an angle iron for staying the plate at that
end of the boiler. An end view of the boiler, showing
its attachment to the boiler, is shown in Fig. 391, the
axle is curved at A, the brackets B being bolted to the
boiler, which rests on a band, having at C nuts, which
secure the band to the brackets B B.

Fig 389.

Fig. 389 is a side elevation of the Frick Company,s
Portable Engine on wheels, and fig. 390 is a sectional
view of the boiler. The construction of the engine
corresponds with that already shown with reference to
the traction engine. The boiler is of the locomotive
pattern, the water surrounding the fire-box, except at
the furnace and ash-pit doors. The stays from the fire-

Fig. 392 represents a semi-portable engine, by the
Lidgerwood Manufacturing Company.

The boiler is mounted upon an iron base, forming a
foundation for the engine, and in which the crank-shaft
bearings are situated. A single guide-bar placed above
the cross-head is used, and a Pickering governor.

Portable engines usually have plain D slide valves.

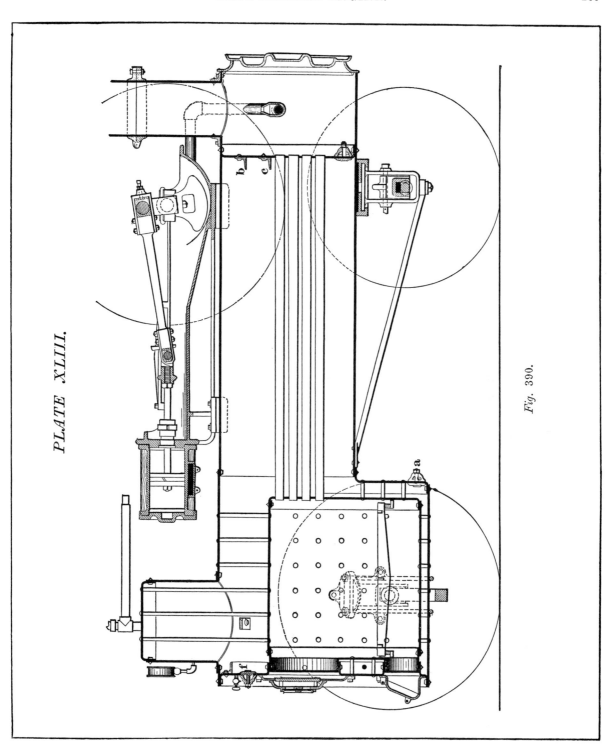

PLATE XLIII.

Fig. 390.

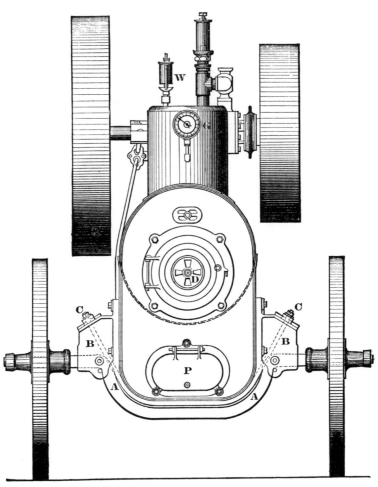

Fig. 391.

End View of the Frick Portable Engine.

Colwell's Engine For Sugar Mills.

Fig. 363 is a representative of the class of beam engine used upon sugar estates for driving the cane-crushing rolls, the driving pinion being shown on the crank-shaft. The link motion is moved for different points of cut-off by a segmental rack, and worm, as shown, and is counter-balanced by a weight suspended to one end of the rack. A common D valve is employed, and a throttling governor. Motion from the link-block rod to the valve-rod is conveyed through a rock-shaft.

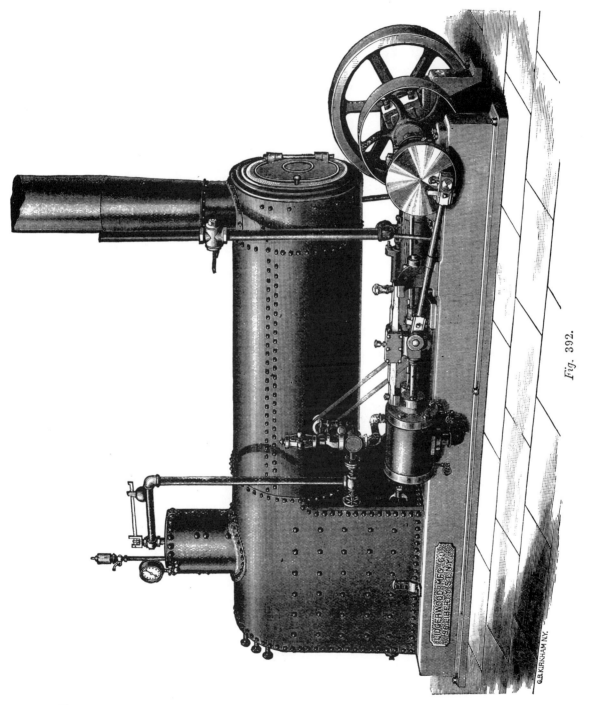

Fig. 392.

Fig. 393.

Colwell's Engine for Sugar Mills.

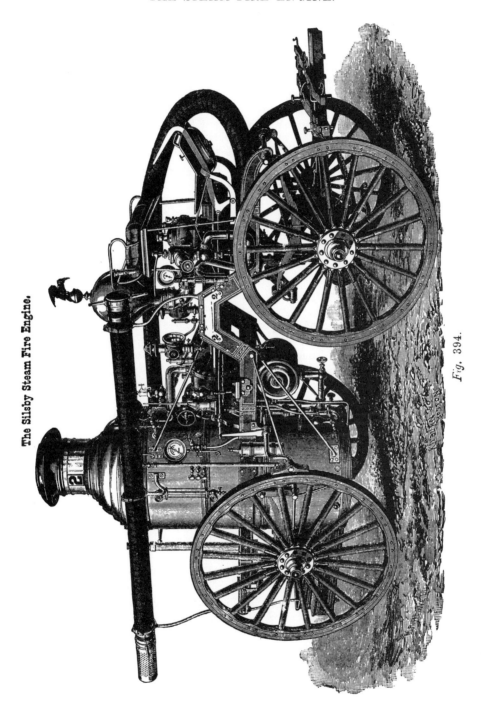

The Silsby Steam Fire Engine.

Fig. 394.

THE STEAM FIRE ENGINE.

Steam fire engines may be divided into two principal classes, those with rotary, and those with reciprocating pumps. When reciprocating pumps are used, the steam and pump pistons are usually on one rod, and between the two is a yoke, such as was shown in Fig. 71, this yoke being employed to drive the fly-wheel.

The steam cylinders usually have plain D slide valves the steam following for three-quarters or more, of the stroke. The pump valves are usually flat rubber discs with spiral springs behind them.

The air chambers of the pumps are so constructed, as to let the water they contain be disturbed as little as possible by the delivery water from the pump, because the water gradually absorbs the air from the chamber, and the presence of the air tends to break up the water column after it has left the hose nozzle. In some cases, the air chambers are, for the above purpose, provided with long necks in which the water may lie, as nearly as may be, undisturbed. In others, a pipe extends up within the chamber, thus separating the water and air.

A delivery pressure greater than the steam pressure is obtained by making the steam piston of larger diameter than the pump piston.

When a rotary pump is used, a delivery pressure greater than the steam pressure may be obtained in two ways, first by reducing the diameter of the pump, and secondly, by making the pump cylinder shorter than the steam cylinder. A rotary pump possesses the advantage that it requires no valves, and the suction and delivery water is kept in continuous motion, unchecked by the rise and fall of valves. Rotary steam cylinders are less economical of fuel than reciprocating ones, but in a steam fire engine. efficiency is of the greater importance than fuel economy.

The Silsby Steam Fire Engine.

Fig. 394 represents a steam fire engine with rotary steam cylinder and pump, and Fig. 395 a longitudinal section through the engine and boiler, the construction being as follows: What may be termed the fire box, extends downwards from the plate *a*, which supports a series of double tubes that extend down towards the fire. These double tubes are one within the other, the inner tube having a V shaped opening at the bottom, so that the water can pass down through the inner tube and up through the space between the inner and outer tube, the latter being closed at its lower end, and being surrounded by the heat in the fire box.

The fire box has water legs, or in other words, there is water, between its sides and the shell of the boiler. Through the body of the boiler from *a* to *b* are the smoke flues, through which the heat and products of combustion pass to the chimney. The dry pipe through which steam is taken for the engine, extends around the boiler as shown at C C, D being the steam pipe for the rotary engine E. The exhaust pipe is shown at F, extending up to G, where there is an exhaust nozzle whose area of opening may be either opened or closed to regulate the draft for the fire. This nozzle consists of a cone having openings on two sides, and capable of adjustment vertically, in a coned seat at the top of the exhaust pipe. When the cone is raised out of its seat, the exhaust is more free, whereas when it is lowered, the area of exhaust opening is reduced, the steam escapes with a greater velocity and, as it carries with it up the chimney the contents of the smoke box, the draft is forced. The constuction of the engine is seen in the cross-sectional view at H. It consists of two intermeshing cams upon shafts that are geared together by the gearing at J K, in Fig. 395, so that the two cams are driven independently, which relieves the intermeshing parts of strain Each cam has two strips, which have contact with the circumferential bore of the cylinder, to maintain a steam tight fit, and these strips follow up the wear. The construction of the pump is similar, except that each cam has three strips for bearing against the cylinder bore.

The globe valve at N admits a portion of the exhaust steam into L, to heat the feed water, the pipe from the engine extending down into the heater L, and being perforated to distribute the steam through the water. The pipe whose exit is shown at M, is merely an overflow.

There are two methods of feeding the boiler as fol-

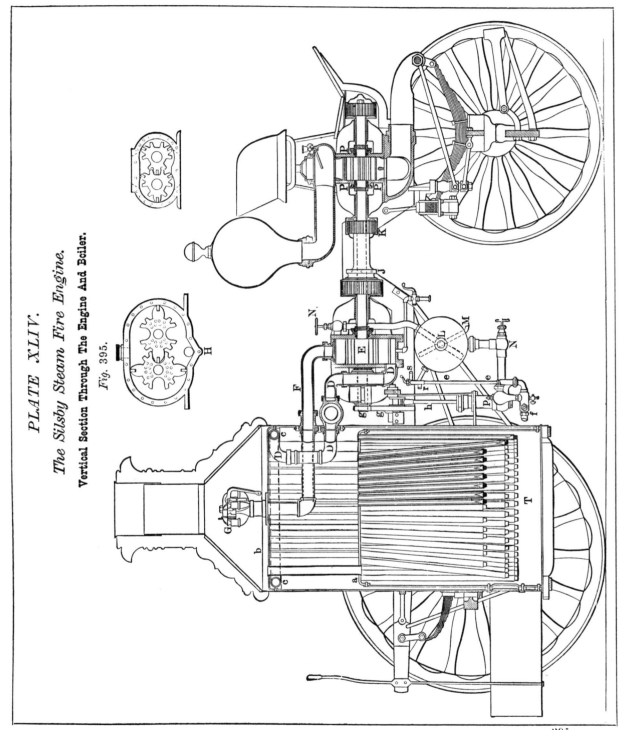

lows: *g g*, are a pair of gears, a crank-pin on the lower
one operating the rod *h* of the pump *p*, which takes
water from the heater L through the suction pipe N,
where there is a globe valve to regulate the amount of
feed. Second, a pipe leads direct from the main pump
P, to S, so that by opening a valve, the water can pass
through pipe *e*, and through *f* into the boiler, without
passing through pump *p*. This method of feeding can
be employed when the pressure in the air chamber

Fig. 396 represents Mundy's hoisting engine, in
which the cylinder is bolted to the side of a frame,
upon which the boiler is mounted. The connecting
rod drives a disc, on whose shaft is a pinion that drives
the hoisting drum or drums, as the case may be, through
a simple or compound train of gearing, according to
the weight the engine is designed to lift.

Fig. 396.

Mundy's Hoisting Engine.

exceeds that in the boiler, and to obtain the required
pressure, the discharge gate of the pump may be partially closed. To prevent the feed pump *p* from freez-
ing in severe weather, a pipe is provided, which may be
used to admit steam from the boiler to the pump. To
promote the draft, a pipe connects from *r* to the ash-pit,
thus allowing the main pump P (from which *r* recieves
water) to wet the ashes.

Fig. 396 is an end view of the frame, showing the
construction of the friction device through which the
power is transferred from the crank disc on the right,
to the gear on the left, and it is seen, that in this example, the crank-shaft operates a spur gear, on whose
side is secured a cone, formed of pieces of wood, whose
end grain is radial from the gear axis.
The end of the drum is provided with a conical seat,

into which the wooden cone may be forced by the lever

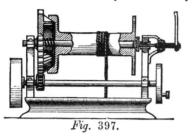

Fig. 397.

Mundy's Friction Drum.

shown at the right hand end of the drum shaft. This lever operates a screw, which moves the drum endways on its shaft, to engage or disengage the cones, as may be required for raising or lowering the load.

Each drum is provided with a ratchet and pawl, as shown, for holding the load independently of the friction cones, or of the piston power.

A steam pump for boiler feeding is shown attached to the boiler. Hoisting engines are provided with flat D slide-valves, and with link motions for reversing purposes.

Fig. 399 shows a single cylinder and single drum

Fig. 398.

The Lidgerwood Hoisting Engine.

hoisting engines, constructed by the Lidgerwood Manufacturing Company, and having a cone friction applied
to the drum. A similar engine is constructed, with a

Fig. 399.

cylinder on each side of the frame, the cranks being at
right angles.

The lowering of the load is effected by partially releasing the grip of the friction cones of the hoisting
drum.

Robertson's Semi-Rotary Engine.

Fig. 399 shows a stationary hoisting engine, with
two cylinders and link motion reversing gear.

In semi-rotary engines the pistons reciprocate in an
arc of a circle, and the great difficulty in this form of
engine has been, that the piston power acts to bend the
piston, and cause it to bind in the cylinder bore.

In Robertson's engine, however, this defect is elimin-
ated, the construction being as follows: In Fig. 400 is
a sectional side view, a a is the cylinder, B B the piston,
c a center piece to which B B is bolted, and which has
journal-bearing in boxes provided upon the cylinder or
frame, to which one end of the connecting-rod F is
pivoted by means of an ordinary crank-pin. G is the
crank, and m the fly-wheel. H is an ordinary eccentric

to operate the slide valve, which is of the three-port
type, h is a piece separating the two chambers a a, the
exhaust port passing through it, and the steam ports
being one on each side of it. E is a gland for packing
the piston. Referring now to Fig. 401, it will be observed that in one casting we have the cylinder, the
frame and the fly-wheel bearings. The bearings upon
which the piston vibrates, are also cast solid upon the
same casting. The pistons B are firmly bolted to the
middle piece carrying the crank-pin, the object being to
facilitate turning the pistons in the lathe.

To secure, beyond peradventure, the pistons upon the
middle piece, they are let or recessed into it.

The pistons, it will be observed, do not fit against
the sides of the cylinder, except over a small projecting
piece at the top, which serves as an abutment for the
packing, the space for which is shown between it and
the end of the gland E.

The space between the pistons and the cylinder bore
is made to be as small as possible, and represents the
clearance found in the passages incidental to three-port-
ed valves. The pistons B B are solid castings, whose
strength insures them against spring. The packing,
which keeps them steam tight, represents the steam
packing of an ordinary piston, the gland and its packing being dispensed with. A feature of this plan is,
that any piston leak becomes apparent at once, and the
nuts being on the outside, the packing may be tight-
ened and the leak stopped immediately, without stopping the engine. Steam is admitted alternately through
the ports on the right and left hand of the pistons.
The eccentric rod is hooked to the rocker arm I, the
latter being provided with a handle, whereby to operate the valve by hand when starting the engine. K is
the valve stem, and L the steam chest.

The Rotary Engine.

A rotary engine is one in which the piston revolves
with the shaft, hence the length of the stroke is the
path of the piston, around the cylinder bore. In some
forms of rotary engines, cam shaped pistons, such as
was shown in Fig. 395, are employed. In others, the
piston head carries small sliding pieces, which are

Fig. 400.

Fig. 401.

Robertson's Semi-Rotary Engine.

moved in and out, to pass an abutment by cam motions while in others, the abutment itself is moved, to allow the piston to pass.

A rotary engine possesses the advantage that there is no reversal of motion of the piston, etc., at the ends of the stroke, and of comparative inexpensiveness to make.

402, which is a longitudinal section on a vertical plane paralell with the shaft, and Fig. 403 an end elevation, with the cylinder cover removed, The cylinder is divided by a central partition A. On the shaft B are two pistons C, which pass through abutment rings E.

These rings fit, at their ends, into recesses or grooves provided in the partition A, and in the cylinder covers.

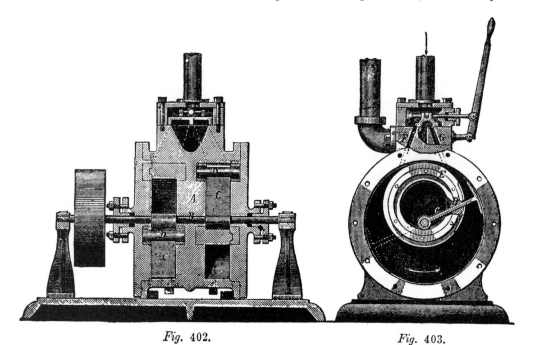

Fig. 402.   *Fig.* 403.

Rotary Engine.

On the other hand, however, the steam pressure acts to press the shaft against the sides of the bearing. The wear of the piston is greatest at the circumference of the cylinder bore, and less as the shaft is approached, which renders it difficult to maintain the piston steamtight. Furthermore, the piston area is small in proportion to the length of the stroke, hence the loss of heat from cylinder radiation is great.

From these causes, rotary engines are less economical than reciprocating or rotative engines, and therefore find their field of usefulness confined to cases in which economy of fuel is not of primary importance

A representative of rotary engines is given in Fig.

The ring is disposed eccentrically to the shaft, and as at its highest point it is in contact with the cylinder between the ports F and G, it forms a constant abut ment for the steam. The latter entering between this abutment and the piston, acts directly upon the piston, which being merely a lever arm as regards the shaft, of course turns the same, traveling in the direction of the arrow. In passing the abutment part of the ring, the flukes fit into a recess, so that the contact between the abutment and cylinder is always maintained. The reversing gear, by which steam is admitted to either port, by means of a common D valve, is operated by the hand lever shown.

Fig. 404.

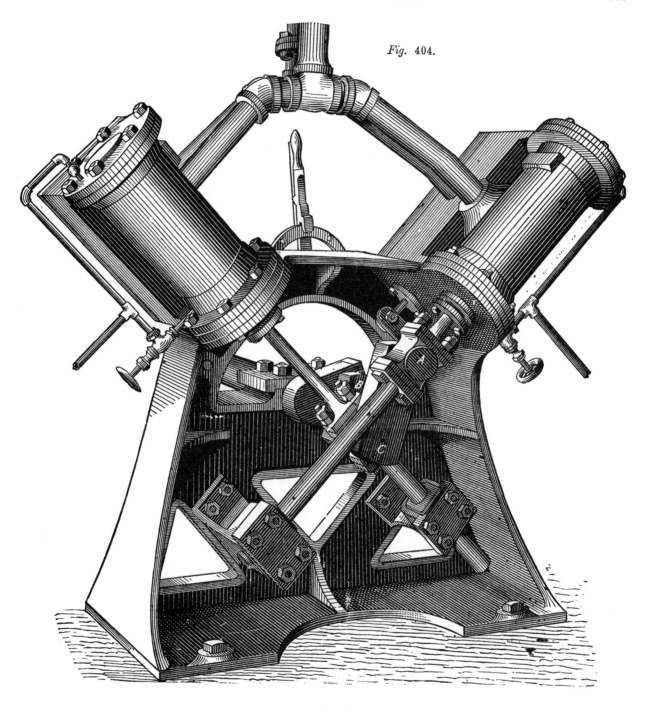

Figs. 403 and 404 represent an Engine in which the cylinders are set at a right angle, and the piston-rods being prolonged past the crank-pins, and passing through guides, no guide bars are necessary.

Fig. 403 is a front, and fig. 404 a plan view of the

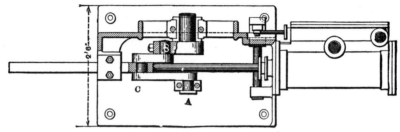

Fig. 405.

engine. A is a crank-pin fast in the crank C, which carries a crank-pin for the other piston-rod. This crank-pin has journal bearing in the crank cheek B, the latter having a crank-pin with journal bearing in

during each engine revolution in the cranks D and C.

As the crank-pins A and C must always move at right angles to each other, it is evident that the long arm of the floating crank has a very peculiar motion. Thus, starting from the positions A, B, and C in the diagram, fig. 406, during the first quarter revolution, it lies at varying angles at the left of the vertical line. During the next quarter it inclines in the opposite direction, and to the right of the vertical line, while during

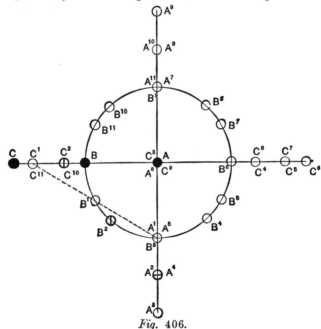

Fig. 406.

the crank D, which is fast upon the main shaft of the engine. The combined lengths of B and D equal the length of crank C, which is half that of the piston-stroke.

It will be observed that crank B is what may be termed a floating or free crank, its two pins revolving

the third quarter the end that was previously at the bottom reaches the top, and continues so during the rest of the stroke. This floating crank seems to revolve in a direction opposite to that of the main shaft. Its position may be found at any position of the piston, by means of the diagram Fig. 406. Suppose, for example, that the parts are in the position shown in Fig. 405, the three crank cheeks or arms being then all in line. The black dots in Fig. 406 will then represent the crank-pins, A being the pin attached to the outer piston-rod, the black dot at B representing the inner pin, whose path of motion is the circle, and the black dot at C representing the crank-pin, for the other piston-rod.

It is obvious that the pin B will always be somewhere

crank-pins C B, fig. 406), position B[1]. Again suppose the outer crank-pin (A fig. 406) moves from A[1] to A[2], and the middle pin (C fig. 404) will have moved to position C[2], and the inner pin (B fig. 404) to position B[2], and so on through all the other numbered letters. similar numbers indicating simultaneous positions. Twelve positions are plotted on the diagram, and it is easy to follow the parts throughout the stroke.

It is obvious, that the cylinders may be placed in

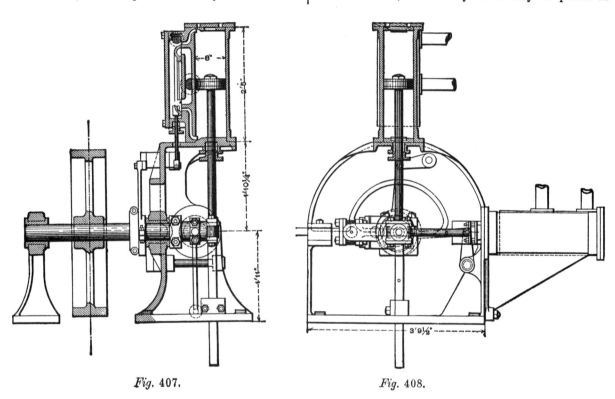

Fig. 407.   *Fig. 408.*

on the circle, which represents its path of motion, and to find its position for any given piston position, all we require to do is to set a pair of compasses to the radius of crank-pins A C, and another pair to the radius of crank-pins B C, and we may trace the motion as follows: Suppose the outer crank-pin moves from A to A[1], and from A[1] (with the radius of crank-pins A C), we may mark at C[1] the corresponding position of the crank-pin C. To find the position crank-pin B will have moved to, we mark from C[1] (with a radius equal to that of the

any required positions, so long as their bores are at a right angle, and also that in place of two cylinders, four may be used, two taking the place of the guides shown in the illustrations.

Figs 407 and 408 show two cylinders, one vertical and the other horizontal, and it is obvious, that we may trace out the paths of motion, as given above, by drawing the center lines to represent the axial lines of the pistons. The objection to this class of engine is the excessive friction and wear of the crank-pins.

Engines for piercing holes in rock, for receiving explosives, are called, in general terms, *Rock-Drills.* A rock drill consists of a cylinder mounted upon an adjustable frame, that permits the cylinder to be set in the direction in which the hole is to be drilled.

The cylinder is fed to the cut, and the rate of feed is so regulated, that the piston-stroke is maintained as nearly as possible in the same part or length of the cylinder bore.

The valve is sometimes operated by tappets in connection with the steam or compressed air that drives the rock-drill, and at others by the piston admitting steam that operates the valve. Steam is used to drive the drill when it is used in the open air, and compressed air when the drilling is to be done in tunnels or similarly confined places, or when the boiler for generating steam could not be placed near enough to the engine to prevent great loss from condensation in the steam pipe.

The Ingersoll Eclipse Rock-Drill.

In fig. 409 is shown the Ingersoll rock-drill mounted upon a tripod, for surface work. As the tripod legs are pivoted at their upper ends, it is obvious that they may be moved to bring the drill into the required position, the weights serving to anchor them.

The drill is attached to the piston-rod, as shown, and the feed is put on by operating the feed screw, which moves the cylinder down a slide provided upon the frame.

A sectional view of the cylinder is given in fig. 410, in which it is seen that it is provided with steam and exhaust ports of the usual pattern.

The two dotted circles F F' represent open passages in the cylinder, which are connected with the exhaust port E, and hence the interior of the cylinder between F F' is at all times open to the atmosphere. Now observe the two passages D D': These are small brass tubes opening a passage from the space in the steam chest at each end of the valves, to the interior of the cylinder within the length between F and F': Hence, if there were nothing in the cylinder to shut the passage D and D', each end of the valve would be open to the exhaust; but we have the piston B moving back and forth in the cylinder, and having a stroke from X to Y. This piston has a long bearing in the cylinder, broken in its center by the annular space S S': This space is a constriction in the diameter of the piston, and makes an open space or chamber all around it. The length of this space is such that, wherever the piston may be in the cylinder, this space is at all times open to one of the passages D D', and hence to one of the holes F F', which leads by way of the exhaust port E, to the open air. S S', therefore, is an exhaust chamber carried up and down with the piston. This exhaust chamber can never be open to both of the passages at D D' at the same time. When the piston is on the up stroke it is open to one of these passages, and when on the down stroke, to the other. The valve V is spool shaped, and has a hole through its longitudinal axis, through which passes the bolt T, which serves to guide the valve in its motion back and forth, and which, by means of a spline, prevents its rolling on its seat. In the bottom of the steam chest there are two cored passages, connecting the tubes D and D' with the ends R' and R of the valve. These passages cross each other, so that R is connected with D', and R' with D.

In the figure the piston has completed the up stroke; the valve has been reversed; and the drill is ready to strike a blow. Suppose the steam to be admitted through the chest to the valve at a point—say O. As the spaces at O, N and N' are in one, the steam will encircle the valve, bearing it down upon its seat through the excess of pressure at O. Escaping over the top of the valve flange it will also occupy R'; this being connected with D, and D being closed by the lower piston head, there is no outlet. Now R being connected with D', and as D' is now open to the piston exhaust chamber, the space behind the valve flange at R is free to the exhaust, and hence the steam pressure in R' holds the valve close at R, so long as D' is open to the piston exhaust passage. Therefore the valve must remain in its present position until the piston moves. The port P being open to the live steam chamber in the valve, and

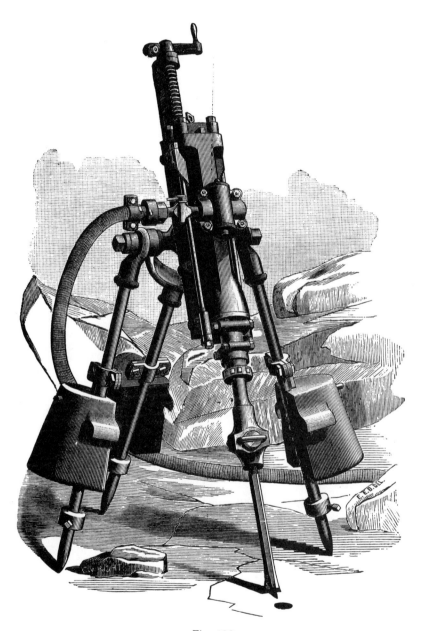

Fig. 409.

the port P to the exhaust, the steam passes through P'
into the cylinder at M, and pressing upon the back of
the piston drives it down.   As the piston moves down
this piston exhaust passage is open to D, and at the
same instant D' is shut off by the upper piston head.
The result is that D is suddenly opened to the atmosphere, and the chamber R' being connected with it, is
exhausted.   The live steam around the valve rushes
towards this exhaust opening, carrying the valve with it
and pressing it against the upper head of the chest at

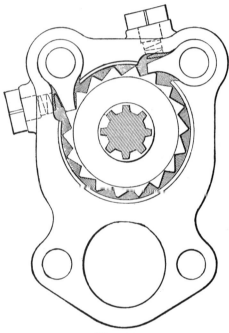

Fig. 411.

R', thus the valve is reversed, the machine exhausts, and
the motion of the piston is reversed.

We here have an intermittent and reciprocative
action of piston and valve, one being dependent upon
and regulated by the other, yet each is separate, and
removed from the other, and without direct mechanical
connection.   The valve of this feature in Rock Drilling Machinery is evident.   Where a piston is made to
strike rock at the rate of three hundred blows a minute,
it is important that it should move freely in its cylinder
and that it should strike nothing but the rock.

By simply feeding down the cylinder, the piston will

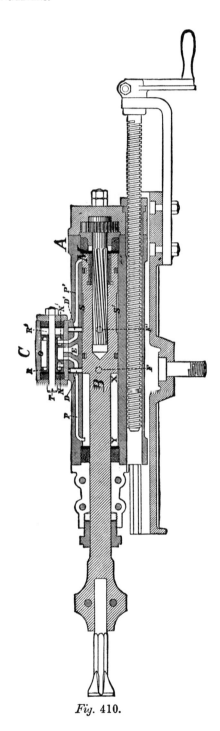

Fig. 410.

work entirely in the upper part, cutting off so soon as the blow is delivered, and increasing its stroke as the hole is driven. This is important, especially in starting or pointing holes.

Fig. 411 illustrates the manner in which the drill is revolved a certain amount during each upward piston-stroke. Within the upper end of the piston is a bar with twist grooves, and fitting to the same is a nut; fast in the head of the piston is a ratchet wheel having two pawls, which permit of the ratchet wheel revolving in one direction only. When the piston makes an upward stroke, the spirals cause it to make a part of a revolution, and the pawls fall into the ratchet teeth, preventing the spiral bar from turning back to the same place.

By using two pawls the teeth of the ratchet may be made twice as coarse, and therefore twice as strong for a given pitch of ratchet, because the ratchet wheel needs to move bnt half the pitch of the ratchet, in order to permit one or the other of the pawls to engage.

The piston diameter varies from 1¾ inches to 5 inches in diameter, the larger sizes being used for deep holes.

It is found that vertical holes may be drilled one-quarter faster than horizontal, the rate of feed being diminished in proportion as the rock is either hard or seamy.

The Ingersoll Air Compressor.

The construction of the Ingersoll air compressor for driving rock-drills is shown in Figs. 412, 413, 414 and 415.

It consists of a steam and air cylinder, whose pistons are in line, and connect to a cross-head C, operating in the guides G, the connecting-rods being connected outside the guides. It is obvious, that the pressure in the air cylinder increases as the air piston approaches the end of the stroke, at which time the steam piston has the least pressure, because of the cut-off and release occurring before it completes its stroke.

This necessitates the employment of heavy fly-wheels W, which store up the power received during the ear-
39

lier part of the stroke, and deliver it back again at the later part.

The air cylinder is provided with a pump, by which water is forced into the cylinder and mingled with the air, thus keeping it cool and preventing its expansion from the heat, that would otherwise be generated during its compression, it being obvious that such heating and expansion would increase the load on the engine, and diminish the quantity of air compressed.

The power of the engine is proportioned to the amount of its load, by means of a Meyer's cut-off valve, the eccentrics operating rock-shafts C C, which, at their upper ends, operate the valve rods, that for the cut-off valve being seen at *f*.

The cut-off valves are moved to vary the point of cut-off by a right and left hand screw, operated by a hand wheel shown at J.

To maintain the required pressure in the receiver in which the compressed air is stored, and to maintain the pressure uniform, the following construction is employed: The cross-head C drives a lever A, which is pivoted at one end, and which operates the pump rod *r*, the pump being on the side of the air cylinder. The pump discharges alternately into each end of the cylinder through discharge pipes, which may be placed in communication or separated by a valve.

The suction pipe is also provided with a valve, and the mechanism is such, that when the air pressure in the receiver begins to increase above that required, a piston closes the suction pipe and opens communication between the two delivery passages, so that the water already in the air cylinder passes through the delivery pipes, from one side to the other of the air piston. At the same time a valve in the admission pipe of the steam cylinder is closed, thus diminishing the steam supply, and therefore the engine speed.

A further feature of this part of the mechanism as applied to the larger sizes of compressors, is that the compressed air in the receiver may be utilized for starting the engine, the construction being as follows: Fig. 414 is a side view of the air cylinder shown removed from the engine, and broken away at one end to expose the valves and the delivery passage *d*, the regulating mechanism being shown in section; and fig. 415 is an end view of the same.

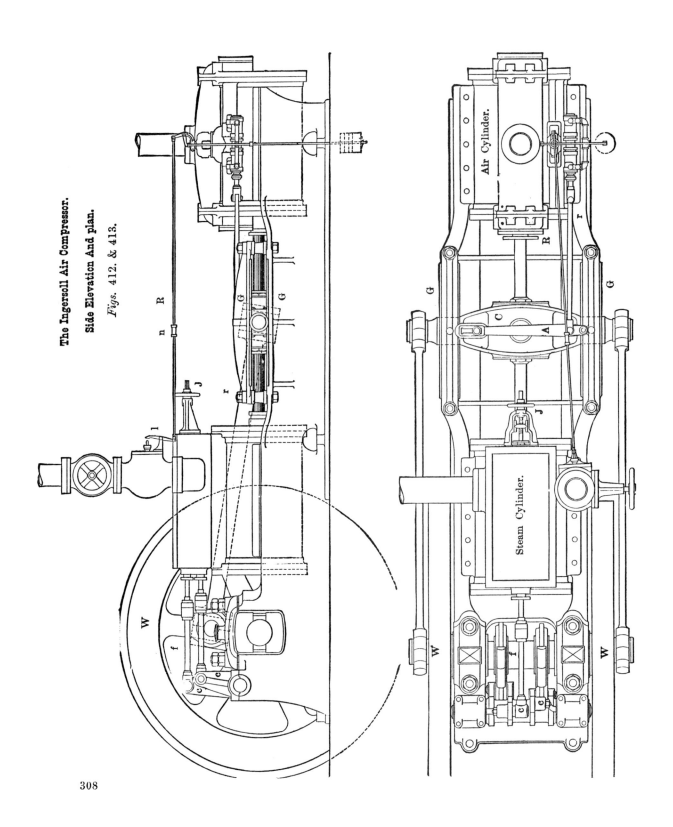

The Ingersoll Air Compressor.

Side Elevation And plan.

Figs. 412. & 413.

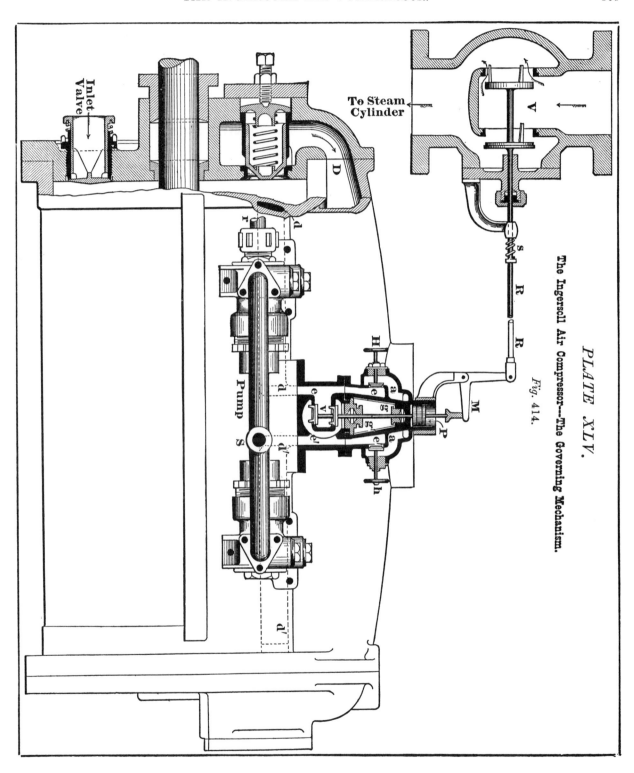

PLATE XLV.

The Ingersoll Air Compressor—The Governing Mechanism.

Fig. 414.

PLATE XLVI.

The Ingersoll Air Compressor. ---- The Governing Mechanism.

Fig. 415.

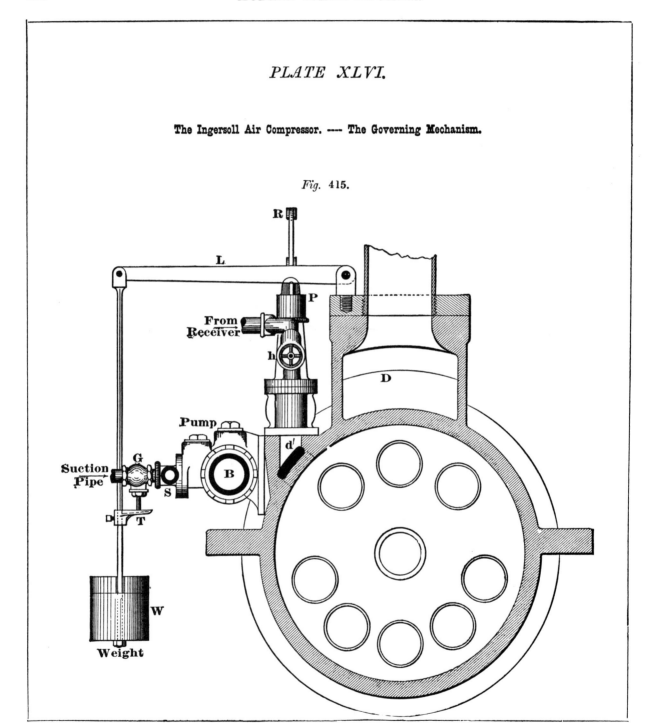

Referring to these two views, a pipe from the receiver admits the compressed air into the chamber *a a*, and also beneath the small piston P, which is held down by the weight shown attached to the end of lever L, hence this corresponds to a common safety valve, except that the raising of the piston does not permit the compressed air to escape. The weight is adjusted in amount to suit the pressure to be maintained in the receiver. Now suppose that from a rock-drill being stopped, or from some other ordinary cause, the pressure in the receiver begins to increase, and piston P will be raised, and this will lift the valve *v*, which is on the same rod as piston P. When valve *v* is raised the two delivery passages *d d'* of the pump (which both communicate with the chambers *e e'*) are in open communication, and the water on the delivery side of the pump passes, as the air piston reciprocates, to and fro in the delivery passages *d' d*, from one end to the other of the air cylinder. From the time that the piston P begins to close, the water supply to the pump begins to diminish, because the tappets T on the rod suspending the weight, meets the valve stem of globe valve G and closes that valve.

Simultaneously, with this action a second one occurs, inasmuch as that the lever L, as it rises, meets a bell crank M, which operates the rod R, and this closes the steam governor valve V, dimishing the supply of steam to the steam cylinder. All these regulating devices, therefore, are operated by the motion of the piston P, which is determined by the pressure in the receiver, and is, therefore, automatic.

Now suppose the engine to have been stopped, its steam supply being shut off, and to start it again the steam is turned on again, and the compressed air in the receiver may be utilized to assist in starting, by the following construction. The annular chamber *a a*, being in open communication with the receiver, is filled with compressed air, which is excluded from the chambers *e e'*, by valves, whose operating hand wheels are shown at H and *h* respectively. If, however, the air piston is at the right hand end of the cylinder, valve *h* may be opened, admitting the compressed air into chamber *e'*, from which it will pass through *d'* into the right hand end of the air cylinder, and thus aid in propelling the piston.

Similarly if the piston is at the other end of the cylinder, valve H may be operated to admit air from *a* through *e* to *d*, to propel the air piston, the valve being closed as soon as the engine is fairly started. Obviously, in operating either of these valves, the direction of engine revolution is to be considered, since it determines which valve must be opened; at *g* and *g'* are glands to close communication between the small cylinder P and the chamber *e'*.

The Knowles Steam Pump.

Fig 416 is a general and fig. 417 a sectional view of the Knowles steam pump, whose construction is as follows: The steam and water pistons are in line, and are connected by a piston-rod, to which is attached an arm A. Upon this arm is a roller having contact with the bottom edge of the rocker bar T, which is pivoted at its center, this bottom edge being curved, so that as the arm A traverses, it alternately lifts one or the other end of the rocker bar.

At B is a link attached to a short arm on the rod D, so that as A traverses to and fro, the vertical vibration its roller gives to rocker arm T, causes B to slightly revolve the rod D of the chest piston, and this causes the chest piston to operate the main B shaped valve, by reason of the following construction: At each end of the steam chest piston, is a small port, one of which is shown at the right hand end of the chest piston. Near each end of the seat of the steam chest piston is a small steam and exhaust port, which are opened and closed by motion of the chest piston, and alternately admit steam to, and exhaust it from the ends of the chest piston, and thus operate it endwise.

At *v* and *v'* are tappets that while not absolutely essential to the valve motion, serve as safeguards to move the chest piston, should the rotary motion of the chest fail to do so.

Referring now to the water cylinder, the two lower valves are for the suction, and the two upper for the delivery, as the pump piston is shown moving to the left, the right hand side is drawing, hence the right hand suction valve is open, while the left hand delivery

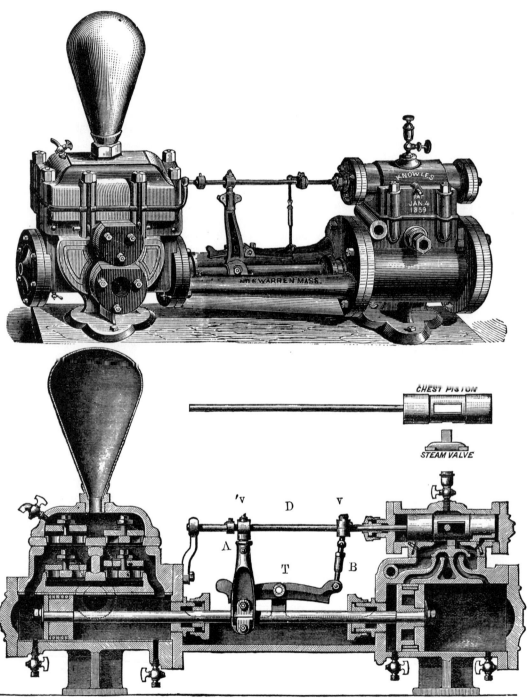

CHEST PISTON

STEAM VALVE

ʼv　　　　D　　　v

Λ

T　　　　B

Figs. 416. & 417.

valve is also open, to permit the escape of the water on the left hand side of the water piston. The left hand suction valve is closed, and held closed by a spiral spring on its back, as well as by the pressure of the

operated from the piston-rod of the other, as was fully explained with reference to fig. 373 of the Worthington Compound Condensing Engine.

When the circumstances render it necessary to use the

Fig. 418.

delivery, while the right hand delivery valve is similarly held closed by its spring, and by the pressure of the delivery, as well as by the suction.

The Worthington Steam Pump.

A general view of a Worthington Steam Pump is in fig. 418. It consists of a pair of steam cylinders and pumps connected together, so as to form virtually one machine. The valve gear for one steam cylinder is

steam expansively, compound steam cylinders are employed, the arrangement being shown in fig. 419. The valve motion is the same as before, except that the valve rod T passes through the low pressure steam chest, to the high pressure steam chest, both valves being worked by one rod. The exhaust from the high pressure cylinder passes through the pipes shown direct to the low pressure steam chest.

Figs. 420 and 421 show the construction as applied to pumps for brewery purposes, in which thick liquids require to be pumped. This suction chamber is at S, and a piston, instead of the usual plunger, is employed,

Fig. 419.

so as to reduce the clearance to a minimum, the lower pair of valves v' v^2, are obviously for the suction, and the upper pair v^3 and v^4 for the delivery. The suction chamber S is common to both pumps, and as one or the other of the suction and of the delivery valves are always open, the suction and delivery streams are continuous.

The steam valve motion is constructed as follows:

move until it covers the exhaust port e, the steam remaining in the cylinder being compressed, and finally causing the piston to pause at the end of its stroke. During this pause the suction valve v' and the delivery valve v^3 will quietly seat themselves without inducing reverse currents in either the suction or delivery pipes, as occurs when the valves close quickly. The admis-

Fig. 420.

The Worthington Pump as Applied to Brewery Purposes.

Two arms A and B, receive motion from the respective piston-rods, the arm on the left hand engine operating the valve for the right hand one. Thus arm A operates a rock-shaft B, connected to the valve rod R, whose block v, that fits in the valve, has a certain amount of lost motion or play. Now suppose motion to occur in the direction of the arrows, and the steam piston P will

sion of steam for the next stroke of piston P, is so regulated that it first permits of this pause, and then drives the piston on the return stroke. The lost motion between the blocks v and the valve, is so regulated in amount, that one piston takes up the motion just before the other pauses, hence a steady and continuous flow of water is maintained.

40

The Gordon & Maxwell Isochronal Pumping Engine.

Fig. 421 represents the Gordon and Maxwell Isochronal steam pump, in which a cataract is employed to govern and equalize the pump motion. Live steam is admitted by the valve F to the chamber E E', in which

piston G being fast upon the rod D″ of the valve moving piston.

When the steam piston B is at the end of its stroke, and the ports are in the position shown in the figure, the valve F has admitted steam which acts against the end face at E′ of the valve moving piston D, and as the port a′ is closed, and the piston B and cataract cylinder at rest, the valve moving piston will move to the left, and operate the main valve to open port a′.

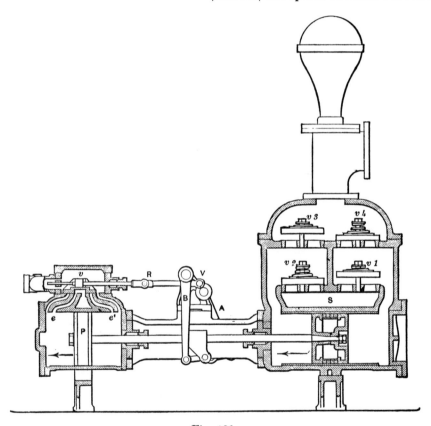

Fig. 421.

Sectional View of the Worthington Pump as Applied to Brewery Purposes.

the piston valve D D′ operates. This valve moving piston operates the valve for the ports a a′ of the main cylinder A. The cataract cylinder H is filled with oil, and is given a positive motion through the medium of arm J, link L″, lever I, and rod I′, the cataract or oil

Between the point at which the valve piston begins to move, and that at which the valve opens the port a′ a certain period of time must elapse, during which the piston B pauses, which gives time for the delivery valves of the pump to close quietly. This period of

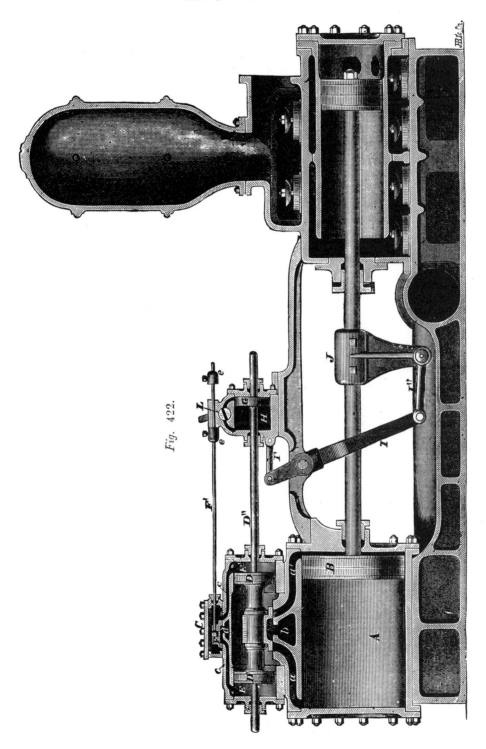

Fig. 422.

rest is caused by the main valve having lap, as shown in the figure, where it is seen that the valve moving piston D will move the main valve to the amount of the steam lap, before port a' begins to open. Suppose a' to be open and B to move to the left, and the cataract cylinder H will be moved to the right, while its piston will stand still. The amount of pressure at E′ acting to move the piston valve D to the left, will remain constant so long as the valve F admits steam to E′. But the amount of counteracting pressure on the face of the oil or cataract piston G, will vary with the effort of the cataract or oil cylinder H, to move to the right, and when this effort places, upon the face of the oil piston, a pressure exceeding that of the steam pres-sure on D, then the piston G will move to the right (in the same direction as H) carrying the piston and main valve over the main steam port, and reducing, or cutting off, as the case may be, the supply of steam to piston B. Now suppose that from some cause or other, the resistance to the pump plunger is suddenly relaxed, or diminished, and piston F will accelerate its speed. This effort will be communicated to the oil cylinder H; but the diminished opening at I will not permit the oil to pass freely through to the other side of G, hence the pressure in H will cause G to move to the right in unison with the oil cylinder, thus moving D to the right, reducing the steam supply to port a', and proportioning it to the resistance offered to the pump piston.

INDEX

319